W9-BEZ-462

SUN, EARTH AND SKY

SECOND EDITION

Kenneth R. Lang

Second Edition

Kenneth R. Lang
Department of Physics and Astronomy
Robinson Hall
Tufts University
Medford, MA, 02155
USA
ken.lang@tufts.edu

Cover illustration: An extreme ultraviolet image of the Sun with a huge, handle-shaped prominence (*top right*). It was taken on 14 September 1999 at a wavelength of 30.4 nanometers with the Extreme-ultraviolet Imaging Telescope, abbreviated EIT, aboard the *SOlar and Heliospheric Observatory,* abbreviated *SOHO.* Prominences are huge clouds of relatively cool dense plasma suspended in the Sun's hot, thin corona. At times, they can erupt, escaping the Sun's atmosphere. Emission in this spectral line shows the upper chromosphere at a temperature of about 60,000 K, or 60,000 degrees kelvin. Every feature in the image traces magnetic field structure. The hottest areas appear almost white, while the darker red areas indicate cooler temperatures. (Courtesy of the *SOHO* EIT consortium. *SOHO* is a project of international collaboration between ESA and NASA.)

Title page illustration: A large, soft X-ray cusp structure (*lower right*) is detected after a coronal mass ejection on 25 January 1992. The cusp, seen edge-on at the top of the arch, is the place where the oppositely directed magnetic fields, threading the two legs of the arch, are stretched out and brought together. Several similar images have been taken with the Soft X-ray Telescope (SXT) aboard *Yohkoh,* showing that magnetic reconnection is a common method of energizing solar explosions. (Courtesy of Loren W. Acton, NASA, ISAS, the Lockheed-Martin Solar and Astrophysics Laboratory, the National Astronomical Observatory of Japan, and the University of Tokyo.)

Library of Congress Control Number: 2005936335

ISBN-10: 0-387-30456-8 e-ISBN: 0-387-33365-7
ISBN-13: 978-0387-30456-4

Printed on acid-free paper.

Printed in Singapore. (Apex/KYO)

9 8 7 6 5 4 3 2 1

springer.com

Dedicated to Julia Sarah Lang

Here comes the Sun
Here comes the Sun and I say
It's alright
Little darling, it's been a long, cold, lonely winter.
Little darling it feels like years since it's been here.
Here comes the Sun. Here comes the Sun and I say
It's alright.
Little darling, the smiles returning to their faces.
. . .
Sun, Sun, Sun, Here it comes.

George Harrison

Preface to the Second Edition

The ***Second Edition*** of the popular ***Sun, Earth and Sky*** has been updated by providing comprehensive accounts of the most recent discoveries made by five modern solar spacecraft since the publication of the ***First Edition*** a decade ago. The *SOlar and Heliospheric Observatory,* abbreviated *SOHO,* launched on 2 December 1995, has extended our gaze from the visible solar disk to within the hidden solar interior and out in all directions through the Sun's tenuous million-degree atmosphere and solar winds. *SOHO's* recent insights have been complemented and extended by other solar spacecraft, including *Ulysses,* which sampled the solar wind during two complete circuits around the Sun, and *Yohkoh,* which completed a decade of observations of solar X-rays. The *Transition Region And Coronal Explorer,* abbreviated *TRACE,* which was launched on 2 April 1998, has provided years of high-resolution, extreme-ultraviolet observations related to the heating of the million-degree outer atmosphere, and its ubiquitous magnetic loops, as well as energetic explosions on the Sun known as solar flares. The *Ramaty High Energy Solar Spectroscopic Imager,* or *RHESSI* for short, launched on 5 February 2002, has revealed new aspects of solar flares, observing them at gamma-ray wavelengths.

Whenever possible we have used new images of the Sun, from the *RHESSI, SOHO,* and *TRACE* spacecraft, many of them never published in book form before. When relevant, line drawings have also been updated. And we have retained works of art that begin each chapter, providing another fascinating perspective of our life-sustaining Sun.

During the past decade, solar astronomers have provided answers to several outstanding mysteries of the Sun, and given us new clues to numerous open questions that remain.

Years of *SOHO* observations of solar oscillations have been combined to look deep inside the Sun, determining its internal rotation to the greatest possible depths and establishing the location of the solar dynamo that amplifies and maintains the Sun's magnetic fields. The detailed mechanisms for creating solar magnetism are nevertheless still lacking.

In the past decade, breakthrough observations with the massive subterranean neutrino detector, the Sudbury Neutrino Observatory, have solved the Solar Neutrino Problem, revealing new physics of the ghostly neutrino, which has a tiny mass and changes form as it moves from the Sun to Earth.

Our ***Second Edition*** also presents several plausible mechanisms that might heat the corona to million-degree temperatures, hundreds of times hotter than the underlying visible solar disk. And it describes the location of energy release for solar flares within the low corona. The exact methods of continuously heating the corona and accelerating particles to enormous energies during solar flares are still open questions.

This ***Second Edition*** of ***Sun, Earth and Sky*** includes answers to the age-old questions of why the Sun shines, and how long it has been shining and will continue to shine, providing life-sustaining heat and light to Earth. And we also describe what the Sun is made of, and where its ingredients came from.

The volume also contains extended discussions of Space Weather, whose importance has grown during the past decade. Our technological society has become increasingly vulnerable to storms from the Sun that can kill unprotected astronauts, disrupt global radio signals, and disable satellites used for communication, navigation, and military reconnaissance and surveillance. We therefore describe methods of forecasting Space Weather, as well as taking evasive action for their threat on the ground or in space.

Recent signs indicate that global warming by human emissions of heat-trapping gases is increasing. The evidence comes from direct measurements of rising surface air temperatures and subsurface ocean temperatures, and from phenomena such as retreating glaciers, increases in average global sea levels, and changes to many physical and biological systems. The book describes these warning signs, the likely future consequences of global warming, and efforts to do something about it, including the *Kyoto Protocol* that went into effect in 2005 without the participation of the United States.

An annotated list of Further Reading is included, for books published between 1995 and 2005, and a Directory of Web Sites includes links to obtain information about total eclipses of the Sun, solar observatories and spacecraft, and space weather. An extensive Glossary completes this ***Second Edition*** of ***Sun, Earth and Sky***. Its preparation was made possible through funding by NASA Grant NAG5–11605, from NASA's Earth-Sun System Program, and NNG05GB00G with NASA's Applied Information Systems Research Program.

Kenneth R. Lang
Tufts University *and* Anguilla, B.W.I.
January 1st, 2006.

Preface to the First Edition

This book was written for my daughter, Julia; her love and comfort have helped sustain me. It is a pleasure to watch Julia noticing details of everything around her. Children and other curious people perceive worlds that are invisible to most of us.

Here we describe some of these natural unseen worlds, from the hidden heart of the Sun to our transparent air. They have been discovered and explored through the space-age extension of our senses.

A mere half-century ago astronomers were able to view the Cosmos only in visible light. Modern technology has now widened the range of our perception to include the invisible realms of subatomic particles, magnetic fields, radio waves, ultraviolet radiation and X-rays. They are broadening and sharpening our vision of the Sun, and providing a more complete description of the Earth's environment. Thus, a marvelous new Cosmos is now unfolding and opening up, as new instruments give us the eyes to see the invisible and hands to touch what cannot be felt.

Giant radio telescopes now tune in and listen to the Sun, even on a rainy day. Satellite-borne telescopes, such as the one aboard the *Yohkoh,* or "sunbeam", satellite, view our daytime star above the absorbing atmosphere, obtaining detailed X-ray images of an unseen Sun. Space probes also directly measure the invisible subatomic particles and magnetic fields in space; for instance, the *Ulysses* spacecraft has sampled the region above the Sun's poles for the first time, and the venerable *Voyager 1* and *2* space probes may have found the hidden edge of the Solar System.

My colleagues have been very generous in providing their favorite pictures taken from ground or space. Numerous diagrams are included, each chosen for the new insight it offers. Every chapter begins with a work of modern art, illustrating how artists depict the mystical and supernatural aspects of the Sun or the subtle variations in illumination caused by changing sunlight. All of these images provide new perspectives on the Sun that warms our soul, and lights and heats our days, and on our marvelous planet Earth that is teeming with an abundance of life.

This book describes a captivating voyage of discovery, recording more than a century of extraordinary accomplishments. Our voyage begins deep inside the Sun where nuclear reactions occur. Here particles of anti-matter, produced during nuclear fusion, collide with their material counterparts, annihilating each other and disappearing in a puff of pure radiative energy.

Neutrinos are also created in the solar core; they pass effortlessly through both the Sun and Earth. Billions of the ghostly neutrinos are passing right through you every second. Massive subterranean neutrino detectors enable us to peer inside the Sun's energy-generating core, but the neutrino count always comes up short. Either the Sun does not shine the way we think it ought to, or our basic understanding of neutrinos is incomplete. Recent investigations suggest that the neutrinos have an identity crisis, transforming into a currently undetectable form.

Today we can peel back the outer layers of the Sun, and glimpse inside by observing its widespread throbbing motions. Visible oscillations caused by sounds trapped within the Sun can be deciphered to reveal its internal constitution. This procedure, called helioseismology, has been used to establish the Sun's internal rotation rate.

We then consider the Sun as a magnetic star. Its visible surface is pitted with dark, cool regions, called sunspots, where intense magnetism partially chokes off the outward flow of heat and energy. The sunspots tend to gather together in bipolar groups linked by magnetic loops that shape, mold and constrain the outer atmosphere of the Sun. The number of sunspots changes from a maximum to a minimum and back to a maximum every 11 years or so; most forms of solar activity vary in step with this magnetic sunspot cycle.

As our voyage continues, we discover that the sharp visible edge of the Sun is an illusion; a tenuous, hot, million-degree gas, called the corona, envelops it. This unseen world, detected at X-ray or radio wavelengths, is never still. Such observations show that the apparently serene Sun is continuously changing. It seethes and writhes in tune with the Sun's magnetism, creating an ever-changing invisible realm with no permanent features.

The *Yohkoh* X-ray telescope has shown that bright, thin magnetized structures, called coronal loops, are in a constant state of agitation, always varying in brightness and structure on all detectable spatial and temporal scales. Dark X-ray regions, called coronal holes, also change in shape and form, like everything else on the restless Sun. A high-speed solar wind squirts out of the holes and rushes past the planets, continuously blowing the Sun away and sweeping interstellar matter aside to form the heliosphere. The heating of the million-degree corona, which expands out to form this solar gale, is one of the great-unsolved mysteries of the Sun.

This book next considers violent solar phenomena that are detected at invisible wavelengths and are synchronized with the sunspot cycle of magnetic activity. In minutes, powerful eruptions, called solar flares, release magnetic energy equivalent to billions of nuclear explosions and raise the temperature of Earth-sized regions to about ten million degrees. Magnetic bubbles, called coronal mass ejections, expand as they propagate outward from the Sun to rapidly rival it in size; their associated shocks accelerate and propel vast quantities of high-speed particles ahead of them.

Our account then turns toward our home planet, Earth, where the Sun's light and heat permit life to flourish. Robot spacecraft have shown that the space outside our atmosphere is not empty! It is swarming with hot, invisible pieces of the Sun.

The Earth's magnetic field shields us from the eternal solar gale, but the gusty, variable wind buffets our magnetic domain and sometimes penetrates within it. Charged particles that have infiltrated the Earth's magnetic defense can be stored in nearby reservoirs such as the Van Allen radiation belts. Spacecraft have recently released chemicals that can illuminate the space near Earth. Other satellites have found a new radiation belt that contains the ashes of stars other than the Sun.

This voyage continues with a description of the multi-colored auroras, or the northern and southern lights. Solar electrons that apparently enter through the Earth's back door, in the magnetotail, are energized locally within our magnetic realm and are guided into the polar atmosphere where they light it up like a cosmic neon sign.

Unpredictable impulsive eruptions on the Sun produce outbursts of charged particles and energetic radiation that can touch our lives. Intense radiation from a powerful solar flare travels to the Earth in just eight minutes, altering its outer atmosphere, disrupting long-distance radio communications, and affecting satellite orbits. Very energetic particles arrive at the Earth within an hour or less; they can endanger unprotected astronauts or destroy satellite electronics. Solar mass ejections can travel to our planet in one to four days, resulting in strong geomagnetic storms with accompanying auroras and electrical power blackouts. All of these effects, which are tuned to the rhythm of the Sun's 11-year magnetic activity cycle, are of such vital importance that national centers employ space weather forecasters and continuously monitor the Sun from ground and space to warn of threatening solar activity.

The Earth is wrapped in a thin membrane of air that ventilates, protects and incubates us. It acts as a one-way filter, allowing sunlight through to warm the surface but preventing the escape of some of the heat into outer space. Without this "natural" greenhouse effect, the oceans would freeze and life as we know it would not exist. Long ago, when the Sun was faint, an enhanced greenhouse effect probably kept the young Earth warm enough to sustain life. Then, as the Sun became more luminous, the terrestrial greenhouse must have been turned down, perhaps by life itself.

The book next shows how the Sun's steady warmth and brightness are illusory; no portion of the spectrum of the Sun's radiative output is invariant. Recent spacecraft measurements have shown that the Sun's total radiation fades and brightens in step with changing activity levels. It doesn't change by much, only by about 0.1 percent over the 11-year sunspot cycle, but the Sun's invisible ultraviolet and X-ray radiation are up to one hundred times more variable than the visible output. Fluctuations in the Sun's visible and invisible radiation can potentially alter global surface temperatures and influence terrestrial climate and weather, alter the planet's ozone layer, and heat and expand the Earth's upper atmosphere.

To completely assess environmental damage by humans to date, and to fully understand how the environment may respond to further human activity, requires an understanding of solar influences on our planet. We must look beyond and outside the Earth, to the inconstant Sun as an agent of terrestrial change. It can both lessen and compound ozone depletion or global warming by amounts that are now comparable to those produced by atmospheric pollutants.

This book therefore next focuses attention on the Earth's protective ozone layer, that is both modulated from above by the Sun's variable ultraviolet output and threatened from below by man-made chemicals. The ozone layer protects us from the Sun's lethal ultraviolet rays, and progress has been made in outlawing the ozone-destroying chemicals. Our ability to reliably determine the future recovery of the ozone layer will depend on adequate knowledge of how it is damaged or restored by the Sun.

Our voyage then continues with a discussion of the Earth's varying temperature. Large natural fluctuations in the record of global temperature changes mask our ability to clearly detect warming caused by human activity, and numerous complexities limit the certainty of computer models used to forecast future global warming. Strong correlations suggest that the 11-year solar activity cycle may be linked to both the Earth's surface temperature and terrestrial weather.

The "unnatural" greenhouse warming might eventually break through the temperature record, if we keep on dumping waste gases into the air at the present rate. The probable consequences of overheating the Earth as the result of human activity are therefore next examined. They suggest that we should curb the build-up of heat-trapping gases despite the great uncertainties about their current effects.

Yet, it is the Sun that energizes our climate and weather. During the past million years, the climate has been dominated by the recurrent, periodic ice ages, which are mainly explained by changes in the amount and distribution of sunlight on the Earth. Variations in the Earth's orbit and axial tilt slowly alter the distances and angles at which sunlight strikes the Earth, thereby controlling the ponderous ebb and flow of the great continental glaciers.

Smaller, more frequent climate fluctuations are superimposed on the grand swings of the glacial/interglacial cycles; these minor ice ages may result from variations in the activity of the Sun itself. For instance, during the latter half of the seventeenth century the sunspot cycle effectively disappeared; this long period of solar inactivity coincided with unusually cold spells in the Earth's northern hemisphere. Observations of Sun-like variable stars indicate that small, persistent variations in the solar energy output could produce extended periods of global cooling or warming. So, a prudent society will benefit by keeping a close watch on the Sun, the ultimate source of all light and heat on the Earth.

I am grateful to numerous experts and friends who have read individual chapters and commented on their accuracy and completeness. They have greatly improved the manuscript, while also providing encouragement. They include Loren Acton, John Bahcall, Dave Bohlin, Ron Bracewell, Raymond Bradley, Ed Cliver, Nancy Crooker, Brian Dennis, Peter Foukal, Mona Hagyard, David Hathaway, Gary Heckman, Mark Hodor, Bob Howard, Jim Kennedy, Jeff Kuhn, Judith Lean, Bill Livingston, John Mariska, Bill Moomaw, Gene Parker, Art Poland, Peter Sturrock, Einar Tandberg-Hanssen, Jean-Claude Vial, Bill Wagner and Wesley Warren. None of them is responsible for any remaining mistakes in the text!

Special thanks go to my entire family – my wife, Marcella, and my three children Julia, David and Marina.

Kenneth R. Lang
Medford, February 1995

Contents

PREFACE TO THE SECOND EDITION vii
PREFACE TO THE FIRST EDITION ix
ICARUS xvi

CHAPTER 1 GOOD DAY, SUNSHINE 2
1.1 The Rising Sun 3
1.2 Fire of Life 5
1.3 Sunlight 7
1.4 Daytime Star 8
1.5 Cosmic Laboratory 10
1.6 Ingredients of the Sun 11
1.7 Children of the Cosmos 16
1.8 Describing the Radiation 16
1.9 Invisible Fires 19

CHAPTER 2 ENERGIZING THE SUN 22
2.1 Awesome Power, Enormous Times 23
2.2 The Sun's Central Pressure Cooker 25
2.3 Nuclear Fusion, Anti-Matter and Hydrogen Burning 26
2.4 Diluting the Radiation 32
2.5 Convection and Granulation 33
2.6 The Sun's Remote Past and Distant Future 36

CHAPTER 3 GHOSTLIKE NEUTRINOS 40
3.1 The Elusive Neutrino 41
3.2 Neutrinos from the Sun 44
3.3 Detecting Almost Nothing 47
3.4 Solving the Solar Neutrino Problem 52

CHAPTER 4 TAKING THE PULSE OF THE SUN 62
4.1 Trapped Sounds 63

4.2 Ode to the Sun 68
4.3 Looking Inside the Sun 71
4.4 Breaking the Symmetry 73
4.5 Internal Flows 76

CHAPTER 5 A MAGNETIC STAR 80
5.1 Magnetic Fields in the Visible Photosphere 81
5.2 The Solar Chromosphere and its Magnetism 88
5.3 Bipolar Sunspots, Magnetic Loops and Active Regions 94
5.4 Cycles of Magnetic Activity 98
5.5 Internal Dynamo 101

CHAPTER 6 AN UNSEEN WORLD OF PERPETUAL CHANGE 106
6.1 The Sun's Visible Edge Is an Illusion 107
6.2 The Million-Degree Corona 113
6.3 Coronal Heating 116
6.4 Closed Coronal Loops and Open Coronal Holes 119
6.5 The Eternal Solar Wind 123
6.6 *Ulysses, SOHO* and the Origin of the Sun's Winds 128
6.7 The Distant Frontier 134

CHAPTER 7 THE VIOLENT SUN 136
7.1 Energetic Solar Activity 137
7.2 Solar Flares 140
7.3 Flare Radiation from Energetic Accelerated Particles 142
7.4 Energizing Solar Flares 148
7.5 Erupting Prominences 152
7.6 Coronal Mass Ejections 156
7.7 Predicting Explosions on the Sun 159

CHAPTER 8 ENERGIZING SPACE 164
8.1 The Ingredients of Space 165
8.2 Earth's Magnetic Cocoon 166
8.3 Penetrating Earth's Magnetic Defense 170
8.4 Storing Invisible Particles Within Earth's Magnetosphere 171
8.5 Northern and Southern Lights 174
8.6 Geomagnetic Storms 181
8.7 Space Weather 183

CHAPTER 9 TRANSFORMING THE EARTH'S LIFE-SUSTAINING ATMOSPHERE 192
9.1 Fragile Planet Earth – The View From Space 193
9.2 Clear Skies and Stormy Weather 193
9.3 The Life-Saving Greenhouse Effect 199
9.4 The Earth's Changing Atmosphere and Oceans 201
9.5 Our Sun-Layered Atmosphere 203
9.6 The Vanishing Ozone 207
9.7 The Inconstant Sun 212
9.8 The Heat Is On 217
9.9 The Ice is Coming 227

QUOTATION REFERENCES 233
GLOSSARY 237
DIRECTORY OF WEB SITES 269
FURTHER READING 271
AUTHOR INDEX 273
SUBJECT INDEX 277

Icarus

The French painter Henri Matisse (1869–1954) thought that happiness is derived from never being a prisoner of anything, including success or style, and he represented freedom from imprisonment in his 1947 book *Jazz* with this cut-out entitled *Icarus,* who seems to be pushing against the downward pull of gravity, trying to break away, soaring into a bright blue sky, and set the human spirit free. Icarus' red heart symbolizes love, which can make one soar, run and rejoice. Its liberating and all-consuming nature can make one see the world anew. (Courtesy of Succession Matisse, Paris.)

SUN, EARTH AND SKY

SECOND EDITION

Chapter One

Good Day, Sunshine

WILD GEESE IN SUNLIGHT These geese are flying south in the northern winter, following the Sun's warmth. In this V-shaped pattern of flight, the lead bird deflects currents of air and makes flying easier for those that follow in its wake. The Earth's magnetic field similarly deflects the Sun's wind. (Courtesy of James Tallon.)

1.1 THE RISING SUN

From earliest times, the Sun has been revered and held in awe. For the Greeks of Aristotle's time, sunlight epitomized the fire in the four basic elements – earth, air, fire and water – from which all things arose. Ancient solar observatories, dedicated to the divine Sun-god Ra, can still be found in Luxor, that enchanting city by the Nile; giant Egyptian obelisks, erected thousands of years ago in Luxor and Heliopolis (City of the Sun), now cast their shadows in sundial fashion across parts of Paris, London, and Rome.

According to this incantation from Ptolemaic Egypt:

> Opening his two eyes, [Ra, the Sun god] cast light on Egypt, he separated night from day. The gods came forth from his mouth and mankind from his eyes. All things took their birth from him.[1]

And in the Old Testament's *Book of Genesis,* we find that the Earth was initially a vast waste, covered by darkness, until God said "Let there be light" and the Sun separated day from night.

Since the time of the ancient Persian prophet Zarathustra (about 1300 BC, Greek Zoroaster), we have associated light with good, beauty, truth and wisdom, in sharp contrast with the dark forces of evil. The war between good and evil in the *Dead Sea Scrolls* is depicted as a battle of the Sons of Light against the Sons of Darkness. Dante's divine journey took him from the dark forest to the radiance of paradise, and today we have the evil darkness of Darth Vader in *Star Wars.*

The Maya, Toltec and Aztec of Central America had a host of Sun gods; the Aztecs regularly fed the hearts of sacrificial victims to the Sun to strengthen it on its daily journey. Shintoism, a religion based on Sun worship, has continued for thousands of years in Japan, the land of the rising Sun. Today you can celebrate sunrise with Hindu worshipers on the terraced banks, or ghats, along the Ganges River at Benares, India's holiest city.

Nowadays, fire symbolically lights the darkness in many of our rituals, including the torch of the Olympic games, and candlelight vigils or dimmed lights that bring focus to tragic events and times of crisis. In everyday life, most of us feel happier on bright days than on gloomy ones, so cheerful people have a "sunny" disposition while an unhappy day is a "dark" one. And throughout the world, oiled Sun-worshippers lie on tranquil beaches, letting the summer Sun warm their bodies and give them strength.

The German romantic painter Caspar David Friedrich (1774–1840) used sunrise to portray a spiritual relationship with nature (Fig. 1.1), comparing the "radiating beams of light" in one of his paintings to "the image of the eternal life-giving Father." Sunlight seems to dominate, consume and absorb everything in the paintings of the British artist Joseph Mallord William Turner (1775–1851), who depicted tiny figures dwarfed by the power, beauty and violence of the physical world. When viewing one of his apocalyptic visions, the spectator can become engulfed and lost in the colored light of the sky and sea (Fig. 1.2). The artist's dying words were "The Sun is God."

Examples of artists' perspectives on the Sun are provided at the beginning of every chapter in this book, each chosen for its artistic value and for the new insights

FIG. 1.1 Woman in morning Sun This portrayal of the glowing sunrise by the German artist Caspar David Friedrich (1774–1840) seems to have a transcendental, mystical quality. The painter once compared the "radiating beams of light" in one of his paintings to "the image of the eternal life-giving Father." (Courtesy of Museum Folkwang, Essen.)

it offers. Here you will find "another light, a stronger Sun" portrayed by the Dutch painter Vincent Van Gogh (1853–1890), who used thick brush stokes of blazing, brilliant pigment, as dense as honey, to portray a powerful, yellow Sun that blazes forth with an almost supernatural radiance. The French artist Claude Monet's (1840–1926) painting of sunrise is included – the one that inaugurated the impressionist movement of painting. He used entire sequences of paintings to depict the subtle changes that varying sunlight causes in our perception of objects, such as haystacks or the cathedral at Rouen.

The chapter frontispieces also include the works of the Spanish painter Joan Miró (1893–1983), who portrayed the powerful red disk of the Sun that caresses our limbs and brings us joy, or links us to the stars beyond. In other instances, we reproduce works that separate the Sun from any reference to the Earth or sky; they show that the Sun can be an intense source of pleasure and beauty by itself.

Writers have also been captivated by the light of the Sun, from the American author Ralph Waldo Emerson (1803–1882), who wrote that pure light was "the reappearance of the original soul," to the German philosopher Friedrich Nietzsche (1844–1900) who wrote in *Thus Spoke Zarathustra:*

> The Moon's love affair has come to an end!
> Just look! There it stands; pale and dejected – before the dawn!

FIG. 1.2 Regulus In this painting by the British artist Joseph Mallord William Turner (1775–1851), every object is in a fiery, misty state. Brilliant yellow rays of light come down from a central, all-powerful Sun, absorbing and consuming everything else. The picture is named after the Roman general Regulus who was punished for his betrayal of the Carthaginians by having his eyelids cut off, and being blinded by the glare of the Sun. Regulus, who is apparently absent from the scene, has been identified with the spectator, staring into the blinding Sun. (Courtesy of the Tate Gallery, London.)

For already it is coming, the glowing Sun –
its love of the Earth is coming!
All Sun-love is innocence and creative desire!
Just look how it comes impatiently over the sea!
Do you not feel the thirst and hot breath of its love?[2]

The Sun warms our soul, and lights and heats our days! Today's astronomers may describe the Sun, and our dependence upon it, in greater scientific detail than artists or writers, but that in no way diminishes their sense of awe for the life-sustaining, even mystical power of the Sun.

1.2 FIRE OF LIFE

The Sun is our powerhouse. It energizes our planet, warming the ground and lighting our days. It is solar heat that powers the winds and cycles water from sea to rain, the source of our weather and arbiter of our climate. And the Sun is the source of the energy that sustains life.

Without its heat and light, all life would quickly vanish from the surface of our planet. And the Sun provides – directly or indirectly – most of our energy. Green plants absorb sunlight where it strikes chlorophyll, giving them the energy to break water molecules apart and energize photosynthesis; plants thereby use the Sun's energy to live and grow, giving off oxygen as a byproduct (Fig. 1.3). Eating these plants nourishes animals. And long-dead, compressed plants provide the petroleum, coal or natural gas that energizes the lights in your house or powers the car you drive.

The Earth glides through space at just the right distance from the Sun for life to thrive on our planet's surface, while other planets freeze or fry. We sit in the comfort

FIG. 1.3 Sunflowers The Sun sustains all living creatures and plants on Earth. Green plants absorb sunlight, giving them the energy to break water molecules apart and energize photosynthesis. Plants thereby use the Sun's energy to live and grow, giving off oxygen as a byproduct. (Courtesy of Charles E. Rodgers.)

zone. Any closer and the oceans would boil away. Further out the Earth would be a frozen wasteland.

We receive just enough energy from the Sun to keep most of our water liquid, which is a requirement for life, as we know it. In comparison, the surface of Venus, just slightly closer to the Sun, is hot enough to melt lead; further away from the Sun, the planet Mars is now frozen into a global ice age. It cannot now rain on Mars, and any liquid water released on its surface will either evaporate or freeze into ice. Turn off the Sun's powerhouse, and in just a few months we could all be under ice.

1.3 SUNLIGHT

Occasionally the mixture of colors in a beam of sunlight is spread before our eyes, as when raindrops act like tiny prisms, bending white sunlight into its separate colors and giving us a rainbow (Fig. 1.4). Our eyes and brain translate the visible sunlight into colors. From long to short waves, they correspond to red, orange, yellow, green, blue and violet. Plants appear green, for example, because they absorb red sunlight and reflect the green portion of the Sun's light.

However, your world might be colored somewhat differently from someone else's. There are subtle differences in the exact shade of color we perceive, depending on the molecules in our eye's detection system. So, even people with normal eyesight do not always see eye to eye.

FIG. 1.4 Light painting This picture was made by using crystals to liberate the spectral colors in visible sunlight, refracting them directly onto a photographic plate. It was obtained in the rarefied atmosphere atop Hawaii's Mauna Kea volcano, where many of the world's best telescopes are located. (Courtesy of Eric J. Pittman, Victoria, British Colombia.)

The most intense radiation of the Sun is emitted at the visible wavelengths of colored light, and our atmosphere permits them to reach the ground. If our eyes were not so sensitive to visible sunlight, we could not identify objects or move around on the Earth's surface. Thus, the sensitivity of our eyes is matched to the tasks of vision.

The Sun also emits invisible radiation, with less intensity than the visible sort. In 1800, for example, the German-born English astronomer William Herschel (1738–1822) discovered invisible radiant heat, or infrared radiation, by noticing a rise in temperature when a thermometer is placed beyond the red end of the visible spectrum of sunlight.

We all glow in the dark, emitting infrared radiation. You can't see anyone's infrared heat radiation, it's outside your range of vision, but you can feel it. In contrast, rattlesnakes have infrared-sensitive eyes that enable them to see the heat radiated by animals at night, and the military uses infrared technology to sense and locate the heat generated by the enemy in the total dark. Night-vision gobbles with infrared sensors are an example.

The Sun also emits invisible ultraviolet radiation, radio waves and X-rays, which differ in the length of their waves. X-rays are very short, much smaller than the ultraviolet whose waves are just a little smaller than those of blue sunlight, and radio waves that are very long. The properties of these different types of the Sun's radiation are described in Section 1.8.

1.4 DAYTIME STAR

All stars are suns, kin to our own daytime star. Indeed, the Sun is just one of about one hundred billion stars in our Galaxy, the Milky Way, and countless billions of galaxies stretch out in the seemingly boundless Universe. But the Sun is a special star; it is our only daytime star! Nothing else in the Universe is so critically important to us. As the Victorian English poet Francis William Bourdillon (1852–1921) wrote:

> The night has a thousand eyes,
> And the day but one;
> Yet the light of the bright world dies,
> With the dying Sun.
>
> The mind has a thousand eyes,
> And the heart but one:
> Yet the light of a whole life dies
> When love is done[3]

The Sun is a quarter million times closer to us than the next nearest star. Because of this closeness, the Sun is about a hundred billion times brighter than any other star. The Sun's brilliance provides ample light for the most exacting studies of its chemical constituents, magnetic fields, and oscillations. This blessing can also be a curse, for the Sun's heat can melt mirrors or burn up electronic equipment when focused to high intensity. For this reason, special mirror configurations are used to reduce the concentration of visible sunlight, while still producing large images that contain fine detail (Fig. 1.5).

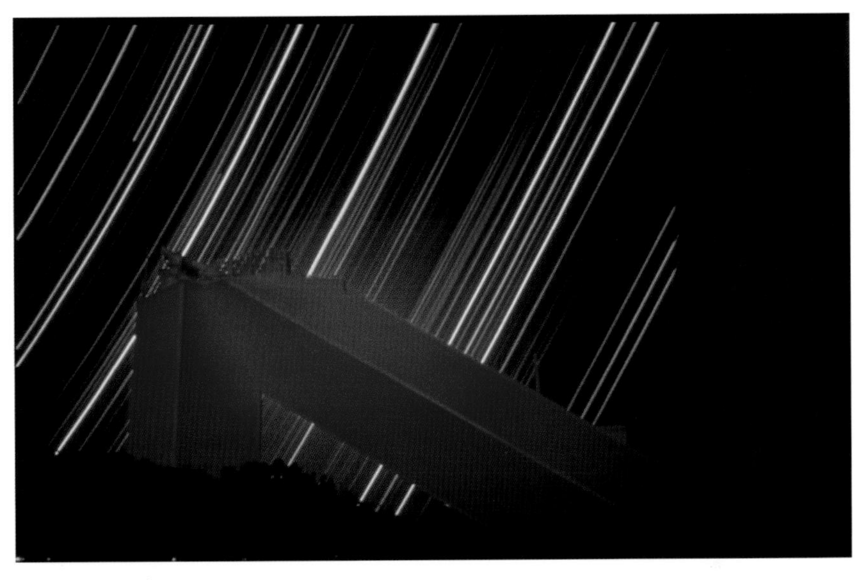

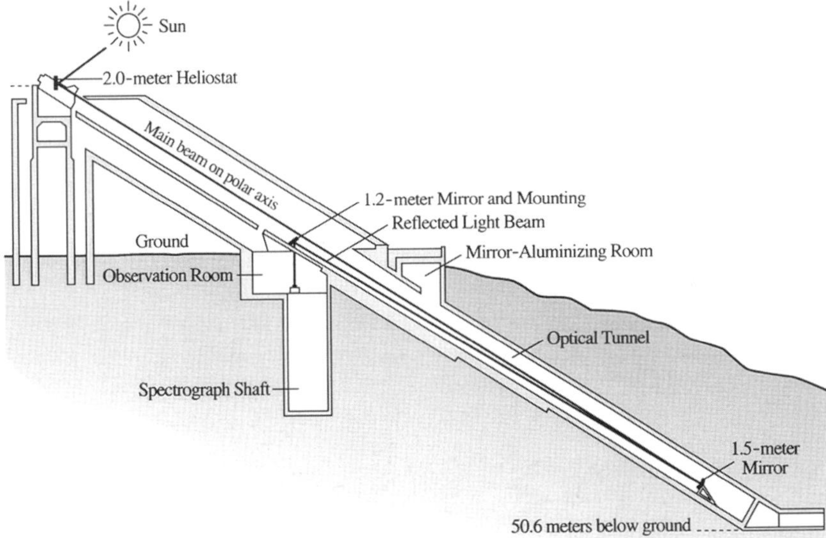

FIG. 1.5 Eyes on the Sun Scattered sunlight colors the McMath solar telescope a stunning red, while stars mark trails across the evening sky *(top)*. A moveable heliostat, perched atop this telescope, follows the Sun and directs its light downward through the long fixed shaft of the telescope (*bottom*). A figured mirror at the shaft bottom reflects and focuses the sunlight toward the observation room. The shaft's axis is parallel to the rotation axis of the Earth, and about three fifths of it is underground. It is kept cool by pumping cold water through tubes in the exterior skin, thereby reducing turbulence in the air inside and keeping the Sun's image steady. (Courtesy of William C. Livingston, NOAO.)

The Sun's proximity allows a level of detailed examination unique among stars. While most other stars appear only as unresolved spots of light in the best telescopes, the Sun reveals its features in exquisite detail. Most ground-based optical telescopes can resolve structures on the Sun's visible disk that are about 750 kilometers across, about the distance from Boston to Washington, D.C. and about three-quarters the size of France; that is comparable to seeing the details on a coin from one kilometer away.

Yet, the resolution of ground-based telescopes is limited by turbulence in the Earth's atmosphere; it reduces the clarity of the Sun's image at visible wavelengths. Similar variations cause the stars to twinkle at night. The best visible images with even finer detail can be obtained using adaptive optics that correct for the changing atmosphere, or from the unique vantage point of outer space using satellite-borne telescopes unencumbered by the limits of our atmosphere.

The other stars are so far away that their surfaces remain unresolved with even the largest telescopes. The Sun therefore permits examination of physical phenomena and processes that cannot be seen in detail on other stars. Furthermore, the Sun's basic properties provide benchmarks and boundary conditions for the study of stellar structure and evolution.

So, all astronomers do not work in the dark. Many of them closely scrutinize our daytime star, deciphering some of the most fundamental secrets of nature.

1.5 COSMIC LABORATORY

The Sun can be a site to test physical theories under conditions not readily attainable in terrestrial laboratories. For example, in contrast to our material world, the Sun's core also contains small quantities of short-lived anti-matter. When subatomic matter and anti-matter collide, they destroy each other, releasing pure radiative energy. We can also detect the process during explosions on the visible solar disk, which briefly become hotter than the center of the Sun.

Other particles made deep inside our home star pass effortlessly through both the Sun and the Earth. Recent observations of these ghostlike neutrinos have helped us understand the subatomic realm at levels beyond the reach of today's most powerful particle accelerators, providing new insight to a theory that might someday unify all the forces of nature.

From afar, the Sun seems to be calm, serene, and unchanging, a steadily shining beacon in the sky; but detailed observations reveal an active, ever-changing Sun. Violent storms and explosive eruptions create gusts in its steady flow of heat, particles and light. The Sun therefore provides us with a unique, high-resolution perspective of the perpetual change and violent activity that characterize much of the Universe.

Thus, we now understand the Sun as a unique star, one so close that it serves as a cosmic laboratory for understanding the physical processes that govern all the other stars, as well as the entire Universe. Everything we learn about the Sun has implications throughout the Cosmos, including planet Earth. As examples, observations of the Sun's visible radiation unlocked the chemistry of the Universe, and investigations of the Sun's internal furnace paved the way to nuclear energy.

1.6 INGREDIENTS OF THE SUN

Celestial objects are composed, like the Earth and we ourselves, of individual particles of matter called atoms. But the atoms consist largely of seemingly empty space, just as the room you may be sitting in appears mostly empty. A tiny, heavy, positively charged nucleus lies at the heart of an atom, surrounded by a cloud of relatively minute, negatively charged electrons that occupy most of an atom's space and govern its chemical behavior.

In the early 20th century, the New Zealand-born British physicist Ernest Rutherford (1871–1937) showed that radioactivity is produced by the disintegration of atoms, and discovered that they emit energetic alpha particles, which consist of helium nuclei; he was awarded the 1908 Nobel Prize in Chemistry for these achievements. By using helium ions to bombard atoms, Rutherford was able to announce in 1911 that most of the mass of an atom is concentrated in a nucleus that is 100,000 times smaller than the atom and has a positive charge balanced by the negative charge of surrounding electrons.

Nearly a decade later, in 1920, Rutherford announced that the massive nuclei of all atoms are composed of hydrogen nuclei, which he named protons. He also postulated the existence of an uncharged nuclear particle, later called the neutron, which was required to help hold the nucleus together and keep it from dispersing as the protons repelled each other. After an eleven-year search, the English physicist James Chadwick (1891–1974) discovered the neutron, in 1932, receiving the 1935 Nobel Prize in Physics for the feat. So, the nucleus of an atom is composed of positively charged protons and neutral particles, called neutrons; both about 1,840 times heavier than the electron.

The simplest and lightest atom consists of a single electron circling around a nucleus composed of a single proton without any neutrons; this is an atom of hydrogen. The nucleus of helium, another abundant light atom, contains two neutrons and two protons, and the helium atom therefore has two electrons.

The atomic ingredients of the Sun can be inferred from dark absorption lines, which are found superimposed on the colors of sunlight (Fig. 1.6). They look like a dark line when the Sun's radiation intensity is displayed as a function of wavelength; such a display is called a spectrum. The term Fraunhofer absorption line is also used, recognizing the German astronomer Joseph von Fraunhofer (1787–1826). By directing the incoming sunlight through a slit, and then dispersing it with a prism, Fraunhofer was able to overcome the blurring of colors from different parts of the Sun's disk, discovering numerous dark absorption lines. By 1814 he had detected and catalogued more than 300 of them, assigning Roman letters to the most prominent.

The Sun is so bright that its light can be spread out into very small wavelength intervals with enough intensity to be detected. The instrument used to make and record such a spectrum is called a spectrograph, a composite word consisting of *spectro* for "spectrum" and *graph* for "record". The spectrograph spreads out the wavelengths into different locations, as a rainbow or prism does. Nowadays it is the grooves of a diffraction grating that reflect sunlight into different locations according to color or wavelength. And you can see such a colored display by looking at a compact disk.

FIG. 1.6 Visible solar spectrum A spectrograph has spread out the visible portion of the Sun's radiation into its spectral components, displaying radiation intensity as a function of wavelength. When we pass from long wavelengths to shorter ones (*left* to *right, top* to *bottom*), the spectrum ranges from red through orange, yellow, green, blue and violet. Dark gaps in the spectrum, called Fraunhofer absorption lines, are due to absorption by atoms in the Sun. The wavelengths of these absorption lines can be used to identify the elements in the Sun, and the relative darkness of the lines helps establish the relative abundance of these elements. (Courtesy of National Solar Observatory, Sacramento Peak, NOAO.)

When a cool, tenuous gas is placed in front of a hot, dense one, atoms in the cool gas absorb radiation at specific wavelengths, thereby producing dark absorption lines. And when a tenuous gas stands alone and is heated to incandescence, emission lines are produced that shine at precisely the same wavelengths as the dark ones. The Sun's dark absorption lines and bright emission lines carry messages from inside the atom, and help us determine its internal behavior.

Adjacent lines of the hydrogen atom exhibit a strange regularity – they systematically crowd together and become stronger at shorter wavelengths. The Swiss mathematics teacher Johann Balmer (1825–1898) published an equation that describes the regular spacing of the wavelengths of the four lines of hydrogen detected in the spectrum of visible sunlight, and they are still known as Balmer lines. The strongest one, with a red color, is also called the hydrogen alpha line.

In 1913, the Danish physicist Niels Bohr (1885–1962) explained Balmer's equation by an atomic model, now known as the Bohr atom, in which the single electron in a hydrogen atom revolves about the nuclear proton in specific orbits with definite, quantized values of energy. An electron only emits or absorbs radiation when jumping between these allowed orbits, each jump being associated with a specific energy and a single wavelength, like one pure note. If an electron jumps from a low-energy orbit to a high-energy one, it absorbs radiation at this wavelength; radiation is emitted at exactly the same wavelength when the electron jumps the opposite way. This unique wavelength is related to the difference between the two orbital energies. Bohr was awarded the 1912 Nobel Prize in Physics for his investigations of the structure of atoms and the radiation emanating from them.

Since only quantized orbits are allowed, spectral lines are only produced at specific wavelengths that characterize or identify the atom. An atom or molecule can absorb or emit a particular type of sunlight only if it resonates to that light's energy. As it turns out, the resonating wavelengths or energies of each atom are unique – they fingerprint an element, encode its internal structure and identify the ingredients of the Sun. In addition, spectral lines yield information about the Sun's temperature, density, motion and magnetism.

Each element, and only that element, produces a unique set of absorption or emission lines. The presence of these spectral signatures can therefore be used to specify the chemical ingredients of the Sun (Fig. 1.7). The abundance calculations depend upon both measurements of the solar lines and on properties of the elements detected in the terrestrial laboratory. The lightest element, hydrogen, is the most abundant element in the Sun and most other stars (Focus 1.1). Altogether, 92.1 percent of the atoms in the Sun are hydrogen atoms, 7.8 percent are helium atoms, and all the other heavier

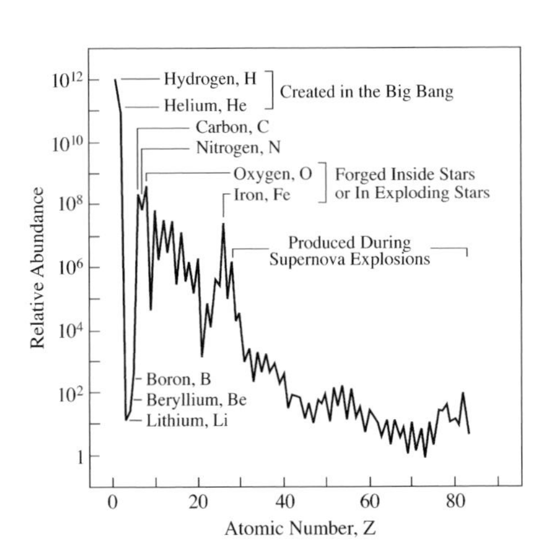

FIG. 1.7 Abundance and origin of the elements in the Sun The relative abundance of the elements in the solar photosphere, plotted as a function of their atomic number, Z. The abundance is specified on a logarithmic scale and normalized to a value a million, million, or 1.0×10^{12}, for hydrogen. Hydrogen, the lightest and most abundant element in the Sun, was formed about 14 billion years ago in the immediate aftermath of the Big Bang that led to the expanding Universe. Most of the helium now in the Sun was also created then. All the elements heavier than helium were synthesized in the interiors of stars that no longer shine, and subsequently wafted or blasted into interstellar space where the Sun originated. Carbon, nitrogen, oxygen and iron, were created over long time intervals during successive nuclear burning stages in former stars, and also during the explosive death of massive stars. Elements heavier than iron were produced by neutron capture reactions during the supernova explosions of massive stars that lived and died before the Sun was born. The atomic number, Z, is the number of protons in the nucleus, or roughly half the atomic weight. The elements shown, He, C, N, O and Fe, have Z = 2, 6, 7, 8 and 26, with atomic weights of 4, 12, 14, 16, and 56, since each nucleus contains as many neutrons as protons with about the same weight. Hydrogen has one proton and no neutrons in its nucleus. The exponential decline of abundance with increasing atomic number and weight can be explained by the rarity of stars that have evolved to later stages of life. (Data courtesy of Nicolas Grevesse.)

FOCUS 1.1

Composition of the Stars

In the mid-nineteenth century, the German physicist Gustav Kirchhoff (1824–1887) discovered a method for determining the ingredients of the stars. Working with the German chemist Robert Bunsen (1811–1899), Kirchhoff showed that every chemical element, when heated to incandescence, emits brightly colored spectral signatures, or emission lines, whose unique wavelengths coincide with those of the dark absorption lines in the Sun's spectrum.

By comparing the Sun's absorption lines with emission lines of elements vaporized in the laboratory, Kirchhoff identified in the solar atmosphere several elements known on Earth, including sodium, calcium and iron. This suggested that stars are composed of terrestrial elements that are vaporized at the high stellar temperatures, and it unlocked the chemistry of the Universe. As Bunsen wrote in 1859:

> At the moment I am occupied by an investigation with Kirchhoff, which does not allow us to sleep. Kirchhoff has made a totally unexpected discovery, inasmuch as he has found out the cause for the dark lines in the solar spectrum and can produce these lines artificially intensified both in the solar spectrum and in the continuous spectrum of a flame, their position being identical with that of Fraunhofer's lines. Hence the path is opened for the determination of the chemical composition of the Sun and the fixed stars.[4]

In a brilliant doctoral dissertation, published in 1925, the American astronomer Cecilia H. Payne (1900–1979) showed that the atmospheres of virtually every luminous, middle-aged star have the same ingredients. Miss Payne, later Payne-Gaposchkin, eventually became the first female Professor in the Faculty of Arts and Sciences at Harvard University, where she had studied. Her calculations also indicated that hydrogen is by far the most abundant element in the Sun and most other stars. But she could not believe that the composition of stars differed so enormously from that of the Earth, where hydrogen is rarely found, so she mistrusted her understanding of the hydrogen atom. Prominent astronomers of the time also did not think that hydrogen was the main ingredient of the Sun and other stars, and this may have played a role in her considerations.

We now know that hydrogen is the most abundant element in the Universe, and that there was nothing wrong with Miss Payne's calculations. The Earth just does not have sufficient gravity to retain hydrogen in its atmosphere. Any hydrogen gas that our planet might have once had must have evaporated away while the Earth was forming and has long since escaped.

Subsequent observations have shown that very old stars have practically no elements other than hydrogen and helium; these stars have probably existed since our Galaxy formed. Middle-aged stars like the Sun contain heavier elements. They must have formed from the ashes of previous generations of stars that have fused lighter elements into heavier ones.

The Sun is mainly composed of light elements, hydrogen and helium, which are terrestrially rare, whereas the Earth is primarily made out of heavy elements that are relatively uncommon in the Sun (Table 1.1). Hydrogen is about one million times more abundant than iron in the Sun, but iron is one of the main constituents of the Earth, which cannot even retain hydrogen gas in its atmosphere.

elements make up only 0.1 percent. In contrast, the main ingredients of the rocky Earth are the heavier elements like silicon and iron, and this explains the Earth's higher mass density – about four times that of the Sun, which is only about as dense as water.

Helium, the second-most abundant element in the Sun, is so rare on Earth that it was first discovered on the Sun – by the French astronomer Pierre Jules Janssen

TABLE 1.1 The twenty most abundant elements in the solar photosphere

Element	Symbol	Atomic Number	Abundance[a] (logarithmic)	Date of Discovery on Earth
Hydrogen	H	1	12.00	1766
Helium	He	2	[10.93 ± 0.01]	1895[b]
Carbon	C	6	8.39 ± 0.05	(ancient)
Nitrogen	N	7	7.78 ± 0.06	1772
Oxygen	O	8	8.66 ± 0.05	1774
Neon	Ne	10	[7.84 ± 0.06]	1898
Sodium	Na	11	6.17 ± 0.04	1807
Magnesium	Mg	12	7.53 ± 0.09	1755
Aluminum	Al	13	6.37 ± 0.06	1827
Silicon	Si	14	7.51 ± 0.04	1823
Phosphorus	P	15	5.36 ± 0.04	1669
Sulfur	S	16	7.14 ± 0.05	(ancient)
Chlorine	Cl	17	5.50 ± 0.30	1774
Argon	Ar	18	[6.18 ± 0.08]	1894
Potassium	K	19	5.08 ± 0.07	1807
Calcium	Ca	20	6.31 ± 0.04	1808
Chromium	Cr	24	5.64 ± 0.10	1797
Manganese	Mn	25	5.39 ± 0.03	1774
Iron	Fe	26	7.45 ± 0.05	(ancient)
Nickel	Ni	28	6.23 ± 0.04	1751

[a] Logarithm of the abundance in the solar photosphere, normalized to hydrogen = 12.00, or an abundance of 1.00×10^{12}. Indirect solar estimates are marked with [ ]. The data are courtesy of Nicolas Grevesse, Université de Liège.

[b] Helium was discovered on the Sun in 1868, but it was not found on Earth until 1895.

(1824–1907) and the British astronomer Joseph Norman Lockyer (1826–1920) as emission lines observed during the solar eclipse of 18 August 1868. Since it seemed to be only found on the Sun, Lockyer named it after the Greek Sun god, Helios, who daily traveled across the sky in a chariot of fire drawn by four swift horses. In 1895, while analyzing a gas given off by a heated uranium mineral, the Scottish chemist William Ramsay (1852–1916) found the spectral signature of helium, thereby isolating it on the solid Earth 27 years after its discovery in the Sun.

Today, helium is used on Earth in a variety of ways, including inflating party balloons and in its liquid state keeping sensitive electronic equipment cold. But there isn't much helium left on the Earth, and we are in danger of running out of it soon.

1.7 CHILDREN OF THE COSMOS

We are made of the same atoms as the Sun. Our bodies, like the Sun, have more hydrogen atoms than any other, but we are composed of a somewhat larger proportion of heavier elements like carbon, nitrogen, and oxygen.

But do not discount the other stars. We are all true children of the stars, partially composed of materials that were forged within ancient stars before the Sun was born. All of the elements heavier than helium were generated long, long ago and far, far away in the nuclear crucibles of other stars, which lit up the night sky and were extinguished before the Solar System came into being.

These stars used up their internal fuel and spewed out their cosmic ashes with explosive force, ejecting the heavier elements into interstellar space. From this recycled material, the Sun, the Earth and we ourselves were formed. So, the carbon in your molecules, the calcium in your teeth, the oxygen in your water and the iron that reddens your blood all came from the interiors of other stars, long since exploded back into space in the death throes of these stars.

But all the hydrogen in the Earth's water and in your body, as well as all of the hydrogen in the stars and most of their internal helium, was synthesized about 14 billion years ago, in the Big Bang that jump-started the expanding Universe. We are thus children of both the stars that exploded during past eons and the Big Bang at the beginning of time.

1.8 DESCRIBING THE RADIATION

The Sun continuously radiates energy that spreads throughout space. This radiation is called "electromagnetic" because it propagates by the interplay of oscillating electrical and magnetic fields in space. Electromagnetic waves all travel through empty space at the same constant speed – the velocity of light. This velocity is usually denoted by the lower case letter c, and it has a value of roughly 300,000 kilometers per second – a more exact value is 299,793 kilometers per second. No energy can be transported more swiftly than this speed of light.

Sunlight has no way of marking time, and it can persist forever. As long as a ray of sunlight passes through empty space and encounters no atoms or electrons it will survive unchanged. Radiation emitted from the Sun today might therefore travel for all time in vacuous space, bringing its message forward to the end of the Universe.

Some of the radiation streaming away from the Sun is nevertheless intercepted at Earth, where astronomers describe it in terms of its wavelength, frequency, or energy. When light propagates from one place to another, it often seems to behave like waves or ripples on a pond (Fig. 1.8). The light waves have a characteristic wavelength, the separation between adjacent wave crests.

Different types of electromagnetic radiation differ in their wavelength (Fig. 1.9), although they propagate at the same speed. The electromagnetic waves entering your eye and those picked up by your radio antenna or used to X-ray your bones are similar except in relation to their wavelength.

X-rays are much smaller than an atom, with wavelengths that are between 10^{-11} and 10^{-8} meters long, and because of this small size, cosmic X-rays are totally absorbed

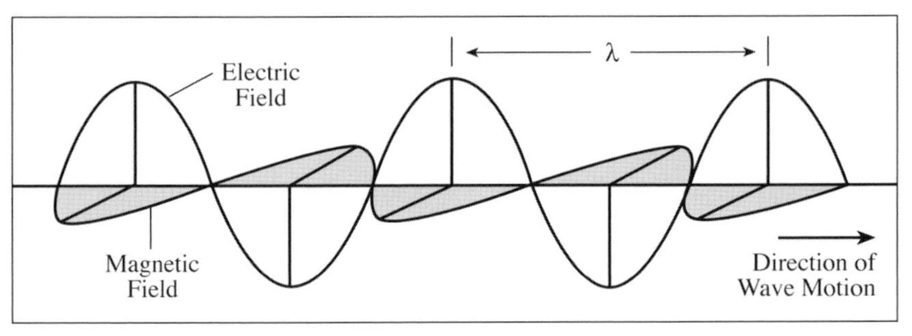

FIG. 1.8 Electromagnetic waves All forms of radiation consist of electric and magnetic fields that oscillate at right angles to each other and to the direction of travel. They move through empty space at the velocity of light. The separation between adjacent wave crests is called the wavelength of the radiation and is often designated by the lower case Greek letter lambda or λ.

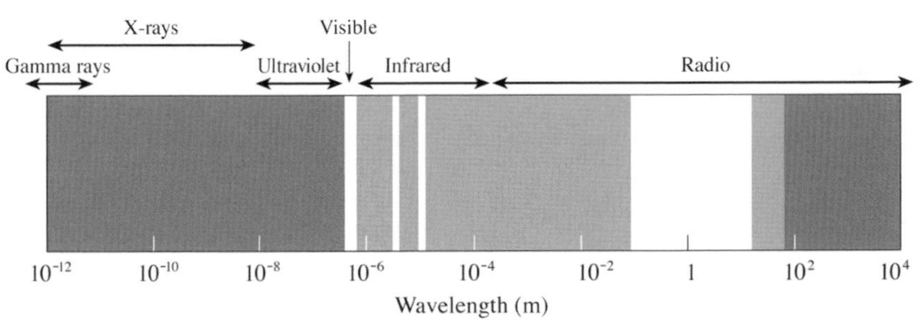

FIG. 1.9 Electromagnetic spectrum Radiation from the Sun and other cosmic objects is emitted at wavelengths from less than 10^{-12} meters to greater than 10^{4} meters. The visible spectrum is a very small portion of the entire range of wavelengths. The lighter the shading, the greater the transparency of the Earth's atmosphere. Solar radiation only penetrates to the Earth's surface at visible and radio wavelengths, respectively denoted by the narrow and broad white areas. Electromagnetic radiation at short X-ray and ultraviolet wavelengths, represented by the dark areas at the left, is absorbed in our air, so the Sun is now observed in these spectral regions from above the atmosphere in Earth-orbiting satellites.

in our atmosphere, never reaching the ground. The wavelength of ultraviolet radiation, which is also absorbed in our air, is just a bit longer, between 10^{-8} and 10^{-7} meters, with extreme ultraviolet radiation lying in the short wavelength part of this range. In contrast radio waves are between 0.001 and 30 meters long. So radio waves can be as big as you are tall, or even as large as a house or skyscraper, too long to enter the eye and not energetic enough to affect vision.

Just as a source of sound can vary in pitch, or wavelength, depending on its motion, the wavelength of electromagnetic radiation shifts when the object emitting or reflecting the radiation moves with respect to the observer (Fig. 1.10). This is called the Doppler effect, after the Austrian physicist Christian Johann Doppler (1803–1853), who discovered it in 1842. If the motion is toward the observer, the Doppler effect

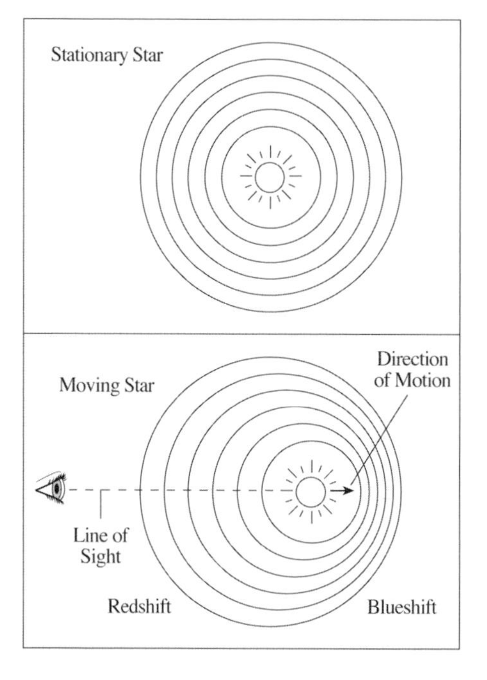

FIG. 1.10 Doppler effect A stationary star (*top*) emits regularly spaced light waves that get stretched out or scrunched up if the star moves (*bottom*). Here we show a star moving away (*bottom right*) from the observer (*bottom left*). The stretching of light waves that occurs when the star moves away from an observer along the line of sight is called a redshift, because red light waves are relatively long visible light waves; the compression of light waves that occurs when the star moves along the line of sight toward an observer is called a blueshift, because blue light waves are relatively short. The wavelength change, from the stationary to moving condition, is called the Doppler shift, and its size provides a measurement of radial velocity, or the velocity of the component of the star's motion along the line of sight. The Doppler effect is named after the Austrian physicist Christian Doppler (1803–1853), who first considered it in 1842.

shifts the radiation to shorter wavelengths, and when the motion is away the wavelength becomes longer. Such an effect is responsible for the change in the pitch of a passing ambulance siren. Because everything in the Universe moves, the Doppler effect is a very important tool for astronomers; it measures the velocity of that motion along the line of sight to the observer.

Sometimes radiation is described by its frequency instead of its wavelength. Radio stations are, for example, denoted by their call letters and the frequency of their broadcasts, usually in units of a million cycles per second, or megahertz, for FM broadcasts.

The frequency of a wave is the number of crests passing a stationary observer each second; the frequency therefore tells us how fast the radiation oscillates, or moves up and down. The product of wavelength and frequency equals the velocity of light, so when the wavelength increases the frequency decreases and *vice versa.*

When light is absorbed or emitted by atoms, it behaves like packages of energy, called photons, which can be created or destroyed. The photons are produced whenever a material object emits electromagnetic radiation, and they are consumed when matter absorbs radiation. Radiation therefore disappears and ceases to exist when absorbed by matter. But energy is neither created nor destroyed; it is just removed from the radiation.

Moreover, each elemental atom can only absorb and emit photon energy in specific amounts. This is a consequence of the unique arrangement of electrons in each atom, and the pattern of photon energy emitted and absorbed can therefore be used to identify the atom.

At the atomic level, the natural unit of energy is the electron volt, abbreviated eV. One electron volt is the energy an electron gains when it passes across the terminals of a 1-volt battery. A photon of visible light has an energy of about two electron volts, or 2 eV. Much higher energies are associated with nuclear processes; they are often specified in units of millions of electron volts, or MeV for short. A somewhat lower unit of energy is a thousand electron volts, called the kilo-electron volt or keV; it is often used to describe X-ray radiation.

The interaction of each type of radiation with matter depends on the energy of its photons, and from the standpoint of the astrophysicist this is the most important property distinguishing one type of radiation from another. In fact, astronomers often describe energetic radiation, such as X-rays or gamma rays, in terms of its energy rather than its wavelength or frequency.

Photon energy is inversely proportional to the wavelength and directly proportional to the frequency. Radiation with a shorter wavelength or a higher frequency therefore corresponds to photons with higher energy. Radio photons have relatively long wavelengths and low frequencies, so they have less energy than the short-wavelength, high frequency X-ray radiation. The low energies of radio photons cannot easily excite the atoms of our atmosphere, so these photons easily pass through the air, even in stormy weather. In contrast, X-rays are totally absorbed when traveling just a short distance through the atmosphere. The energetic X-rays produced by machines here on Earth pass right through your skin, muscles, or teeth. It also takes much less energy to broadcast a radio signal over short distances than to take an X-ray of a broken bone.

The energy of stellar radiation at a given wavelength can be related to the thermal energy, or the temperature, of the emitting gas. Hot stars tend to be bluer in color, for example, and colder stars are redder. This is because the most intense emission occurs at a radiation frequency and energy that increase with the temperature of the star's visible disk. In other words, the emitted power peaks out at a wavelength that varies inversely with the temperature, and this applies to all gaseous objects in the Cosmos. Thus, the cold, dark spaces between the stars radiate most intensely at the longer, invisible radio wavelengths, while a hot, million-degree gas emits most of its energy at short X-ray wavelengths that are also invisible.

1.9 INVISIBLE FIRES

There is more to the Sun than meets the eye! In addition to visible light, there are invisible gamma ray, X-ray, ultraviolet, infrared and radio waves. The whole solar spectrum extends from short gamma rays: that are comparable to the size of an atom's nucleus, to long radio waves that are as broad as a mountain; and the Sun is so bright that it can be examined with precision in every spectral region. Observations at these invisible wavelengths have indeed broadened and sharpened our vision of the Sun.

However, our atmosphere effectively blocks most forms of invisible radiation including ultraviolet and X-ray radiation. Radio waves are the only kind of invisible radiation that is not absorbed in the Earth's atmosphere, so radio astronomy provided the first new window on the Sun.

FIG. 1.11 Very Large Array Each of the 27 radio telescopes of the Very Large Array, abbreviated VLA, measures 25 meters in diameter, or about the size of house, and weighs 235 tons (2.35×10^5 kilograms). These telescopes are placed along the arms of a Y-shaped array on a desert near Socorro, New Mexico, and interconnected electronically to provide a total of 351 pairs of telescopes. The telescopes can be rolled along tracks to change their configuration and create a radio zoom lens. When the telescopes are pushed to the outer ends of each arm and their output combined in a computer, the VLA creates a radio telescope with a diameter as large as 34 kilometers and an angular resolution that can be smaller than 1 second of arc. (Courtesy of NRAO, AUI and NSF.)

Astronomers use conventional radio telescopes to observe the Sun (Fig. 1.11); but radio telescopes do not really look at the Sun, they listen to it. Such telescopes usually have a metallic, dish-shaped, or parabolic, reflector that focuses the radio waves at a receiver. The long, straight antenna on your automobile or home radio similarly intercepts radio signals. Moreover, the Sun is the brightest, noisiest radio object in the sky, and because the atmosphere does not distort radio signals we can observe the radio Sun on a cloudy day, just as your home radio works even when it rains or snows outside.

To look at the Sun through windows other than the radio or visible ones, we must loft telescopes above the atmosphere. This was done first by using balloons and sounding rockets, and then with satellites that orbit the Earth above the atmosphere. Satellite-borne telescopes now view the Sun at ultraviolet and X-ray wavelengths, above the Earth's absorbing atmosphere at places where night can be brief or non-existent.

Thus, astronomers now have new ways to extract previously unobtainable information about the Sun. They are aided by new telescopes and sophisticated computers that gather in an increasing wealth of unsuspected information. Computerized telescopes now operate, from the ground and in orbit, in each of the invisible domains of the electromagnetic spectrum, creating images that provide new insight to the Sun.

Much of this book describes the invisible Sun, an unseen world of perpetual change and cosmic violence that lies outside the visible solar disk. And as the title *Sun, Earth and Sky* suggests, our book also describes the Sun's interaction with planet Earth, mainly through invisible radiation and tiny, energetic particles that cannot be seen. It involves a global, space-age perspective that looks up at the Sun and down at the Earth, at both visible and invisible wavelengths, or directly samples the space outside Earth with orbiting satellites.

But before we begin our journey through these largely invisible realms, there is an equally fascinating world that lies hidden below everything we can see on the Sun; clever techniques are required to perceive this unseen interior of the Sun.

Chapter Two

Energizing the Sun

IMPRESSION OF THE RISING SUN This portrayal of sunrise by the French artist Claude Monet (1840–1926) gave the Impressionist Movement its name. (Courtesy of the Musée Marmottan, Paris.)

2.1 AWESOME POWER, ENORMOUS TIMES

The Sun is extremely massive, containing 99.9 percent of the total mass of the Solar System. The gravitational pull of the Sun's large mass controls the movements of everything in its vicinity. That is the reason the planets belong to the Solar System; their motion is dominated by the Sun.

Our home star is also very big, with a radius of about 109 times that of the Earth. And such a large, hot object must shine brightly.

The Sun's luminosity is diluted by the square of the distance as it spreads out into the increasing volume of space. By measuring the amount of solar radiation intercepted by the Earth, at a mean distance of about 150 million kilometers from the Sun, we can extrapolate back to infer the total energy emitted by the Sun every second. It radiates a power of 380 million billion billion watts, or 3.8×10^{26} watts, which is known as the absolute luminosity of the Sun and designated by the symbol $L_{\odot}$ (Table 2.1).

So the Sun is relentlessly losing its energy, radiating it away at an enormous rate. In just one second, the Sun emits more energy than humans have used since the beginning of civilization. Its' fire is too brilliant to be perpetually sustained; after all, nothing can stay hot forever, and all things wear out with time.

Why does the Sun stay hot, and how long has it been shining? A normal fire of the Sun's intensity would soon burn out. That is, no ordinary fire can maintain the Sun's steady supply of heat for long periods of time. If the Sun were composed entirely of coal, with enough oxygen to sustain combustion, it would be burned away and totally consumed in a few thousand years.

In the mid-nineteenth century, the German physicist and physiologist Hermann von Helmholtz (1821–1894) proposed that the origin of the Sun's radiated energy is the gravitational contraction of its large mass. If the Sun were gradually shrinking, the infalling matter would heat the solar gases to incandescence, just as the air inside a tire pump warms when it is compressed; in more scientific terms, the Sun's gravitational energy would be slowly converted into the kinetic energy of motion and heat the Sun up. Helmholtz also made the first precise formulation of the principle of conservation of energy, in which the total energy of a system and its surroundings remain constant even if it may be changed from one form of energy to another.

The Irish physicist William Thomson (1824–1907), later Lord Kelvin, then showed that the Sun could have illuminated the Earth at its present rate for about 100 million years by slowly contracting. In his article entitled "On the Age of the Sun's Heat", published in 1862, Thomson wrote:

> It seems, therefore, on the whole most probable that the Sun has not illuminated the Earth for 100 million years, and almost certain that he has not done so for 500 million years. As for the future, we may say, with equal certainty, that inhabitants of the Earth cannot continue to enjoy the light and heat essential to their life, for many million years longer, unless sources now unknown to us are prepared in the great storehouse of creation.[5]

Thomson incidentally developed the scale of temperature that starts from absolute zero – the temperature at which atoms and molecules cease to move and have no kinetic energy; it is now known as the Kelvin temperature scale and is widely used

TABLE 2.1 The Sun's Physical Properties[a]

Property	Value	
Mean distance, AU	$1.495\,978\,7 \times 10^{11}$ m, and about 150 million km	
(from radar measurements of the distance to Venus and Kepler's third law)		
Light travel time from Sun to Earth	499 seconds	
Radius, $R_{\odot}$	6.955×10^{8} meters (109 Earth radii)	
(from distance and angular extent)		
Volume	1.412×10^{27} m^{3} (1.3 million Earths)	
Mass, $M_{\odot}$	1.989×10^{30} kg (332,946 Earth masses)	
(from distance and Earth's orbital period using Kepler's third law)		
Escape velocity at photosphere	6.178×10^{5} m s^{-1}	
Mean density	1,409 kg m^{-3}	
Solar constant, $f_{\odot}$	1,366 J s^{-1} m^{-2} = 1,366 W m^{-2}	
Luminosity, $L_{\odot}$	3.854×10^{26} J s^{-1} = 3.854×10^{26} W	
(from solar constant, distance and Stefan-Boltzmann law)		
Principal chemical constituents *(from analysis of Fraunhofer lines)*	(By Number Of Atoms)	(By Mass Fraction)
Hydrogen	92.1 percent	X = 70.68 percent
Helium	7.8 percent	Y = 27.43 percent
All others	0.1 percent	Z = 1.89 percent
Age	4.6 billion years	
(from age of oldest meteorites)		
Density (center)	151,300 kg m^{-3}	
Pressure (center)	2.334×10^{11} bars	
Pressure (photosphere)	0.0001 bar	
Temperature (center)	1.56×10^{7} K	
Temperature (photosphere)	5,780 K	
Temperature (chromosphere)	6×10^{3} to 2×10^{4} K	
Temperature (transition region)	2×10^{4} to 2×10^{6} K	
Temperature (corona)	2×10^{6} to 3×10^{6} K	
Rotation Period (equator)	26.8 days	
Rotation Period (30° latitude)	28.2 days	
Rotation Period (60° latitude)	30.8 days	
Magnetic Field (sunspots)	0.1 to 0.4 T = 1×10^{3} to 4×10^{3} G	
(from Zeeman effect)		
Magnetic Field (polar)	0.001 T = 10 G	

[a] Mass density is given in kilograms per cubic meter, denoted kg m^{-3}; the density of water is 1,000 kg m^{-3}. The unit of pressure is bars, where 1.013 bar is the pressure of the Earth's atmosphere at sea level. The unit of luminosity is joule per second, power is often expressed in watts, where 1.0 watt = 1.0 Joule per second.

by astronomers. The unit for this scale is written kelvin, without a capital K, or just denoted by a capital K. Water freezes at 273 K and boils at 373 K, and to convert to degrees Celsius, abbreviated by C, just subtract 273, or $C = K - 273$. The conversion to degrees Fahrenheit, denoted by F, is more complicated, with $F = (9K/5) - 459.4$.

During the ensuing decades, radioactivity was discovered, leading to the realization that the Earth's rocks are older than Thomson's value for the age of the Sun's heat. This paradox was resolved when scientists discovered that nuclear fusion powers the Sun, providing the ultimate source of stellar energy.

Most of the matter on Earth is completely stable, but some atoms are unstable. Such radioactive atoms, like uranium, spontaneously change form when their nucleus hurls out energetic particles, radiates energy and relaxes to a less energetic state, forming a lighter, stable atom, like lead, in the process. This nuclear transformation can be used to determine the age of the rocks on the Earth's surface.

The radioactive dating method is something like determining how long a log has been burning by measuring the amount of ash and waiting a while to determine how rapidly the ash is being produced. Except you do not need to know the total amount of radioactive ash. The abundance ratio of stable decay atoms to their unstable parents, such as the relative amounts of lead and uranium, can be used with the known rate of radioactive decay to determine the time that has elapsed since the rocks were formed. This technique indicates that the oldest known rocks in the Solar System, from the Moon and meteorites, were formed 4.6 billion years ago when we think the Sun originated together with the planets and their moons.

Fossils of primitive creatures are found etched in rocks more than 3.5 billion years old, so the Sun was apparently warm enough to sustain life back then. Unusual powers must be at work to make the Sun shine so hotly for so long. Indeed, the only known process that can fuel the Sun's fire at the presently observed rates for billions of years involves nuclear fusion in the Sun's hot, dense core.

2.2 THE SUN'S CENTRAL PRESSURE COOKER

Most of the Earth is solid, and we can therefore walk on its surface. By contrast, all of the Sun is a gas, and it has no surface.

Under the extreme conditions within the Sun, the gaseous atoms lose their identity! The atoms move rapidly here and there, colliding with each other at high speeds; the violent force of these collisions is enough to fragment the atoms into their constituent pieces. The interior of the Sun therefore consists mainly of the nuclei of hydrogen atoms, called protons, and unattached electrons that have been torn off the atoms by innumerable collisions and set free to move throughout the Sun.

Negatively charged electrons neutralize the positively charged protons, so the mixture of electrons and protons, called plasma, has no net charge. But every particle in the Sun's plasma is charged, and therefore electrically conducting. And like any electrically charged object, the solar material generates magnetic fields when it moves.

The entire Sun is nothing but a giant, hot ball of plasma. Plasma has been called the fourth state of matter to distinguish it from the gaseous, liquid, and solid ones. A candle flame is plasma, as are all the stars in the Universe.

With their electrons gone, the hydrogen nuclei, or protons, can be packed together much more tightly than normal atoms. This is because nuclei are about 100,000 times smaller than the atoms they normally occupy. The bare nuclei can be squeezed together within the empty space of former atoms.

To understand the Sun's interior, imagine a hundred mattresses stacked into a pile. The mattresses at the bottom must support those above, so they will be squeezed thin. Those at the top have little weight to carry, and they retain their original thickness. The nuclei at the center of the Sun are similarly squeezed into a smaller volume by the overlying material, so they become hotter and more densely concentrated.

Deep down inside, within the dense, central core, the Sun's temperature has risen to 15.6 million kelvin, and the gas density is greater than 10 times that of solid lead. Such extreme central conditions were recognized as long ago as 1870 when Jonathan Homer Lane (1819–1880), an American astrophysicist at the U.S. Patent Office, assumed that gas pressure supports the weight of the Sun. As the result of such crowding, the nuclei in the Sun's center collide more frequently with higher speeds than elsewhere in the Sun, and push more vigorously outward. This pushing is called gas pressure, and it is the force that keeps the Sun from collapsing.

At the center of the Sun, the gas pressure needed to resist the weight of the overlying gas is 233 billion times the pressure of our atmosphere at sea level. The high-speed motions and collisions of particles with temperatures of 15.6 million degrees provide the enormous central pressure.

At greater distances from the center, there is less overlying material to support and the compression is less, so the plasma gets thinner and cooler (Fig. 2.1). Halfway from the center of the Sun to the surface, the density is the same as that of water, and about nine tenths of the distance from the center to the Sun's apparent edge, we find material as tenuous as the transparent air that we breathe on Earth.

At the visible solar disk, the rarefied gas is about one thousand times less dense than our air, the pressure is less than that beneath the foot of a spider, and the temperature has fallen to 5,780 kelvin. Any hot gas with the radius of the Sun and a disk temperature of 5,780 kelvin will emit the Sun's luminosity.

2.3 NUCLEAR FUSION, ANTI-MATTER AND HYDROGEN BURNING

The extraordinary conditions within the center of the Sun provided one clue to the mysterious process that keeps the Sun hot and makes it shine. Other important evidence was accumulated at the Cavendish Laboratory at Cambridge University in England, where the New Zealand-born British physicist Ernest Rutherford (1871–1937) showed, in 1920, that the massive nuclei of all atoms are composed of hydrogen nuclei, which he named protons. In the previous year the English chemist Francis W. Aston (1877–1945) invented the mass spectrograph and used it to show that the mass of the helium nucleus is slightly less than the sum of the masses of the four hydrogen nuclei, or protons, which enter into it.

At the same time that Rutherford and Astron were discovering the inner secrets of the atoms, the British astronomer, Arthur Stanley Eddington (1882–1944), also at Cambridge University, was trying to understand the internal workings of the Sun and other stars. Eddington, an avid reader of mystery novels, once likened the process to analyzing the clues in a crime. He knew that certain elements can be transformed into other ones in the terrestrial laboratory, and reasoned that stars are the crucibles in which the elements are made. He further realized that such stellar alchemy would release energy, arguing that hydrogen is transformed into helium inside stars, with the resultant mass difference released as energy to power the Sun.

Eddington could therefore lay the foundation for solving the Sun's energy crisis, concluding in a paper entitled "The Internal Constitution of the Stars" written in 1920 that:

> What is possible in the Cavendish Laboratory may not be too difficult in the Sun.... The reservoir [of a star's energy] can scarcely be other than the subatomic energy.... There is sufficient subatomic energy in the Sun to maintain its output of heat for 15 billion years.[6]

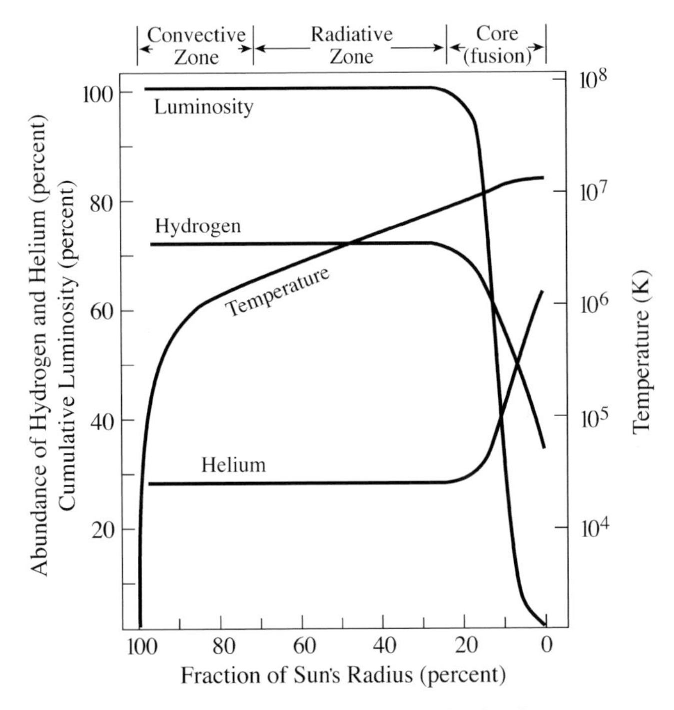

FIG. 2.1 Internal compression The Sun's luminosity, temperature, and composition all vary with depth in its interior, from the Sun's visible disk (*left*) to the center of the Sun (*right*). The nuclei are squeezed into a smaller volume by the pressure of the material above, becoming hotter and more densely concentrated at greater depths. At the Sun's center, the temperature has reached 15.6 million kelvin, and the pressure is 233 billion times that of the Earth's atmosphere at sea level. Nuclear fusion in the Sun's energy-generating core synthesizes helium from hydrogen, so this region contains more helium and less hydrogen than the primordial amounts detected in the light of the visible Sun.

In the same article, he continued with the prescient statement that:

> If indeed, the subatomic energy in the stars is being freely used to maintain their great furnaces, it seems to bring a little nearer to fulfillment our dream of controlling this latent power for the well-being of the human race – or for its suicide.[7]

Great ideas have a curious way of surfacing in different places at about the same time. In an essay entitled "Atomes et Lumière, or Atoms and Light", the French physicist Jean Baptiste Perrin (1870–1942) argued that "radioactive" transformation of the elements could maintain the Sun's luminous output at its present rate for several billion years, or perhaps several dozen billion years, and that the mass lost during the transformation of four hydrogen nuclei into one helium nucleus would supply energy.[8]

It was probably Eddington who convinced most astronomers that subatomic (that is, nuclear) energy must fuel the stars. During the ensuing decade it was realized that the lightest known element, hydrogen, is the most abundant element in the Sun, so hydrogen nuclei, or protons, must play the dominant role in nuclear reactions within our daytime star. Protons must somehow be fused together, forming helium nuclei, but the details were lacking. After all, Rutherford had only shown that atomic nuclei contain protons the same year as Eddington's historic article, and the neutron wasn't discovered until 1932.

Physicists were nevertheless convinced that protons could not react with each other inside the Sun. The proton is positively charged and positive charges repel each other. The force of repulsion between like charges becomes larger and larger as they are brought closer and closer. And even at the enormous central temperature of the Sun, the protons did not seem to have enough energy to overcome this electrical repulsion and merge together. In other words, the Sun's core did not seem hot enough to permit protons to fuse together.

But Eddington was certain that subatomic energy fueled the stars, and in the mid-1920s retorted defiantly that:

> The helium that we handle must have been put together at some time and some place. We do not argue with the critic who urges that the stars are not hot enough for this process; we tell him to go and find a *hotter place*.[9]

As it turned out, Eddington was right and the physicists were wrong.

The paradox was resolved after the Russian physicist George Gamow (1904–1968) explained why the nuclei of radioactive substances are releasing energetic particles. He used the uncertain, probabilistic nature of the very small, adopting the quantum theory to show in 1928 how a subatomic particle could be anywhere, everywhere, and nowhere at all. According to quantum theory, a very small particle does not have a well-defined position, and instead acts like a spread-out thing, existing in a murky state of possibility with a set of probabilities of being in a range of places.

As a result of this location uncertainty, a subatomic particle's sphere of influence is larger than was previously thought. It might be anywhere, although with decreasing probability at regions far from the most likely location. This explains the escape of fast-moving, energetic particles from the nuclei of radioactive atoms like uranium; these particles usually lack the energy to overcome the nuclear barrier, but some of them have a small probability of escaping to the outside world. In this surreal world of subatomic probability, one could relentlessly throw a ball against a wall, watching it bounce back countless times, until eventually the ball would tunnel under the wall, effectively passing through it. As Ahab said in Herman Melville's (1819–1891) *Moby Dick,*

> How can the prisoner reach outside except by thrusting through the wall?[10]

A similar tunneling process, or barrier penetration, occurs the other way around at the center of stars like the Sun. It means that a proton has a very small but finite chance of occasionally moving close enough to another proton to overcome the barrier of repulsion and tunnel through it. Protons therefore sometimes get close enough to fuse together, even though their average energy is well below that required to overcome their electrical repulsion.

But this bizarre tunneling reaction doesn't occur all the time. For fusion to occur, the collision must still be almost exactly head-on, and between exceptionally fast protons. Nuclear reactions therefore proceed very slowly inside the Sun, and it is a good thing. If the temperature were high enough to permit frequent fusion, the Sun would blow up! After all, similar nuclear processes produce the explosive energy in hydrogen bombs.

Unlike a bomb, the temperature-sensitive reactions inside the Sun act like a thermostat, releasing energy in a steady, controlled fashion at exactly the rate needed to keep the Sun in equilibrium. If a star shrinks a little and gets hotter inside, more nuclear energy is generated, making the star expand and restoring it to the original temperature. If the Sun expanded slightly, and became cooler inside, subatomic energy would be released at a slower rate, making the Sun shrink again and restoring equilibrium.

So, we now know that the Sun shines by nuclear fusion, whereby hydrogen nuclei, or protons, fuse together into helium nuclei, also known as alpha particles. The detailed sequence of nuclear reactions is known as the proton-proton chain (Fig. 2.2), since it begins by the fusion of two protons.

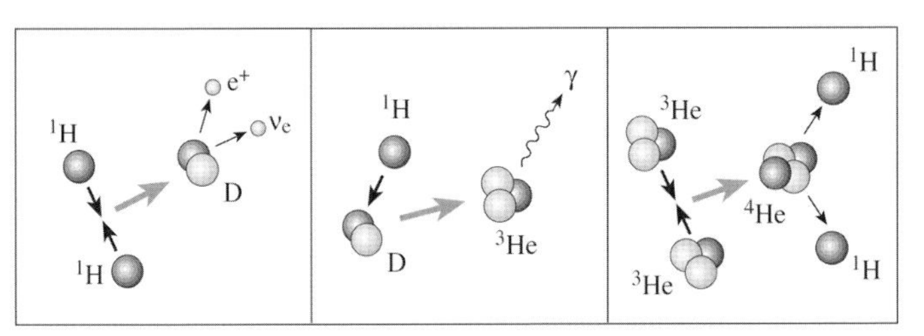

FIG. 2.2 The proton-proton chain The Sun gets its energy when hydrogen nuclei are fused together to form helium nuclei within the solar core. This hydrogen burning is described by a sequence of nuclear fusion reactions called the proton-proton chain. It begins when two protons, here designated by the symbol ^1H, combine to form the nucleus of a deuterium atom, the deuteron that is denoted by D, together with the emission of a positron, e^+, and an electron neutrino, ν_e. Another proton collides with the deuteron to make a light nuclear isotope of helium, ^3He, and then a nucleus of normal heavy helium, ^4He, is formed by the fusion of two light ^3He nuclei, returning two protons to the gas. Overall, this chain successively fuses four protons together to make one helium nucleus. Even in the hot, dense core of the Sun, only rare, fast-moving particles are able to take advantage of the tunnel effect and fuse in this way.

At the suggestion of the German physicist Carl Friedrich von Weizsacker (1912–), the German-born American physicist Hans A. Bethe (1906–2005) investigated the fusion of two protons. And then Gamow, who had defected to the United States, suggested to one of his graduate students, Charles Critchfield (1910–1994), that he calculate the details of the proton-proton reaction. These results were sent to Bethe, who found them correct, and in 1938 the two published a joint paper entitled "The Formation of Deuterons by Proton Combination."

Later that year, Gamow organized a conference in Washington, D.C. to bring astronomers and physicists together to discuss the problem of stellar energy generation. The Swedish astronomer Bengt Strömgren (1908–1987) reported that since the Sun was predominantly hydrogen it would have a central temperature of about 15 million kelvin, rather than 40 million as estimated by Eddington under the assumption that the Sun had about the same chemical composition as the Earth. The lower temperature meant that the calculations of Bethe and Critchfiled correctly predicted the Sun's luminosity, and Bethe, who attended the conference, was able to explain just how the Sun shines by the proton-proton chain (Focus 2.1), while also showing how more massive stars could use carbon as a catalyst in burning hydrogen into helium. In 1967 Bethe was awarded the Nobel Prize in Physics for his discoveries concerning energy production in stars.

Also in 1938, Weizsäcker (1912–) independently investigated the proton-proton chain, and showed how other nuclear reactions could fuel stars that are more massive than the Sun using carbon as a catalyst in the synthesis of helium from hydrogen. Bethe soon went to Los Alamos, New Mexico, to use his knowledge of nuclear physics in support of the construction of the first atomic bomb, and at about the same time Weizsäcker helped the Germans investigate the feasibility of constructing nuclear weapons. The Americans developed the bomb first, bringing an end to World War II (1939–1945).

In the burning of hydrogen, four protons are united, but two of them have to be changed into neutrons. This is because the helium nucleus consists of two protons and two neutrons. Something must be carrying away the charge of the proton, leaving behind a neutron with little change of mass; that mysterious agent is the anti-particle of the electron.

In 1931, Paul Adrien Maurice "P. A. M." Dirac (1902–1984), then at Cambridge University, predicted the existence of anti-matter. For Dirac, mathematical beauty was the most important aspect of any physical law describing nature. He noticed that the equations that describe the electron have two solutions. Only one of them was needed to characterize the electron; the other solution specified a sort of mirror image of the electron – an anti-particle, now called the positron for "positive-electron."

Dirac's trust in the beauty and symmetry of his equations led him to predict:

> A new kind of particle, unknown to experimental physics, having the same mass and opposite charge to an electron.[11]

The American physicist Carl D. Anderson (1905–1991) discovered the then-unknown positron in 1932, when studying high-energy particles from space called cosmic rays; they create positrons and many other subatomic particles when colliding with nuclei in

FOCUS 2.1
Proton-Proton Chain

The Sun shines by a sequence of nuclear reactions, called the proton-proton chain, in which four protons are fused together to form a helium nucleus that contains two protons and two neutrons. Each nuclear transformation releases 25 MeV, or 0.000 000 000 004, or 4×10^{-12}, Joules of energy. This is due to the fact that the mass of the resulting helium nucleus is slightly less (a mere 0.007 or 0.7 percent) than the mass of the four protons that formed it, and the missing mass appears as energy.

The energy content of the lost mass is given by $E = \Delta mc^2$. Because the velocity of light, c, is a very large number, the annihilation of relatively small amounts of mass, Δm, produces large quantities of energy, E. Moreover, that energy is multiplied by the huge number of reactions that occur inside the Sun every second. Roughly 100 trillion trillion trillion, or 10^{38}, helium nuclei are created every second, resulting in a total mass loss of 5 million tons, or 5 billion kilograms, per second, which is enough to keep the Sun shining with its present luminous output of 385.4 million billion billion, or 3.854×10^{26}, watts, where a power of one watt is equal to an energy loss of one Joule every second.

The details of this proton-proton chain were first described by Han A. Bethe (1906–2005) in a paper entitled "Energy Production in Stars," published in 1939. In the first step of the proton-proton chain, two protons, each designated by either ^1H or p, are united to form a deuteron, D, the nucleus of a heavy form of hydrogen known as deuterium. Since a deuteron consists of one proton and one neutron, one of the protons entering the reaction must be transformed into a neutron, emitting a positron, e^+, to carry away the proton's charge, together with a low-energy electron neutrino, ν_e, to balance the energy in the reaction. A positron is the anti-matter particle of the electron. This initiating proton-proton reaction is written:

$$p + p \rightarrow D + e^+ + \nu_e.$$

Each proton inside the Sun is involved in a collision with other protons millions of times in every second, but only exceptionally hot ones are able to tunnel through their electrical repulsion and fuse together. Just one collision in every ten trillion trillion initiates the proton-proton chain.

The electron neutrinos escape from the Sun without reacting with matter, carrying energy away. And every positron is immediately annihilated when colliding with an electron, e^-, producing energetic radiation at short gamma-ray wavelengths, γ. This energy-producing interaction can be written as:

$$e^- + e^+ \rightarrow 2\gamma .$$

The next step follows with little delay. In less than a second the deuteron collides with another proton to form a nucleus of light helium, He^3, and releases yet another gamma ray.

$$D + p \rightarrow He^3 + \gamma.$$

This reaction occurs so easily that deuterium cannot be synthesized inside stars; it is quickly consumed to make heavier elements.

In the final part of the proton-proton chain, two such light helium nuclei meet and fuse together to form a nucleus of normal heavy helium, He^4, returning two protons to the solar gas; this step takes about a million years on average.

$$He^3 + He^3 \rightarrow He^4 + 2p.$$

This normal helium nucleus contains two protons and two neutrons.

A total of six protons are required to produce the two He^3 nuclei that go into this last reaction. Since two protons and a helium nucleus are produced, the net result of the proton-proton chain is:

$$4p \rightarrow H_e^4 + \text{gamma-ray radiation} + 2 \text{ neutrinos}.$$

our outer atmosphere. At the time of his discovery, Anderson was unaware of Dirac's theoretical prediction of the positron. In 1933 Dirac was awarded the Nobel Prize in Physics for his prediction; Anderson received the honor in 1936 for the discovery of the positron, in the same year as the Austrian physicist Victor Hess (1883–1964) who shared the prize for his discovery of cosmic rays.

Once created, anti-matter does not stay around for very long. And it is a good thing, for we occupy a material world, and any anti-matter will self-destruct when it encounters ordinary matter. The singer Madonna expressed it with a different connotation:

> We are living in a material world
> And I am a material girl.[12]

When an electron and positron meet, they annihilate each other and disappear in a puff of energetic radiation. This is how some of the Sun's nuclear energy is converted into radiation.

2.4 DILUTING THE RADIATION

All the Sun's nuclear energy is released deep down inside its high-temperature core, and no energy is created in the cooler regions outside it. The energy-generating core extends to about one quarter of the distance from the center of the Sun to the visible disk, accounting for only 1.6 percent of the Sun's volume. But about half the Sun's mass is packed into its dense core.

Because we cannot see inside the Sun, astronomers combine basic theoretical equations, such as those for equilibrium and energy generation or transport, with observed boundary conditions, such as the Sun's mass and luminous output, to create models of the Sun's internal structure. These models consist of two nested spherical shells that surround the hot, dense core, like Russian dolls (Fig. 2.3). In the innermost shell, called the radiative zone, energy is transported by radiation; it reaches out from the core to 71.3 percent of the distance from the center of the Sun to the visible solar disk. The radiative zone is encompassed by a higher layer known as the convective zone, where turbulent motion, called convection, transports energy.

Even though light is the fastest thing around, radiation does not move quickly from the center of the Sun to its visible surface. The energy made inside the Sun's pressurized core slowly trickles out to finally escape as the light we see.

The solar core is so dense that a single gamma ray produced by nuclear fusion at the center of the Sun cannot move even a fraction of a millimeter before banging into a subatomic particle, where the radiation is scattered or absorbed and re-emitted with less energy. This radiation quickly collides with another particle in the radiative zone, and is eventually re-radiated at yet lower energy. The process continues over and over again countless times, as the radiation moves outward on a haphazard, zigzag path, steadily losing energy at each encounter.

As a result of this continued ricocheting and innumerable collisions in the radiative zone, it takes about 170 thousand years, on the average, for radiation to work its way out from the Sun's core to the bottom of the convective zone. By this time, the radiation has shed so much energy that it emerges on the other side of the convective

FIG. 2.3 Anatomy of the Sun The Sun is powered by the nuclear fusion of hydrogen in its core at a central temperature of 15.6 million kelvin. Energy produced by these fusion reactions is transported outward, first by countless absorptions and emissions within the radiative zone, and then by turbulent motion in the outer convective zone. The visible disk of the Sun, called the photosphere, has a temperature of 5,780 kelvin. Just above the photosphere is the thin chromosphere and a transition region to the rarefied, million-degree outer atmosphere of the Sun. They are represented by an image at the extreme ultraviolet wavelength of 30.38 nanometers wavelength, emitted by singly ionized helium, denoted He II, at a temperature of 60,000 kelvin, taken from the Extreme-ultraviolet Imaging Telescope, abbreviated EIT, aboard the *SOlar and Heliospheric Observatory*, or *SOHO* for short. (Courtesy of the *SOHO* EIT consortium and NASA. *SOHO* is a project of international cooperation between ESA and NASA.)

zone as relatively unenergetic visible light. In contrast, sunlight moves freely through interplanetary space, taking only eight minutes to travel from the Sun to the Earth.

2.5 CONVECTION AND GRANULATION

Radiation streaming out from the Sun's energy-generating core is stopped at the bottom of the convective zone, which occupies the outer 28.7 percent of the solar radius. In this region, the relatively cool, opaque solar gas absorbs great quantities of radiation without re-emitting it. This causes the material to become hotter than it would otherwise be, and the Sun must find another way to release the pent-up energy.

The heated material expands, becomes less dense than the gas in overlying layers, and rises to the visible disk of the Sun in roughly ten days. The hot material then cools by radiating sunlight into space, and sinks back down to become reheated and rise again. The roiling currents of hot and cool gas create a churning, wheeling motion that carries heat from the bottom to the top, like a boiling pot of water or other liquid that is similarly heated from below (Fig. 2.4).

In 1801 the English astronomer William Herschel (1738–1822) noticed that the Sun has a granular appearance. More than a century later, high-resolution photographs revealed that this granulation is composed of closely packed cells having bright centers surrounded by dark lanes. They mark the top of the convective zone.

Recent images of the Sun's white light, or all the colors combined, resolve the granules when taken under conditions of excellent seeing (Fig. 2.5). The mean angular distance between the bright centers of adjacent granules is about 2.0 seconds of arc, corresponding to about 1,500 kilometers at the Sun. That seems very large, but an individual granule is about the smallest thing you can see on the Sun when peering through the Earth's turbulent atmosphere.

There are at least a million granules on the visible solar disk at any moment, exhibiting a non-stationary, overturning motion. The bright center of each granule, or convection cell, is the highest point of a rising column of hot gas. The dark edges

FIG. 2.4 Benard Convection When a gas or liquid is heated from below, convection takes place in vertical cells, known as Benard cells. Warmer material rises at the centers of the cells and cooler material falls around their boundaries. This figure shows the hexagonal convection pattern in a layer of silicone oil that is heated uniformly from below and exposed to ambient air above; light reflected from aluminum flakes shows fluid rising at the center of each cell and descending at the edges. The regular motion, long duration, and polygonal shapes of these cells are somewhat distorted in the Sun's turbulent convective zone. (Courtesy of Manuel G. Velarde and M. Yuste, Universidad Nacional de Educacion a Distancia, Madrid, Spain.)

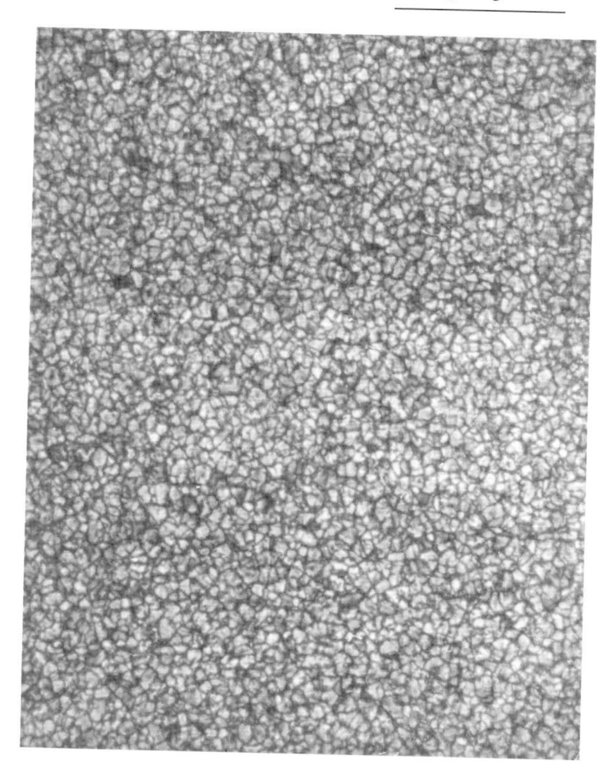

FIG. 2.5 Double, Double Toil and Trouble When optical telescopes zoom in and take a detailed look at the visible Sun, they resolve a strongly-textured granular pattern. Hot granules, each about 1,500 kilometers across, rise at speeds of 500 meters per second, like supersonic bubbles in an immense, boiling cauldron. The rising granules burst apart, liberating their energy, and cool material then sinks downwards along the dark, intergranular lanes. This photograph was taken at 430.8 nanometers with a 1-nanometer interference filter. It has an exceptional angular resolution of 0.2 seconds of arc, or 150 kilometers at the Sun. (Courtesy of Richard Muller and Thierry Roudier, Observatoire du Pic-du-Midi et de Toulouse.)

of each granule are the cooled gas, which sinks because it is denser than the hotter gas. And each individual granule lasts only about 15 minutes before it is replaced by another one, never reappearing in precisely the same location.

The granules are superimposed on a larger cellular pattern, called the supergranulation. Unlike the granules, which move up and down, the supergranulation is detected as a pattern of horizontal flow across the solar disk, and it remains invisible in white-light images that show the granules. By subtracting long- and short-wavelength images of the Sun in 1960, the American physicist Robert Leighton (1919–1997) discovered the supergranulation, a pattern of convection cells that are about 30,000 kilometers across, or about twice the size of the Earth. Roughly 2,500 supergranules are seen on the visible solar disk, each persisting for one or two days (Fig. 2.6).

Material in each supergranule cell rises in the center, moves away from the center with a typical velocity of about 400 meters per second, and sinks down again at the cell boundary. The Sun's magnetic field is carried along with this flow, piling up at the supergranular cell edges and creating a network of concentrated magnetic field.

Modern instruments aboard the *SOlar and Heliospheric Observatory,* abbreviated *SOHO,* have shown that the supergranulation outflow is relatively shallow, disappearing at depths of about 5,000 kilometers or about one sixth of a cell diameter. Moreover, the supergranulation undergoes oscillations and supports waves that move across the Sun, resembling the "wave" of spectators at a baseball, football or soccer game. Since the supergranulation wave moves in the same direction as the Sun rotates, the supergranulation appears to rotate faster around the Sun than the other visible solar gas and magnetic features like sunspots.

The visible disk of the Sun, with its granulation and supergranulation, caps the convective zone and completes our current model of the solar interior. As we shall next see, its ingredients and dimensions change dramatically over the eons, on cosmic time scales.

2.6 THE SUN'S REMOTE PAST AND DISTANT FUTURE

So how long has the Sun been shining, and how long will it keep on doing so? Nothing in the Cosmos is fixed and unchanging, and nothing escapes the ravages of time. Everything moves and evolves, and that includes the seemingly constant and unchanging Sun. It formed together with the planets about 4.6 billion years ago, when a spinning interstellar cloud of dust and gas fell in on itself. The center got denser and denser, until it became so packed, so tight and so hot that protons came together and fused into helium, making the Sun glow. Ever since then, the Sun has slowly grown in luminous intensity with age, a steady, inexorable brightening that is a consequence of the increasing amounts of helium accumulating in the Sun's core.

Although the Sun is consuming itself at a prodigal rate, the loss of material is insignificant in comparison with its total mass. The mass of the Sun is two thousand trillion trillion tons or 2×10^{30} kilograms, about a third of a million times the mass of the Earth. Over the past 4.6 billion years the Sun has consumed only a few hundredths

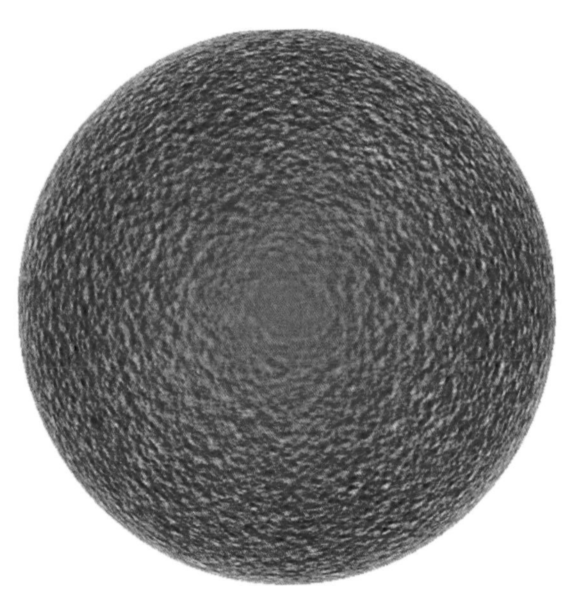

FIG. 2.6 Supergranulation Thousands of large convection cells, the supergranules, are detected in this Dopplergram which shows motion toward and away from the observer, along the line of sight, as light and dark patches, respectively. It was obtained using the Doppler effect of a single spectral line with the Michelson Doppler Imager, abbreviated MDI, instrument on the *SOlar and Heliospheric Observatory*, or *SOHO* for short. The image contains about 2,500 supergranules, each about 30,000 kilometers across. Near the disk center, where the Doppler effect detects radial motion, the supergranulation is hardly visible at all, thus indicating that the velocities are predominantly horizontal. Supergranules flow horizontally outward from their centers with a typical velocity of 400 meters per second. (Courtesy of the *SOHO* MDI/SOI consortium. *SOHO* is a project of international cooperation between ESA and NASA.)

of one percent of its original mass. The reason is essentially because very little mass (just 0.007 or 0.7 percent) is annihilated in forming a helium nucleus.

A more significant concern is the depletion of the Sun's hydrogen fuel within its nuclear furnace. Since thermonuclear reactions are limited to the hot dense core, the Sun will eventually run out of hydrogen – in about 7 billion years. The Sun will then expand to engulf the closest planet, Mercury.

As the hydrogen in the Sun's center is slowly depleted over time, and steadily replaced by heavier helium, the core must keep on producing enough pressure to keep the Sun from collapsing in on itself. And the only way to maintain the pressure and keep on supporting the weight of overlying material is to increase the central temperature. As a result of the slow rise in temperature, the rate of nuclear fusion gradually increases and the Sun inexorably brightens.

This means that the Sun was significantly dimmer in the remote past, and the Earth should have been noticeably colder then, but this does not agree with geological evidence. Assuming an unchanging terrestrial atmosphere, with the same composition and reflecting properties as today, the decreased solar luminosity would have caused the Earth's global surface temperature to drop below the freezing point of water about 2 billion years ago. The oceans would have been frozen solid, there would be no liquid water, and the entire planet would have been locked into a global ice age something like Mars is now.

Yet, sedimentary rocks, which must have been deposited in liquid water, date from 3.8 billion years ago. There is fossil evidence in those rocks for living things at about that time. Thus, for billions of years the Earth's surface temperature was not very different from today, and conditions have remained hospitable for life on Earth throughout most of the planet's history.

The discrepancy between the Earth's warm climatic record and an initially dimmer Sun has come to be known as the faint-young-Sun paradox. It can be resolved if the Earth's primitive atmosphere contained about a thousand times more carbon dioxide than it does now. Greater amounts of carbon dioxide would enable the early atmosphere to trap greater amounts of heat near the Earth's surface, warming it by the greenhouse effect. That would keep the oceans from freezing.

Over time the Sun grew brighter and hotter. The Earth could only maintain a temperate climate by turning down its greenhouse effect as the Sun turned up the heat. Our planet's atmosphere, rocks, oceans, and perhaps life itself, apparently combined to decrease the amount of carbon dioxide over time, keeping the Earth's temperature steady as the Sun slowly brightened.

But the Sun cannot shine forever, for it will eventually use up the hydrogen fuel in its core. Although it has converted only a trivial part of its original mass into energy, the Sun has processed a substantial 37 percent of its core hydrogen into helium during the past 4.6 billion years. And the Sun will have used up all its available core hydrogen in another 7 billion years. It will then balloon into a red giant star with a dramatic increase in size and a powerful rise in luminosity (Fig. 2.7).

Although the outer solar atmosphere will cool and redden, the Sun will also expand and move much closer to the Earth. Mercury will become little more than a memory, being pulled in and swallowed by the swollen Sun. The giant Sun will be 2,300 times

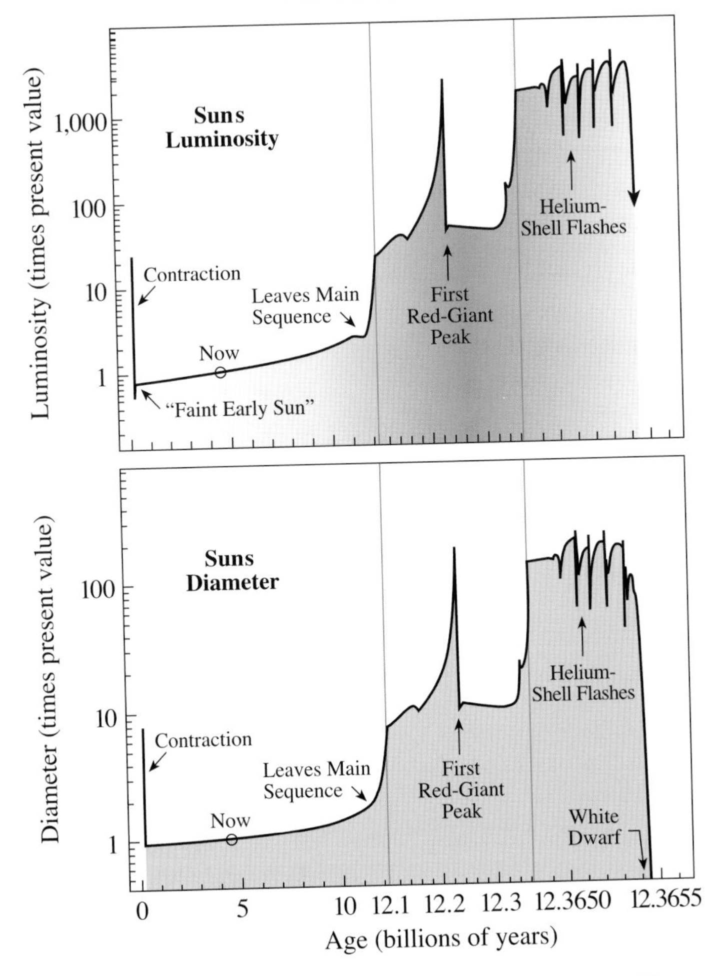

FIG. 2.7 The Sun's fate In about eight billion years the Sun will become much brighter (*top*) and larger (*bottom*). The time scale has been expanded near the end of the Sun's life to show relatively rapid changes. (Courtesy of I-Juliana Sackmann and Arnold I. Boothroyd.)

brighter than it is now, resulting in a substantial rise in temperature throughout the Solar System. It will become hot enough to melt the Earth's surface.

Moreover, terrestrial life will be wiped out well before then. In just 1 billion years from now the Sun will have brightened by 10 percent. Calculations suggest that the Earth's oceans will then evaporate at a rapid rate, resulting in a hot, dry uninhabitable Earth. And if that doesn't do us in, any Earthly life is doomed to fry in about 3 billion years from now. The Sun will then be hot enough to boil the Earth's oceans away, leaving the planet a burned-out cinder, a dead and sterile place.

Meanwhile, during the apocalyptic period of planetary destruction, the core of the Sun will continue to contract until the central temperature is hot enough to ignite helium burning – at about a hundred million kelvin. But this conversion of helium into carbon does not last very long, compared to the Sun's 12 billion years of hydrogen burning. In about 35 million years, the core helium will have been used up, and there will be no heat left to hold up the Sun. In a last desperate gasp of activity, the Sun will shed the outer layers of gas to produce an expanding "planetary" nebula around the star, while the core collapses into a white dwarf star.

By this time, intense solar winds will have stripped the Sun down to about half its original mass, and gravitational collapse will squeeze that into an insignificant cinder about the size of the present-day Earth. Nuclear reactions will then be a thing of the past, and there will be nothing left to warm the Sun or planets. The former Sun will just gradually cool down and fade away into old age, plunging all of the planets into a deep freeze.

Such events are in the very distant future, of course. For now, the Sun provides us with an up-close laboratory of the subatomic particles known as neutrinos. Astronomers have discovered fundamental properties of the neutrinos by observing those emitted from the Sun's energy-generating core.

Chapter Three

Ghostlike Neutrinos

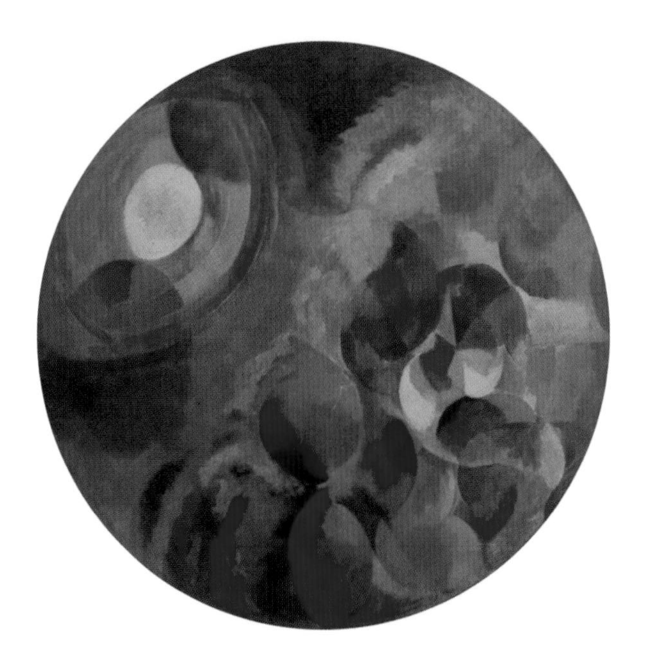

SIMULTANEOUS CONTRASTS: SUN AND MOON In this work the French cubist painter Robert Delaunay (1885–1941) has portrayed the Sun with no reference to the earthly world. The Sun is seen as a source of pure light and color that can be appreciated in its own right without reference to the sunlight that illuminates terrestrial objects. (Courtesy of the Museum of Modern Art, New York. Mrs. Simon Guggenheim Fund, Oil on canvas, 53 inches diameter.)

3.1 THE ELUSIVE NEUTRINO

Neutrinos, or little neutral ones, are very close to being nothing at all. They are tiny, invisible packets of energy with no electric charge and almost no mass, traveling at nearly the velocity of light. These subatomic particles are so insubstantial, and interact so weakly with matter, that they streak through almost everything in their path, like ghosts that move right through walls. Unlike light or any other form of radiation, the neutrinos can move nearly unimpeded through any amount of material, even the entire Universe.

Each second, trillions upon trillions of neutrinos that were produced inside the Sun pass right through the Earth without even noticing it is there. The indestructible neutrinos interact so rarely with the material world that almost nothing ever happens to them. Billions of ghostly neutrinos from the Sun are passing right through you every second, whether you are inside a building or outdoors, or whether it is day or night, and without your body noticing them, or them noticing your body.

The neutrinos are the true ghost riders of the Universe. As American writer John Updike (1932–) put it:

> Neutrinos, they are very small.
> They have no charge and have no mass
> And do not interact at all.
> The Earth is just a silly ball
> To them, through which they simply pass,
> Like dust maids down a drafty hall.[13]

How do we know that such elusive, insubstantial particles even exist? They are required by a fundamental principle of physics, known as the conservation of energy. According to this rule, the total energy of a system must remain unchanged, unless acted upon by an outside force. We know of no process that disobeys this principle.

Nevertheless, in a type of nuclear decay process, called beta-decay, the nucleus of a radioactive atom disintegrates through the release of a beta particle, now known to be an electron, whose energy is less than that lost by the initial nucleus. Careful measurements failed to turn up the missing energy, which seemed to have vanished into thin air, and this suggested that energy might not be conserved during beta-decay. However, it turned out that a mysterious, invisible particle was spiriting away the missing energy. It was the elusive neutrino, whose existence was postulated more than half a century ago by Wolfgang Pauli (1900–1958), a brilliant Austrian physicist (Fig. 3.1).

Pauli proposed a "desperate way out" of the energy crisis. He speculated that a second, electrically neutral particle, produced at the same time as the electron, carried off the remaining energy. The sum of the energies of both particles remains constant, so the energy books are balanced during beta-decay, and the principle of conservation of energy is saved. As Pauli expressed it in 1933:

> The conservation laws remain valid, the expulsion of beta particles [electrons] being accompanied by a very penetrating [energetic] radiation of neutral [uncharged] particles, which has not been observed so far.[14]

FIG. 3.1 Wolfgang Pauli This Austrian physicist, Wolfgang Pauli (1900–1958), predicted the existence of the neutrino to solve an energy crisis in a type of radioactivity called beta-decay. He thought that the invisible neutrino would never be seen, but it was subsequently discovered as a byproduct of nuclear reactions on the Earth and in the Sun. (Courtesy of the American Institute of Physics Niels Bohr Library, Goudsmit Collection.)

When first discovered, the electrons emitted during beta-decay were called beta particles, to distinguish them from alpha particles (helium nuclei) and gamma rays (high-energy radiation) that are also emitted during radioactive decay processes. From their measured charge and mass, we now know that beta-rays are not rays at all but instead ordinary electrons moving at nearly the velocity of light.

Pauli thought he had done "a terrible thing", for his desperate remedy postulated an invisible particle that could not be detected. Dubbed the neutrino, or "little neutral one" by the Italian physicist Enrico Fermi (1901–1954), the new particle could not be observed with the technology of the day, since the neutrino is electrically neutral, has almost no mass, and moves at nearly the velocity of light. So the neutrinos were removing energy that would never be seen again. (Even in Pauli and Fermi's time, the observed high-energy shape of the emitted electron's energy spectrum indicated that the mass of the neutrino is either zero or very small with respect to the mass of the electron.)

In 1934 Fermi formulated the mathematical theory of beta-decay in a paper that was rejected by the journal *Nature* because "it contained speculations too remote from reality to be of interest to the reader." As beautifully described by Fermi, the decay process occurs when the neutron in a radioactive nucleus transforms into a proton with the simultaneous emission of an energetic electron and a high-speed neutrino. When left alone outside a nucleus, a neutron will, in fact, self-destruct in about 10 minutes into a proton, plus an electron to balance the charge and a neutrino to help remove the energy.

As far as anyone could tell, an atomic nucleus consists only of neutrons and protons, so the electron and neutrino seemed to come out of nowhere. They do not reside within the nucleus and are born at the time of nuclear transformation. No one knew exactly how the neutrinos were created.

How do you observe something that spontaneously appears out of nowhere and interacts only rarely with other matter? Calculations suggested that the probability of a neutrino interacting with matter, so one could see it, is so incredibly small that no one could ever detect it. To see one neutrino, you would have to produce enormous numbers of them at about the same time, and build a very massive detector to increase the chances of catching it. Although almost all of the neutrinos would still pass through any amount of matter unhindered and undetected, a rare collision with other subatomic particles might leave a trace.

Nuclear reactors, first developed in the 1940s, produce large numbers of neutrinos. Such reactors run by a controlled chain reaction in which neutrons bombard uranium nuclei, causing them to split apart and create more neutrons to continue the chain reaction, thereby producing large amounts of energy with an enormous flux of neutrinos in the process. A similar thing occurs in an atomic bomb, except that the chain reaction runs out of control with an explosive release of energy. If you place a very massive detector near a large nuclear reactor, and appropriately shield the detector from extraneous signals, you might just barely observe the telltale sign of the hypothetical neutrino.

The existence of the neutrino was finally proven with Project Poltergeist, an experiment designed by Clyde L. Cowan (1919–1974) and Frederick Reines (1918–1998) of the Los Alamos National Laboratory in New Mexico. They placed a 10-ton (10,000-liter) tank of water next to a powerful nuclear reactor engaged in making plutonium for use in nuclear weapons. After shielding the neutrino trap underground and running it for about 100 days over the course of a year, Reines and Cowan detected a few synchronized flashes of gamma radiation that signaled the interaction of a few neutrinos with the nuclear protons in water.

The neutrinos were not themselves observed, and they never have been. Their presence was inferred by an exceedingly rare interaction. One out of every billion billion, or 10^{18}, neutrinos that passed through the water tank hit a proton, producing the telltale burst of radiation.

In June 1956, Cowan and Reines telegraphed Pauli with the news:

> We are happy to inform you that we have definitely detected neutrinos from fission fragments by observing inverse beta-decay of protons![15]

And Pauli promptly sent them a case of champagne in recognition of their accomplishment. Half a century later, Reines was awarded the 1995 Nobel Prize in Physics for this accomplishment, but his colleague Cowan just did not live long enough to share it. The inverse beta-decay mentioned in the telegram incidentally occurs when a nuclear proton absorbs a neutrino and turns into a neutron, at the same time emitting a positron, the anti-matter counterpart of the electron, which immediately annihilates with an electron and produces the radiation that was detected.

The ghostly neutrino, which most scientists had thought would never be detected, had finally been observed, and thoughts turned to catching neutrinos generated in the hidden heart of the Sun.

3.2 NEUTRINOS FROM THE SUN

Solar neutrinos are produced in profusion by thermonuclear reactions in the Sun's core, removing substantial amounts of energy that is never seen again. When hydrogen is burned into helium, thereby making the Sun shine, four protons are united, but two of them have to be turned into neutrons to make a helium nucleus. Each conversion of a proton into a neutron coincides with the birth of a neutrino; so two neutrinos are created each time a helium nucleus is made.

To make the Sun's energy at its present rate, a hundred trillion trillion trillion, or 10^{38}, helium nuclei are formed each second, and twice that number of neutrinos are released by the Sun every second. Although the Earth intercepts only a small fraction of these, it is still an enormous number. Each second, about 70 billion solar neutrinos fly through every square centimeter of the surface of the Earth facing the Sun, and out through the opposite surface unimpeded.

The Sun thus bathes our planet with a beam of neutrinos that is as steady as sunlight. And unlike radiation inside the Sun, which is diluted and transformed during its 170-thousand-year journey from the Sun's core, neutrinos pass quickly through the massive body of the Sun. To neutrinos, the Sun is essentially transparent, so they bring us a unique message from deep within its hidden interior, telling us what is happening in the center of the Sun right now – or to be more exact, 499 seconds ago, the time required to move at a neutrino's velocity from the center of the Sun to the Earth.

By catching and counting solar neutrinos, we can open the door of the Sun's nuclear furnace and peer inside its energy-generating core. But the feasibility of detecting neutrinos from the Sun depends on exactly what nuclear reactions are involved in making the Sun shine, the internal composition and structure of the Sun, and how it has evolved with time.

Both the amount and energies of the neutrinos produced inside the Sun depend on the elements being fused (Fig. 3.2). The great majority of solar neutrinos, about 82.5 percent, has relatively low energies and is generated during the head-on collision of two protons. In this proton-proton reaction, one of the protons turns into a

FIG. 3.2 Solar neutrino energy spectrum Neutrinos are produced inside the Sun as a byproduct of nuclear fusion reactions in its core, but both the amounts and energies of the neutrinos depend on the element fused and the detailed model of the solar interior. Here we show the energy spectrum of neutrinos predicted by the Standard Solar Model. The largest flux of solar neutrinos is found at low energies; they are produced by the main proton-proton, abbreviated pp, reaction in the Sun's core. Less abundant, high-energy neutrinos are produced by a rare side reaction involving boron-8. The shading denotes the detection range of the gallium, chlorine and water experiments. The gallium experiment can detect the low-energy pp neutrinos, as well as those of higher energy; both the chlorine and water detectors are sensitive to the high-energy boron-8 neutrinos. The neutrino fluxes from continuum sources (pp and boron 8) are given in the units of number per square meter per second per million electron volts (MeV) at one astronomical unit, the Earth's mean distance from the Sun. Neutrinos should also be generated at two specific energies when beryllium-7 captures an electron, and also during a relatively rare proton-electron-proton, abbreviated pep, reaction; their fluxes are given in number per square meter per second. The neutrino energy spectra from reactions with carbon, nitrogen and oxygen, abbreviated C, N and O, are drawn with dashed lines; they account for about 1.8 percent of the conversion of protons into helium within the Sun, while the proton-proton chain produces 98.2 percent of the helium. (Adapted from John N. Bahcall, Aldo M. Serenelli, and Sarbani Basu, *Astrophysical Journal* **621,** L87 2005.)

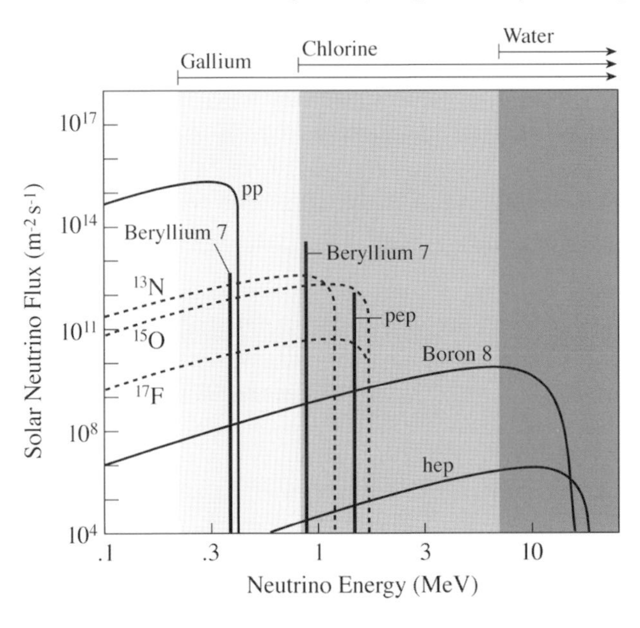

neutron, emitting a low-energy neutrino with energy of no more than 0.42 million electron volts, abbreviated 0.42 MeV.

Scientists have identified several less common neutrino-producing reactions, which have no substantive bearing on the total amount of energy produced in the Sun. The highest-energy neutrinos are produced in one of these rare side reactions. About 15.7 percent of the time, two types of helium nuclei, already produced in the solar core, fuse together to form beryllium-7 and energetic radiation. In very rare cases, about once in every 5,000 completions of the proton-proton chain, a beryllium-7 nucleus fuses with a proton, forming a radioactive boron-8 nucleus that ultimately decays back into two helium nuclei emitting a neutrino in the process. The boron-8 neutrinos can have energies as great as 15 MeV, or about 36 times more energy than the most energetic proton-proton neutrino.

William A. "Willy" Fowler (1911–1995), a nuclear astrophysicist at the California Institute of Technology, was one of the first to realize the ramifications of this rare side reaction; and much later, in 1993, he received the Nobel Prize in Physics for his theoretical and experimental studies of nuclear reactions in the stars and the early Universe.

More energetic neutrinos find it easier to interact with the material world. However, such encounters are still exceedingly rare, and the Sun produces fewer high-energy neutrinos than low-energy ones. So, if you are in the neutrino-catching business, it's roughly a zero sum game as far as how many neutrinos you might detect at different energies, and it won't be very many of them whatever the energy.

The flux of solar neutrinos expected at the Earth is calculated using supercomputers that produce an evolving sequence of theoretical models, culminating in the Standard Solar Model that best describes the Sun's luminous output, size and mass at its present age. Such calculations have been developed and refined over the past four decades, notably by John N. Bahcall (1934–2005) of the Institute for Advanced Study at Princeton, and other astrophysicists throughout the world, such as Sylvaine Turck-Chièze (1951–) at Saclay, France.

The computer models always include three basic assumptions:

(1) Energy is generated by hydrogen-burning reactions in the central core of the Sun, and there is no mixing of material between the core and overlying regions. The nuclear reaction rates depend on the density, temperature and composition of the core, as well as coefficients extrapolated from laboratory experiments.
(2) The outward thermal pressure, due to the energy-producing reactions, just balances the inward pressure due to gravity, thereby keeping the Sun from either collapsing or blowing up.
(3) Energy is transported from the deep interior to the visible solar disk via radiation and convection. The great bulk of energy is carried by radiative transport with an opacity determined from atomic physics calculations. Heavy elements provide a sort of dirt, or opacity, that blocks the flow of radiation, just as a dirty window keeps sunlight from fully illuminating a room.

One begins with a newly formed Sun having a uniform composition, and an element abundance that is observed in the visible solar disk today. The model then imitates the evolution of the Sun to its present age of 4.6 billion years by slowly converting hydrogen into helium within the model core. The central nuclear reactions supply both the radiated luminosity and the local heat (thermal pressure), while also creating neutrinos and producing composition changes in the core. The Sun's current luminosity and size are obtained after 4.6 billion years if about 37 percent of the hydrogen in the core has now been transformed into helium.

3.3 DETECTING ALMOST NOTHING

Once the Standard Solar Model has specified the neutrino flux, the predictions are extended to specific experiments that detect neutrinos of different energies. The neutrino reaction rate with atoms in these detectors is so slow that a special unit has been invented to specify the experiment-specific flux. This Solar Neutrino Unit, abbreviated SNU and pronounced "snew", is equal to one neutrino interaction per second for every trillion trillion trillion, or 10^{36}, atoms. And even then, the predictions are only a few SNU per month for even the largest most-massive, detectors that were first constructed.

Of course, it isn't easy to catch the elusive neutrino. The vast majority of neutrinos pass right through matter, but there is a finite chance that a neutrino will interact with some of it. When this slight chance is multiplied by the prodigious quantities of neutrinos flowing from the Sun, we conclude that a few of them will occasionally strike an atom's nucleus squarely enough to produce a nuclear reaction that signals the presence of an invisible neutrino. The neutrino detector must nevertheless consist of large amounts of material, literally tons of it, to allow interaction with even a tiny fraction of the solar neutrinos and measure their actual numbers.

Unlike a conventional optical telescope, that is placed as high as possible to minimize distortion by the obscuring atmosphere, a neutrino telescope is buried beneath a mountain, or deep within the Earth's rocks inside mines. This is to shield the neutrino detector from deceptive signals caused by cosmic interference. There, beneath tons of rock that only the neutrino can penetrate, detectors can unambiguously measure neutrinos from the Sun. Otherwise, neutrino detectors near the Earth's surface would detect high-energy particles and radiation produced by other energetic particles, called cosmic rays, interacting with the Earth's atmosphere. But neither the primary cosmic rays nor their secondary atmospheric emissions can penetrate thick layers of rock.

Thus, solar neutrino astronomy involves massive, subterranean detectors that look right through the Earth and observe the Sun at night or day. The first such neutrino telescope, constructed in 1967 by Raymond Davis, Jr. (1914–), was a 615-ton tank located 1.5 kilometers underground in the Homestake Gold Mine near Lead, South Dakota (Fig. 3.3). The huge cylindrical tank was filled with 378 thousand liters of cleaning fluid, technically called perchloroethylene or "perc" in the dry-cleaning trade; each molecule of the stain remover consists of two carbon atoms and four chlorine atoms.

Most solar neutrinos passed through the tank unimpeded. Occasionally, however, a neutrino scored a direct hit with the nucleus of a chlorine atom, turning one of its neutrons into a proton, emitting an electron to conserve charge and transforming the chlorine atom into an atom of radioactive argon. Only neutrinos more energetic than 0.814 million electron volts, abbreviated 0.814 MeV, triggered the nuclear conversion. None of the Sun's abundant proton-proton neutrinos have enough energy to cause this transformation, but the Homestake chlorine experiment was sensitive to the much rarer, higher-energy neutrinos produced by the less common boron-8 fusion reactions in the Sun.

The new argon atom rebounds from the encounter with sufficient energy to break free of the perc molecule and enter the surrounding liquid. Because the argon is chemi-

FIG. 3.3 Underground neutrino detector The original solar neutrino detector located 1.5 kilometers underground, in the Homestake Gold Mine near Lead, South Dakota, to filter out strong signals from energetic cosmic particles. The huge cylindrical tank was filled with 378 thousand liters of cleaning fluid. When a high-energy solar neutrino interacted with the nucleus of a chlorine atom in the fluid, radioactive argon was produced, which was extracted to count the solar neutrinos. This experiment operated for more than 25 years, always finding fewer neutrinos than expected from the Standard Solar Model – see Fig. 3.4. (Courtesy of Brookhaven National Laboratory.)

cally inert, it can be culled from the liquid by bubbling helium gas through the tank; and the number of argon atoms recovered in this way measures the incident flux of solar neutrinos.

Every few months Davis and his colleagues flushed the tank with helium, extracting about 15 argon atoms from a tank the size of an Olympic swimming pool. That was a remarkable achievement considering that the tank contains more than a million trillion trillion, or 10^{30}, chlorine atoms. And the scientists persisted for nearly thirty years, like aging hunters tending a trap, capturing a total of just 2,000 neutrinos. This implied that nuclear fusion reactions were indeed providing the Sun's energy, making it shine, which was the sole motivation for carrying out the experiment. But there was a small, unexpected problem with the result.

For nearly two decades, from 1968 to 1987, the Homestake detector always yielded results in conflict with the most accurate theoretical calculations. The final experiment value was 2.55 $\pm$ 0.25 SNU, where the $\pm$ value denotes an uncertainty of one standard deviation. (A standard deviation is a statistical measurement of the uncertainty of a measurement; a definite detection has to be above three standard deviations and preferably above five of them.) In contrast, the most recent theoretical result using the Standard Solar Model predicts that the Homestake detector should have observed a flux 8.5 $\pm$ 1.8 SNU. So the tank full of cleaning fluid captured almost one-third the expected number of neutrinos (Fig. 3.4). The discrepancy between the observed and calculated values is known as the Solar Neutrino Problem.

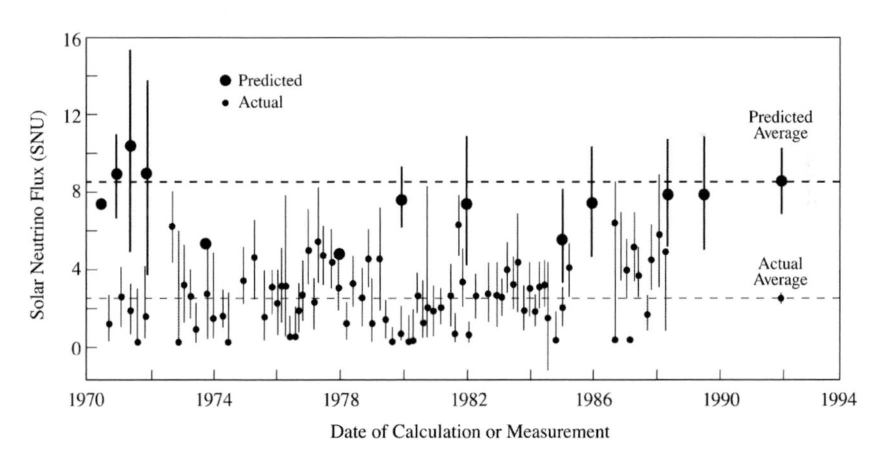

FIG. 3.4 Solar neutrino fluxes Calculated and measured solar neutrino fluxes have consistently disagreed for several decades. The fluxes are measured in solar neutrino units, or SNU, defined as one neutrino interaction per trillion trillion trillion, or 10^{36}, atoms per second. Measurements from the chlorine neutrino detector *(small dots)* give an average solar neutrino flux of 2.55 $\pm$ 0.25 SNU (*lower broken line*), well below theoretical calculations (*large dots*) that predict a flux of 8.5 $\pm$ 1.8 SNU (*upper broken line*) for the Standard Solar Model. Other experiments have also observed a deficit of solar neutrinos, suggesting that either some process prevents neutrinos from being detected or the method by which the Sun shines differs from that predicted by current theoretical models.

In 1987, another giant, underground neutrino detector, called Kamiokande, began to monitor solar neutrinos, confirming the neutrino deficit observed by Davis. This second experiment, located in a mine at Kamioka, Japan, and known as Kamiokande, consisted of a 4,500-ton, or 4.5 million liters, tank of pure water. Nearly a thousand light detectors were placed in the tank's walls to measure signals emitted by electrons knocked free from water molecules by passing neutrinos.

The number of scattered electrons detected by Kamiokande indicated that the flux of high-energy neutrinos is just under half the neutrino flux expected from the Standard Solar Model, or to be exact 0.48 ± 0.07 times the predicted value. An incident neutrino must have an energy of at least 7.5 MeV to produce a detectable recoil electron in water, so Kamiokande had a much higher threshold than even the chlorine experiment, at 0.814 MeV. The dominant source in both experiments is the rare boron-8 neutrinos, so they independently confirm an apparent deficit of high-energy solar neutrinos, although the observed deficit was in different proportions.

When an energetic solar neutrino collides with an electron in the water, the neutrino knocks the electron out of its atomic orbit, pushes it forward in the direction of the incident neutrino, and accelerates the electron to nearly the velocity of light. In water, the electron moves faster than the light it radiates, and as a result the electron produces a cone-shaped pulse of light about its path. The faint blue glow is technically known as Cherenkov radiation, named after the Soviet physicist Pavel A. Cherenkov (1904–1958) who discovered the effect in 1934.

The axis of the light cone gives the electron's direction, which is the direction from which the neutrino arrived. And since the observed electrons were preferentially scattered along the direction of an imaginary line joining the Earth to the Sun, the Kamiokande water experiment confirmed that the neutrinos are indeed produced by nuclear reactions in the Sun's core. After 1000 days of observation, Yoji Totsuka (1942–), speaking for the Kamiokande collaboration led by Masatoshi Koshiba (1926–), could therefore report in 1991 that:

> The directional information tells us that neutrinos are coming from the Sun, [providing] the first [direct] evidence that the fusion processes are taking place in the Sun.[16]

In contrast, the chlorine experiment cannot tell what direction the neutrinos are coming from. However, because the Sun is so close it should dominate the cosmic neutrino input to Earth, and no other known cosmic source could be providing the high-energy neutrinos the chlorine experiment detected for so long.

The Japanese experiment had incidentally already detected, on 23 February 1987, a brief burst of just 11 neutrinos from a rare and distant supernova explosion known as SN 1987A. The total energy emitted by all the undetected neutrinos during the explosion of the massive star is about equal to all the energy emitted by the Sun in its entire 4.6-billion-year history.

The predicted flux of the high-energy boron-8 neutrinos from the Sun has a large uncertainty, about 37 percent, for it depends strongly on the uncertain temperature at the center of the Sun; the amount of these neutrinos varies roughly as the eighteenth power of the core temperature. Scientists therefore developed a method of detecting the more numerous low-energy, proton-proton neutrinos using gallium, a rare and expensive metal used in the red lights of hand calculators and other pieces of electronic equipment. The predicted flux of the proton-proton neutrinos has only a 2 percent uncertainty, primarily because it is relatively insensitive to the core temperature.

When a low-energy neutrino has a rare, head-on collision with the nucleus of a gallium-71 nucleus, containing 31 protons and 40 neutrons, one of its neutrons is changed into a proton, emitting an electron to conserve charge and producing a nucleus of a germanium-71 atom, with 32 protons and 39 neutrons. The energy threshold for this nuclear conversion is 0.23 MeV, so it should be sensitive to a sizeable fraction of the proton-proton neutrino flux, whose energy varies from 0 to 0.422 MeV, as well as the more energetic boron-8 neutrinos. The germanium produced in this way can be chemically separated from the gallium, and counted by its radioactive decay; thereby providing a measure of the flux of incident neutrinos with energies above the 0.23 MeV threshold.

Unfortunately, it takes at least 30 tons of gallium to produce a detectable signal from solar neutrinos. The total world production of gallium was about 10 tons per year, and a ton of gallium costs half a million dollars – a lot more than cleaning fluid! Moreover, a neutrino conversion of a gallium nucleus does not happen very often. Only about one such event is expected to happen each day in 30 tons of gallium. But this is serious business; so two international collaborations spent millions of dollars to place huge tanks of gallium underground.

In 1990, the Soviet-American Gallium Experiment, abbreviated SAGE, began operation at the Baksan Neutrino Observatory located in a long tunnel, some 2 kilometers below the summit of Mount Andyrchi in the northern Caucasus Mountains in Russia. SAGE used 60 tons of gallium metal kept molten in reactor vessels at about 303 kelvin. A second multinational experiment, dubbed GALLEX for GALLium EXperiment, started operating in 1991 within the Gran Sasso Underground Laboratory, located 1.4 kilometers below a peak in the Apennine Mountains of Italy. The GALLEX experiment used 30 tons of gallium in 100 tons of highly concentrated gallium-chloride solution.

The GALLEX collaboration operated between 1991 and 1997, and a new series of measurements was then carried out at the same location by the Gallium Neutrino Observatory, abbreviated GNO, from 1998 to 2002. The combined GALLEX and GNO result was 70.8 ± 4.5 SNU. Between 1990 and 2001, SAGE obtained a very similar measurement of 70.8 ± 5.3 SNU. The two results are well below the predicted 131 ± 10 SNU using the Standard Solar Model for the gallium experiments.

The low-energy neutrinos from the direct fusion of two protons, in the proton-proton reaction, contribute just slightly more than half of the predicted gallium detection rate, or about 70 SNU of the predicted 131 SNU, and the high-energy boron-8 neutrinos contribute the rest of the predicted amount. When combined with the chlorine and the Kamiokande results, which show a deficit of high-energy neutrinos, the gallium experiments must have detected the low-energy proton-proton neutrinos. This was an important result, for it provided confirmation that hydrogen fusion makes the Sun shine.

The results from the first four solar neutrino experiments are given in Table 3.1, where they are also compared to theoretical calculations using the Standard Solar Model. The chlorine experiment detects about one fourth of the expected flux of neutrinos, and

TABLE 3.1 The first solar neutrino experiments

Target	Experiment	Threshold Energy (MeV)	Measured Neutrino Flux (SNU)[a]	Predicted Neutrino Flux (SNU)	Ratio: Measured/ Predicted
Chlorine 37	HOMESTAKE	0.814	2.55 ± 0.25	8.5 ± 1.8	0.30 ± 0.02
Water	KAMIOKANDE	7.5			0.48 ± 0.07
Gallium 71	GALLEX + GNO	0.23	70.8 ± 4.5	131 ± 10	0.54 ± 0.05
Gallium 71	SAGE	0.23	70.8 ± 5.3	131 ± 10	0.54 ± 0.05

[a] Here the uncertainties are one standard deviation. The units of the measured and predicted values for the Kamiokande experiment are 10^{10} m^{-2} s^{-1}, while the Solar Neutrino Unit, or SNU, is used for the other three experiments.

the neutrino-electron scattering experiment about one half the predicted amount. The two gallium experiments also detect fewer neutrinos than predicted. Taken together, all four experiments seem to have confirmed that the solar neutrinos are missing, and that the Solar Neutrino Problem is real.

The importance of the observed solar neutrino deficit was recognized by the award of the 2002 Nobel Prize in Physics to Raymond Davis Jr. (1914–), who pioneered the field of neutrino astrophysics, and to the Japanese physicist Masatoshi Koshiba (1926–) for the discovery of cosmic neutrinos, from both the Sun and the distant exploding star, or supernovae, SN 1987A. Another astrophysicist, the Italian-born American Riccardo Giacconi (1931–) shared the 2002 prize for his pioneering investigations of cosmic X-rays. And since the prize cannot be given to more than three scientists, other significant contributors to neutrino astronomy or the theory of neutrinos could not be recognized.

3.4 SOLVING THE SOLAR NEUTRINO PROBLEM

Standard Solar Model Confirmed

After almost forty years of meticulous measurements and calculations, the neutrino count still came up short! Massive underground detectors always observed fewer neutrinos than theory says they should detect, and the Solar Neutrino Problem was a continuing embarrassment.

But what's all the fuss about anyway? The difference between the Standard Solar Model and observations is only a factor of two or three, or perhaps four. The excitement arises because the stakes are high! Either the Sun does not shine the way we think it ought to, or our basic understanding of neutrinos is in error.

That is, there are two methods of solving the Solar Neutrino Problem. One method is to create a non-standard solar model that modifies our astrophysical description of the Sun and produces the observed number of neutrinos. In the other solution, solar neutrinos are produced in the quantity predicted by the Standard Solar Model, but there is some flaw in our understanding of how sub-atomic particles behave, requiring new properties for the neutrino.

Is there something amiss with our understanding of the internal operations of the Sun? When the Solar Neutrino Problem first arose, some harried astrophysicists pondered deeply and lost sleep at night, coming up with all sorts of explanations. Most of them involved a reduction of the central temperature of the Sun by about 3 percent. If the center of the Sun were about half a million degrees cooler than presently thought, the nuclear reactions would run at a slower rate, producing neutrinos in the observed amounts and resolving the dilemma.

A lower central temperature might result from a rapidly rotating core or a strong central magnetic field that would help hold the Sun up against the inward tug of its own gravity, thereby reducing the pressure and temperature at its center; or the temperature might be reduced by mixing from the Sun's outer layers. Nevertheless, Sylvaine Turck-Chièze (1951–) and her colleagues at Saclay, France, effectively used the oscillations of the Sun as a thermometer to measure the central temperature of the Sun, showing in 2001 that it is very close to the value of 15.6 million kelvin calculated using the Standard Solar Model. In other words, the oscillation data, obtained with instruments aboard the *SOlar and Heliospheric Observatory,* placed constraints on the neutrino flux emitted by nuclear reactions in the Sun's core, showing that the neutrino deficit measured on Earth cannot be explained by adjustments to the solar model calculations.

Even before these temperature measurements, John Bahcall (1934–2005) had teamed up with Hans A. Bethe (1906–2005), who first elucidated the fusion reactions that power the Sun in the 1938, arguing in 1990 that:

> We do not know of any modifications of the astrophysical calculations of the state of the solar interior that could lead to the reconciliation [of the Homestake and Kamiokande observations with theoretical calculations] without requiring new physics for the neutrino.[17]

The Standard Solar Model, as well as our explanations for how other stars shine and evolve, therefore seemed to be on solid ground. So the astrophysicists could sleep at night, and most scientists agreed that the solution to the Solar Neutrino Problem had to result from an incomplete understanding of neutrinos.

Neutrinos Change Form and Oscillate Between Types

Where have all the solar neutrinos gone? The ghostlike neutrinos are transforming themselves into an invisible form during their journey from the center of the Sun, escaping detection by changing character. The reason we can't see some neutrinos is because they are hiding in disguise, having undergone metamorphosis into a different form, like a caterpillar into a butterfly or moth. After all, other particles, such as the neutron, turn into something else when left alone. Of course, there may not be anything more fundamental for a neutrino to change into, but solar neutrinos could turn into even more evasive types of neutrinos, thereby rendering them invisible.

We haven't mentioned it yet, but scientists have learned that there are three separate types, or flavors, of neutrinos, each named after the fundamental, subatomic par-

FOCUS 3.1
Leptons, Quarks and Electroweak Theory

Electrons, muons and tau particles, along with their corresponding neutrinos, are collectively known as *leptons,* the Greek word for "slender", as they are all significantly less massive than most other elementary particles. The leptons are thought to be fundamental particles that do not consist of anything else; they are therefore amongst the basic building blocks out of which the Universe is constructed.

We also now know that protons and neutrons are not themselves fundamental, but consist of smaller particles, the quarks, buried deep within them. The whimsical name quark was taken from James Joyce's (1882–1941) *Finnegan's Wake, Book 2, Episode 4,* including the phrase "Three quarks for Mustar Mark!"

Two kinds of quarks, called up and down, respectively reside within the proton and neutron, and the transmutation of a proton into a neutron involves changing an up quark into a down quark. Every time one quark is changed into another, it produces a neutrino.

Neutrinos occasionally interact with other sub-atomic particles through a force that is at least 1,000 times weaker than the electromagnetic force and 100,000 times feebler than the strong force. The electromagnetic force binds electrons to protons, and the strong force holds protons and neutrons together in the nucleus. The electromagnetic force and the weak force are unified in a single electroweak theory developed by Sheldon Glashow (1932–), Abdus Salam (1926–1996) and Steven Weinberg (1933–); in 1979 they were awarded the Nobel Prize in Physics for this feat, even before the discovery of the electroweak forces that were predicted by their theory. In the standard electroweak theory, the neutrinos are assumed to be completely without mass.

ticle with which it is most likely to interact (Focus 3.1). All of the neutrinos generated inside the Sun are electron neutrinos, designated ν_e; this is the kind that interacts with electrons, denoted e^-. The other two flavors, the muon neutrino, ν_μ, and the tau neutrino, ν_τ, interact with muons, μ, and tau particles, τ, respectively. The muon and tau particles are unfamiliar to most of us because they die shortly after birth. The muon decays into an electron, a muon neutrino and an electron anti-neutrino in just two millionths, or 2×10^{-6}, of a second, and the tau particle disappears just three-tenths of a million-millionth, or 3×10^{-13}, of a second after it is made. The muon was discovered in 1936 as part of the particle fall out generated when a high-energy cosmic-ray particle slams into the Earth's atmosphere. Indirect evidence for the tau neutrino was obtained from the decay of the tau particle, discovered by Martin L. Perl (1927–) and his colleagues in 1975 when using the linear particle accelerator at Stanford University; he shared the 1995 Nobel Prize in Physics for this discovery, with Frederick Reines (1918–1998) for his discovery of the neutrino. More direct evidence for the tau neutrino was detected in 2000, using another particle accelerator at Fermilab, near Chicago.

Neutrinos apparently have an identity crisis! Each type of neutrino is not completely distinct, and the different types can be transformed into each other. In the language of quantum mechanics, neutrinos do not occupy a well-defined state; they instead consist of a combination or mixture of states, each with a specific mass. As neutrinos move through space, the mass states come in and out of phase with each other, so the mixture they form changes with time.

The effect is called neutrino oscillation since the probability of metamorphosis between neutrino types has a sinusoidal, in and out, oscillating dependence on path length. The chameleon-like change in identity is not one way, for a neutrino of one type can change into another kind of neutrino and back again as it moves along. Like the Cheshire cat, the elusive neutrino can appear and disappear.

In 1967 the Italian atomic physicist Bruno Pontecorvo (1913–1993) proposed the idea that one type of neutrino might transform, or oscillate, into another type in the vacuum of space; he was also the first person to propose using a chlorine detector to study neutrinos. And in 1969, Pontecovo and Vladimir Gribov (1930–1997), working in Russia, proposed that the discrepancy between the observed solar neutrinos and theoretical expectations for the Sun could be explained if solar neutrinos switch from electron neutrinos to another type as they travel in the near vacuum of space from Sun to Earth, thereby escaping detection. Almost a decade later, the American physicist Lincoln Wolfenstein (1923–) showed that the neutrinos could oscillate, or change type, more vigorously by interacting with matter, rather than in a vacuum, and in 1985 the Russian physicists, Stanislav P. Mikheyev (1940–) and Alexei Y. Smirnov (1951–) explained how the matter oscillations might explain the Solar Neutrino Problem.

The theory, named the MSW effect after the first letters of the last names of the scientists who developed it, proposed that the electron neutrinos generated in the solar core could change type on their way out of the Sun. This metamorphosis would happen extremely rarely in the vacuum of space, but might be amplified in the dense interior of the Sun. Interactions between the electron neutrinos and the densely packed solar electrons can, when the density is just right, alter the mass state of a neutrino traveling out through the Sun, thereby changing it into a muon neutrino. Once formed, the muon neutrino does not change back into an electron neutrino; it travels out into space and remains invisible to the first solar neutrino detectors.

In order to change from one form to another, neutrinos must have some substance in the first place. They can only pull off their vanishing act if the neutrino, long thought to have no mass, possesses a very small one. Such a tiny neutrino mass has been inferred from measurements of non-solar neutrinos using the high-tech $100 million Super-Kamiokande detector (Fig. 3.5), which replaced the older, nearby Kamiokande instrument in 1996.

Super-Kamiokande can observe both solar electron neutrinos and atmospheric muon neutrinos. The electron neutrinos are created by nuclear fusion at the center of the Sun, while the muon neutrinos are created when fast-moving cosmic rays enter the Earth's atmosphere from outer space. The solar electron neutrinos are distinguished by their relatively low energy, near the 5 MeV threshold of the detector. A high energy of 1,000 MeV is typical of the atmospheric muon neutrino. Neutrinos of higher energy produce a tighter cone of light, so the solar electron neutrino makes a fuzzy, blurred and ragged light pattern, while the atmospheric muon neutrino produces a neat, sharp-edged ring of light.

After monitoring light patterns for more than 500 days, the Super-Kamiokande scientists reported in 1998 that muon neutrinos produced in the atmosphere change type in mid-flight. Subsequent experiments using neutrinos generated on Earth have

FIG. 3.5 Super-Kamiokande This neutrino detector has been built a kilometer underground in a Japanese zinc mine. The huge stainless-steel vessel, 40 meters tall and 40 meters wide, has been filled with 50,000 metric tons (50 million liters) of highly purified water. About 13,000 light sensors, called photo-multiplier tubes, are uniformly arranged on the inner walls of the vessel. The photo-multiplier tubes are so sensitive that they can detect a single photon of light – a light level approximately the same as the light seen on Earth from a candle on the Moon. The light sensors can monitor the entire volume of the pure transparent water for the blue flash of Cherenkov light generated by an electron recoiling from a direct hit by a neutrino. (Courtesy of Yoji Totsuka, Institute for Cosmic Ray Research, University of Tokyo.)

FOCUS 3.2

Terrestrial neutrinos

Although our planet is perpetually bathed with electron neutrinos, produced by nuclear fusion reactions in the Sun's core, other kinds of neutrinos are also produced on Earth by cosmic rays entering the atmosphere, by man-made, high-energy particle accelerators, and by man-made nuclear reactors. They have all been used to demonstrate that neutrinos change type, or flavor, oscillating between types as they travel through material on the Earth.

In 1998 the Japanese Super-Kamiokande group reported the discovery of neutrino oscillations when observing muon neutrinos generated by cosmic rays interacting with the atmosphere. There were roughly twice as many muon neutrinos coming from the atmosphere directly over the Super-Kamiokande detector than those coming up from the other side of the Earth. The muon neutrinos are produced in the atmosphere above every place on our planet, but some of them apparently disappeared while traveling through the Earth to arrive at the detector from below.

The further a neutrino travels, the more time it has to oscillate, and that would account for why there are fewer muon neutrinos arriving from the back side of Earth, some 12,700 kilometers away, than from 40 kilometers up in the atmosphere directly above the detector. Some of the muon neutrinos generated in the atmosphere on the Earth's far side had apparently changed or oscillated into undetected tau neutrinos somewhere along their way through the Earth. The atmospheric neutrino data was subsequently used to show how the expected neutrino oscillations depend on the neutrino energy and the travel distance.

Intense muon neutrino beams generated by a Japanese particle accelerator were then directed through the Earth to the Super-Kamiokande underground neutrino detector, located about 250 kilometers away. The observed deficit of detected muon neutrinos confirmed the disappearing atmospheric muon neutrino result.

American and Japanese scientists next teamed up to construct the $20 million Kamioka Liquid scintillator Anti-Neutrino Detector, abbreviated KamLAND, at the site of the older Kamiokande solar neutrino detector. KamLAND detects electron anti-neutrinos that have traveled through the Earth from 51 nuclear reactors in Japan plus 18 reactors in South Korea. The measurements, reported in 2003 to 2005, show that some of the electron anti-neutrinos are disappearing when traveling to the detector. This work confirmed the solar electron neutrino deficit that had been detected for more than three decades, and provided compelling evidence for neutrino oscillation and mass that had previously been demonstrated by observations of solar neutrinos from the Sudbury Neutrino Observatory, or SNO for short. The combined SNO and KamLAND results have constrained the neutrino mass and oscillation parameters, and indicated that most of the solar neutrino metamorphosis occurs in the Sun's interior.

confirmed the effect (Focus 3.2). They suggest that although all solar neutrinos are born electron neutrinos, they do not stay that way. Nevertheless, the terrestrial neutrinos did not come from the Sun, and are not directly related to nuclear fusion reactions there. So the solution to the Solar Neutrino Problem was not definitely known until 2001, when a new underground solar neutrino detector in Canada, the Sudbury Neutrino Observatory, demonstrated that solar neutrinos are changing type when traveling to the Earth.

The Sudbury Neutrino Observatory

The Sudbury Neutrino Observatory, abbreviated SNO and pronounced "snow", is a collaboration of Canadian, American and British scientists. The detector is located 2 kilometers underground in a working nickel mine near Sudbury, Ontario. Like the previous water detectors, it observes boron-8 solar neutrinos with energies above 7.5 MeV. But unlike Kamiokande or Super-Kamiokande, the SNO detector contains heavy water.

Heavy water is chemically similar to ordinary water, and it doesn't appear or taste any different. In fact, heavy water exits naturally as a constituent of ordinary tap or lake water in a ratio of about one part in 7,000, and expensive chemical and physical processes can separate it.

The hydrogen in heavy water has a nucleus, called a deuteron, which consists of a proton and a neutron. For ordinary water, the hydrogen is about half as light, with a nucleus that contains only a proton and no neutron. And it is the heavier deuteron that makes the Sudbury Neutrino Observatory sensitive to not just one type of neutrino but to all three known varieties of neutrinos.

One thousand tons, or one million liters, of heavy water, with a value of $300 million, was placed in a central spherical cistern with transparent acrylic walls (Fig. 3.6).

FIG. 3.6 Sudbury neutrino observatory The central spherical flask of this neutrino observatory is 12 meters in diameter, and is surrounded by a geodesic array of thousands of light sensors to detect the flash of light from the interaction of a neutrino with the heavy water. (Courtesy of Kevin Lesko, Lawrence Berkeley National Laboratory.)

Since the scientists could not afford its cost, the heavy water was borrowed from Atomic Energy of Canada Limited, which stockpiled it for use in its nuclear power reactors – the heavy water moderates neutrons created by uranium fission in the reactors.

A geodesic array of about 10,000 photo-multiplier tubes surrounds the vessel to detect the flash of light given off by heavy water when it is hit by a neutrino. Both the light sensors and the central tank are enveloped by a 7,800-ton jacket of ordinary water (Fig. 3.7), to shield the heavy water from weak natural radiation, gamma rays and neutrons from the underground rocks. As with the other neutrino detectors, the overlying rock blocks energetic particles generated by cosmic rays.

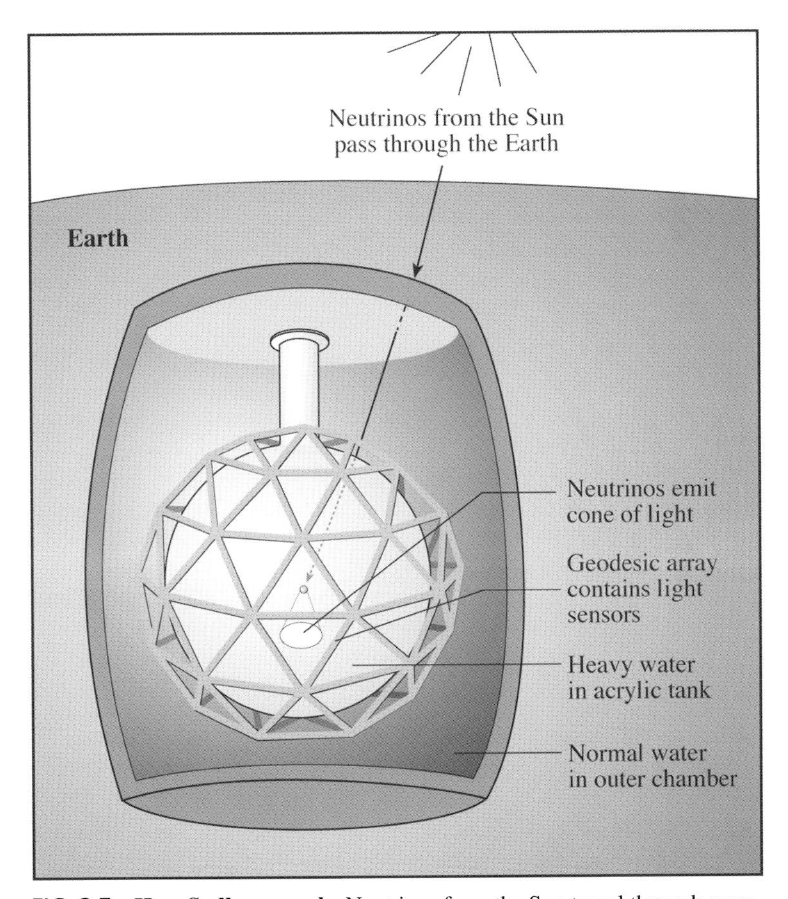

FIG. 3.7 How Sudbury works Neutrinos from the Sun travel through more than two kilometers of rock, entering an acrylic tank containing 1,000 tons (1 million liters) of heavy water. When one of these neutrinos interacts with a water molecule, it produces a flash of light that is detected by a geodesic array of photo-multiplier tubes. Some 7,800 tons (7.8 million liters) of ordinary water surrounding the acrylic tank blocks radiation from the rock, while the overlying rock blocks energetic particles generated by cosmic rays in our atmosphere. The heavy water is sensitive to all three types of neutrinos.

FOCUS 3.3

Neutrino Reactions with Heavy Water

The heavy water in the Sudbury Neutrino Observatory can be used to either detect electron neutrinos from the Sun or to measure all three neutrino types that might interact with the water – the electron, muon, or tau neutrinos.

A rare direct hit by a solar electron neutrino can change the neutron in deuterium to a proton, and the neutrino into an electron, by the "charged current" reaction:

$$\nu_e + D \rightarrow p + p + e^-,$$

where ν_e is the electron neutrino, D is a deuteron, the nucleus of a deuterium atom, and p and e^- respectively denote a proton and an electron. The electron will get most of the neutrino energy since it has a smaller mass than the protons, just as when a gun is fired, the bullet, being lighter, gets most of the energy. The electron will be so energetic that it will be ejected from the deuteron at the speed of light, producing a flash of blue Cherenkov light.

The nucleus of heavy water is also a target for all three flavors of neutrinos. When a neutrino of type x, denoted ν_x, interacts with a deuteron, D, it can break the nucleus into its sub-atomic constituents, the proton, p, and a neutron, denoted by n, by the "neutral current" reaction:

$$\nu_x + D \rightarrow p + n + \nu_x.$$

The neutron that is knocked lose by the impacting neutrino can be readily detected by the gamma rays which are emitted when the neutron is captured by another nucleus; the gamma rays will scatter electrons which produce detectable Cherenkov light. This reaction is equally sensitive to all neutrino types, and adding ordinary table salt to the heavy water enhances its sensitivity to the reaction. The chlorine in the salt increases the neutron capture rate and creates more light.

The Sudbury Neutrino Observatory can be operated in two modes, one sensitive only to electron neutrinos and the other equally sensitive to all three types of neutrinos (Focus 3.3). When the experiment was operated in the first mode, measuring electron neutrinos from the Sun, it made history. In June 2001, Arthur "Art" McDonald (1943–), the SNO Project Director, announced that:

> We now have high confidence that the [Solar Neutrino Problem] discrepancy is not caused by models of the Sun but by changes in the neutrinos themselves as they travel from the core of the Sun to the Earth. Earlier measurements had been unable to provide definitive results showing that this transformation from solar electron neutrinos to other types occurs. The new results from SNO, combined with previous work, now reveal this transformation clearly, and show that the total number of electron neutrinos produced in the Sun is just as predicted by detailed solar models.[18]

By detecting about 10 neutrinos a day, the SNO scientists accurately measured about one-third of the solar electron neutrinos predicted by the Standard Solar Model. The Super-Kamiokande detected about half the predicted number, but it had a small sensitivity to other neutrino types. The difference between the two measurements provided evidence that solar neutrinos oscillate, or change type, when traveling from the Sun, and showed that the total number of neutrinos emitted by the Sun agrees with predictions by the Standard Solar Model. After 33 years, the Solar Neutrino Problem had been solved!

The epochal SNO results continued in succeeding years, when 2 tons of table salt was added to the 1,000 tons of heavy water. In 2003 the group announced that the total number of neutrinos of all types reaching the Earth from the Sun is precisely equal to the number of electron neutrinos produced by thermonuclear reactions in the core of the Sun, but that two-thirds of the electron-type neutrinos were observed to change to muon- or tau-type neutrinos before reaching the Earth. By 2005, the SNO salt-phase results for solar neutrinos had been combined with the KamLAND experiment using neutrinos generated by terrestrial nuclear reactors, providing estimates for the tiny neutrino mass and for other properties of neutrino oscillation.

Chapter Four

Taking the Pulse of the Sun

POLLARD WILLOWS AT SUNSET In this painting, the Dutch artist Vincent Van Gogh (1853–1890) captures the sulfur yellow and pale gold-lemon color of the Sun. The artist literally worshipped the Sun, which for him represented the source of life and wellbeing – a wellbeing that eluded him in his delirium and madness. But even through the iron bars of an insane asylum, Van Gogh never wearied of watching "the Sun rise in its glory." (Courtesy of State Museum Kroller-Müller, Otterlo, The Netherlands.)

4.1 TRAPPED SOUNDS

The Sun is playing a secret melody, which produces a widespread throbbing motion of its surface. The sounds are coursing through the Sun's interior, causing the entire globe, or parts of it, to move in and out, slowly and rhythmically like the regular rise and fall of tides in a bay or of a beating heart (Fig. 4.1). Such radial oscillations are imperceptible to the naked eye; the surface moves a hundred-thousandth (0.000 01) of the solar radius.

The Sun's tiny periodic motions have nevertheless been observed as subtle changes in the wavelength of absorption lines that are formed in the solar gas. When part of the Sun heaves up toward us, the wavelength of the line formed in that region is shortened, introducing a small blueshift in its spectrum; if the region moves away from us, back toward the solar interior, the wavelength is lengthened, introducing a redshift in the spectrum.

These motions are inferred by subtracting an image of the Sun taken in the long-wavelength side of a stationary, or non-moving, absorption line from an image taken in the short-wavelength side of the line. In such a "Dopplergram", an outward motion of a region results in an increase in brightness at that place in the subtracted image, while an inward motion darkens it. The Dopplergrams show that bouncing regions move in and out all over the Sun (Fig. 4.2).

The American astrophysicist Robert B. Leighton (1919–1997) and his students, Robert W. Noyes (1934–) and George W. Simon (1934–), were the first to employ such techniques. They used subtracted images to show that the local vertical motions are not random in their time variation, but instead oscillate with a period of about five minutes. As Leighton announced, at an international conference in 1960, a sequence of Dopplergrams indicated that:

> These vertical motions show a strong oscillatory character, with a period of 296 ± 3 seconds [about five minutes].[19]

The observed five-minute oscillations had a velocity amplitude of about half a kilometer per second, and covered roughly a third of the visible disk at any given time, localized within numerous patches a few thousand kilometers across. Moreover, the repetitive, vertical oscillations did not seem to be continuous, apparently lasting only a few periods before disappearing.

But the Sun does not resonate with a single pure note. The observed oscillations turned out to be due to the combined effect of millions of sound waves traveling in all directions and with a large range of sizes. And it is not just the visible part of the Sun, but the interior as well that is oscillating.

Some of these oscillations can last several weeks to months, and they are not confined to localized regions on the Sun. Careful observations showed that the five-minute oscillations are a global phenomenon, coherent across the entire Sun. Eventually, other solar astronomers demonstrated that the entire Sun is ringing like a bell, with sound waves that resonate within its interior and penetrate to its very core.

In 1961 Franz Daniel "F. D." Kahn (1926–1998) showed that sound waves could be trapped near the visible solar disk, essentially because the temperature increases both

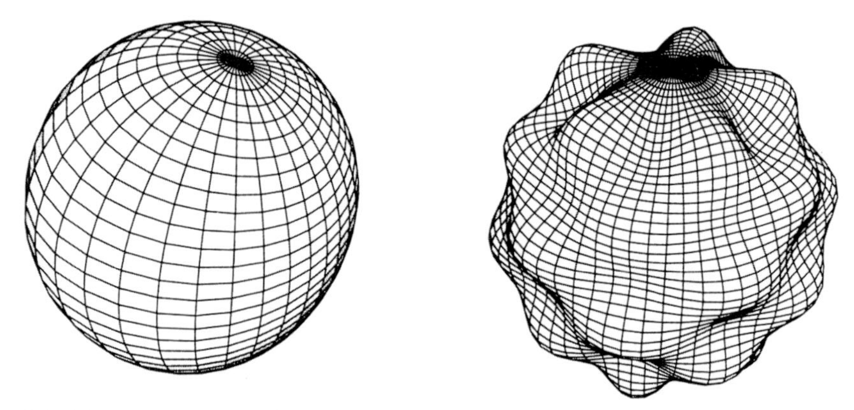

FIG. 4.1 Internal contortions The Sun exhibits over a million shapes produced by its oscillations. Two of these shapes are illustrated here with exaggerated amplitude. (Courtesy of Arthur N. Cox and Randall J. Bos, Los Alamos National Laboratory.)

above and below it. The hotter material acts as a kind of mirror, reflecting sound waves back into the regions they come from. So, the Sun's throbbing motions could be caused by sound waves trapped inside the Sun; on striking the visible disk and rebounding back down, the sound waves cause the gases there to move up and down.

Building upon this result, Leighton and his colleagues concluded in 1962 that:

> The atmosphere may therefore act as a *wave guide* for laterally moving [acoustic] waves, and the observed oscillation may correspond to the lower "cutoff" frequency of the waveguide.[20]

In an extensive observational study of the five-minute oscillations, Edward N. Frazier (1939–), at the University of California, Berkeley, showed, in 1968, that the oscillatory power is concentrated at specific combinations of size and duration, suggesting to him that they are not being pushed in and out by the motion of granules seen on the Sun. Instead, the five-minute oscillations were attributed to sound waves generated by the same convection that produces the up and down granular motion. In his own words:

> The well-known five-minute oscillations are primarily standing resonant acoustic [sound] waves.... [They] are not formed directly from the "piston action" of a convective cell impinging on the stable photosphere, but rather are formed within the convection zone itself.[21]

According to this interpretation, a sound wave resonates within the convective zone, like the plucked string of a guitar or the beat of a drum, effectively standing in one place and growing in power. This resonance effect is somewhat analogous to repeated pushes on a swing. If the pushes occur at the same point in each swing, they can increase the energy of the motion. In the absence of such a resonance, the perturbations would be haphazard and the effect would eventually fade away. When you regularly move water

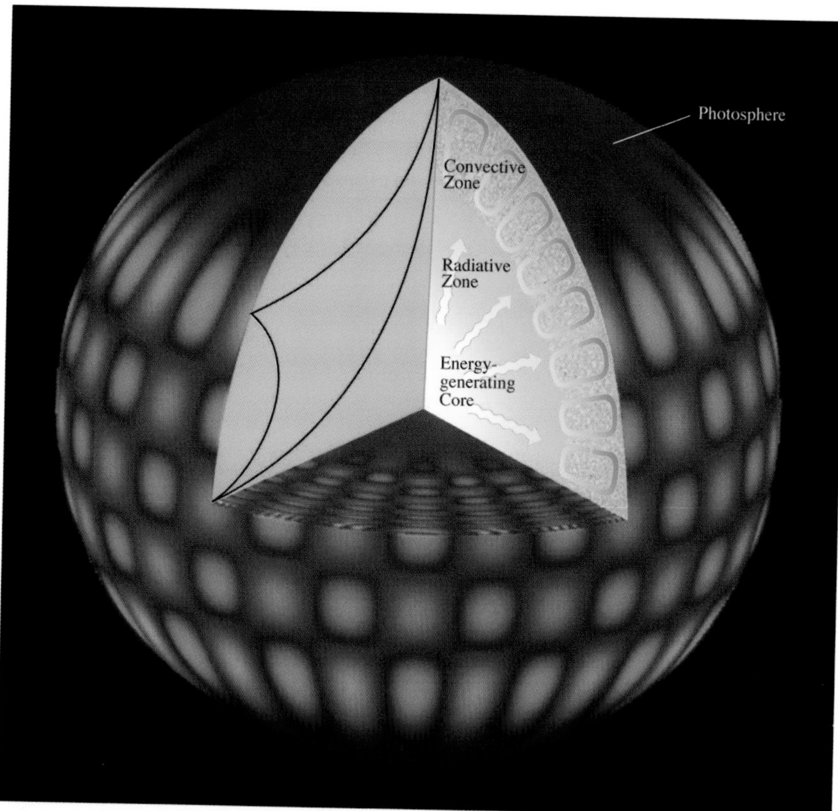

FIG. 4.2 The pulsating Sun Sound waves inside the Sun cause the visible solar disk to move in and out. This heaving motion can be described as the superposition of literally millions of oscillations, including the one shown here for regions pulsing inward (*red regions*) and outward (*blue regions*). The sound waves, whose paths are represented here by black lines inside the cutaway section, resonate through the Sun. They are produced by hot gas churning in the convective zone, which lies above the radiative zone and the Sun's core. (Courtesy of John W. Harvey, National Optical Astronomy Observatories, except cutaway.)

in a bathtub, the waves similarly grow in size, but when you swish it randomly, the water develops a choppy confusion of small waves.

As suggested in 1970 by Roger Ulrich (1942–) at the University of California, Los Angeles, and independently in 1971 by John Leibacher (1941–), then at the Harvard College Observatory, and Robert F. Stein (1935–), then at Brandeis University, the convective zone acts as a resonant cavity, or spherical shell. The sound waves are trapped inside this circular waveguide, and can't get out. They therefore go around and around, bouncing repeatedly against the photosphere like a hamster caught in an exercise wheel, reverberating between the cavity boundaries and driving oscillations in the overlying material.

A real breakthrough came in 1975, when the German astronomer Franz-Ludwig Deubner (1934–) showed that the oscillating power is concentrated into narrow ridges in a spatial-temporal display of horizontal wavelength and period, which meant that they are due to the standing acoustic waves that had been predicted. That is, the vertical motions vary in space and time across the Sun, but only at specific sizes and periods, creating a regular pattern in the apparently random oscillations. For a given size, only certain periods will give a cavity that has the proper depth for a resonant superposition. This pattern is described in terms of narrow bands, or ridges, of enhanced oscillation power when decomposed into a two-dimensional display of size and period, or horizontal wavelength and frequency (Fig. 4.3).

Many of the sound waves eventually fade away without contributing much to the observed motions. Other sound waves are amplified by repeated reflection, like a swing

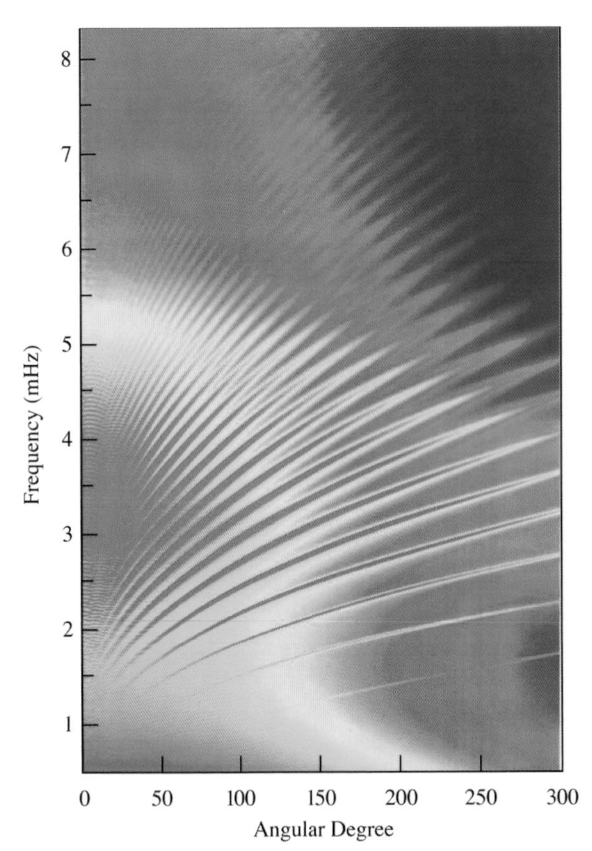

FIG. 4.3 Oscillation period and size Sound waves resonate deep within the Sun, producing oscillations of the photosphere with periods near five minutes, or frequencies near 3 milliHertz, abbreviated 3 mHz and equal to 0.003 cycles per second. Only waves with specific combinations of period or frequency (*left axis*) and size, wavelength, or angular degree (*bottom axis*) resonate within the Sun, producing the fine-tuned "ridges" in the oscillation power shown in this image, obtained from 2 months of data from the Michelson Doppler Imager, abbreviated MDI, instrument aboard the *SOlar and Heliospheric Observatory,* or *SOHO* for short. The angular degree is the inverse of the size or spatial wavelength, and an angular degree of 150 corresponds to waves about 27,000 kilometers in size. The oscillation power is contained within specific combinations of frequency and degree, demonstrating that the oscillations detected in the photosphere are due to internal standing waves confined within resonant cavities. (Courtesy of the *SOHO* MDI/SOI consortium. *SOHO* is a project of international cooperation between ESA and NASA.)

that is pumped at regular intervals. Such a standing wave rises in the same location, over and over again, pushing the visible solar disk in and out at the same places every time it circulates around the Sun. A sound wave's path inside the Sun then forms a regular sequence of loops, like the lace filigree on a napkin or the hoops in a round rug (Fig. 4.4).

When a sound wave angles up to the visible solar gases, it strikes them with a glancing blow, turning around and traveling back into the Sun, like light reflected from a mirror,

FIG. 4.4 Sound paths The trajectories of sound waves are shown in a cross section of the solar interior. The rays are bent inside the Sun, like light within the lens of an eye, and circle the solar interior in spherical shells or resonant cavities. Each shell is bounded at the top by a large, rapid density drop near the photosphere and bounded at the bottom at the inner turning point (*dotted circles*), where the bending rays undergo total internal refraction, owing to the increase in sound speed with depth inside the Sun. How deep a wave penetrates and how far around the Sun it goes before it hits the photosphere depends on the harmonic degree, l. The white curve is for $l = 0$, the blue one for $l = 2$, green for $l = 20$, yellow for $l = 25$ and red for $l = 75$. (Courtesy of Jørgen Christensen-Dalsgaard and Philip H. Scherrer.)

and causing the gas to rise and fall in a ponderous rhythm. Above this level, which is the same for all waves, the sound waves are evanescent and cannot propagate.

The inner turning point, or cavity bottom, depends on the increase of sound speed, or wave velocity, with depth. Because the speed of sound is greater in a hotter gas, it increases in the deeper, hotter layers of the Sun. The deeper part of a wave front traveling obliquely into the Sun moves faster than the shallower part and pulls ahead of it. Gradually the advancing wave is once again headed back up. Sound is similarly refracted down into the cool air above a mountain lake; the hotter, higher air bends the sound downward permitting it to travel great distances across the lake's surface.

So, the increase in sound speed with depth eventually bends a downward-moving sound wave back up, and this inner turning point depends upon the period of the wave. Long-wavelength sound waves penetrate deep into the Sun before they return, while short waves travel through shallower and cooler layers and bounce off the visible solar disk more often. The turn-around depth also depends on physical conditions within the Sun's interior and can be used to determine their radial variation within the Sun.

Each note of the vibrating Sun is similar to the sound wave produced when you tap a crystal glass or strike a doorbell's chime; an outside force makes each sound, which lasts only a short time. Something must therefore ring the solar bell, and set the oscillations in motion. And because any oscillation must eventually lose energy and disappear, the solar oscillations must be continually excited.

The sound, or acoustic, waves that produce the five-minute oscillations are probably generated by vigorous turbulence in the convective zone, where motions at near-sonic speed are expected to be a strong source of acoustic waves. It's somewhat like the deafening roar of a jet aircraft or the loud hissing noise of a boiling pot of water. As the hot convective bubbles rise in the Sun, their motion disturbs the gas they flow through and starts it oscillating. And the violent, turbulent convective motions occur continuously and randomly, so the visible solar disk is always oscillating in tune with its internal sounds.

4.2 ODE TO THE SUN

The Sun reverberates in millions of ways, creating a resonant symphony of notes, each with its own wavelength and path of propagation, trapped within a well-defined spherical shell and bouncing back and forth as it goes around and around like a Ferris wheel. The combined sound has been compared to a resonating gong in a sandstorm, being repeatedly struck with tiny particles and randomly ringing with an incredible din.

At first, this cacophony seemed hopelessly chaotic and complex, but then a hidden order and regular pattern was discovered in the noise. The trapped sound waves combine and reinforce each other to shake the Sun in a regular way, something like cars on a highway that bunch up together, and then disperse as they move along. So the observed up and down motions are in reality the combined effect, or superposition, of millions of separate long-lived notes or vibrations.

Each individual vibration moves the visible solar disk in and out by only a few tens of meters or less, at speeds of about ten centimeters per second. When millions of these vibrations are superimposed, they produce oscillations with peak values as fast as half a kilometer per second, or about five thousand times faster than those of individual vibrations. These large-amplitude combinations are the well-known, five-minute oscillations. They grow and decay as individual vibrations go in and out of phase to combine and disperse and then combine again. Groups of birds or schools of fish will similarly gather together, move apart, and then congregate again.

Upon close examination, it is found that five minutes is not the only oscillation period. The Sun simultaneously resonates with many different periods from 3 minutes to 1 hour. Moreover, the brief duration and small size of the localized five-minute oscillations is an illusion resulting from the combination of many individual components, each persisting for days or longer and reverberating with varying spatial extent. Indeed, it is this longevity that enables astronomers to build up a signal and record single notes within the noisy din.

A major obstacle to studies of the pulsating Sun is the rotating Earth, which keeps us from observing the Sun around the clock and accumulating long records of the solar oscillations. Nightly gaps in the data create background noise that hides all but the strongest oscillations, and the longer sound waves with low frequencies that probe the solar interior to the greatest depths are especially difficult to detect.

Fortunately, continuous uninterrupted observations have been obtained from a joint mission of the European Space Agency, or ESA for short, and NASA, called the *SOlar and Heliospheric Observatory,* abbreviated *SOHO.* It was launched on 2 December 1995, and reached its permanent observing position on 14 February 1996, at a place where the Earth's gravitational pull just equals that of the Sun. *SOHO* therefore orbits the Sun along with the Earth, and does not move around our planet. Previous spacecraft observing the Sun orbited the Earth, which would regularly obstruct their view.

Our atmosphere makes some kinds of solar oscillations difficult to record; especially very short sound waves with exceptionally high frequencies. For these sounds, ground-based observations are something like trying to listen to a Mozart piano concerto when your son is blasting rap music on his boom box. Out in quiet, peaceful and tranquil space, unperturbed by terrestrial interference, *SOHO* has a long, clear undistorted view. It has obtained recordings of both large-scale global oscillations and small-scale ones with unprecedented quality, providing continuous independent velocity measurements over a total of about 800,000 locations on the visible solar disk.

SOHO has kept watch over the Sun for decades. Except, that is, for a three-month interlude when controllers lost contact with the spacecraft (Focus 4.1).

A worldwide network of solar imaging observatories, known as the Global Oscillations Network Group, or GONG for short, has also observed the Sun around the clock. It consists of six identical instruments distributed in longitude around the world, producing a map of photospheric velocities every minute that the Sun is visible at each

FOCUS 4.1

SOHO Lost In Space and Recovered From Oblivion

After more than two years of uninterrupted views of the Sun, and completing its primary mission with unqualified success, *SOHO*'s eyes were abruptly closed. During routine maintenance maneuvers and calibrations on 25 June 1998, the spacecraft spun out of control and engineers could not re-establish radio contact with it.

A group of experts, with the ponderous title "The *SOHO* Mission Interruption Joint ESA/NASA Investigation Board", was assembled to find out what went wrong. Their post-mortem found no fault with the spacecraft itself; a sequence of operational mistakes added up to its loss. A combination of human error and faulty computer command software, which had not been previously used or adequately tested, disabled some of *SOHO*'s stabilizing gyroscopes. Continuous firings of its jet thrusters failed to bring the spacecraft into balance, and instead sent it spinning faster and faster.

For nearly a month, the crippled spacecraft failed to respond to signals sent daily. Scientists feared that *SOHO* might be drifting away from its expected orbit, never to be heard from again. However, the wayward satellite was found by transmitting a powerful radar pulse to it from the world's largest radio telescope located in Arecibo, Puerto Rico. The faint radio echo indicated that *SOHO* was located in the right, predicted part of space, turning relatively slowly at the rate of one revolution every 53 seconds. This gave everyone renewed hope and optimism that radio contact would eventually be established and the spacecraft ultimately recovered.

At this time, *SOHO*'s solar panels were turned edge-on toward the Sun, so its electrical power could not be renewed. Its internal energy had drained away, and the spacecraft was unable to receive or send communications. Power might nevertheless be regained as the panels slowly turned to a more favorable alignment with the Sun during *SOHO*'s annual orbit around the star. Antennas in NASA's Deep Space Network therefore continued to send the satellite wake up messages, asking it to call home.

After six weeks of silence, a feeble and intermittent response was received from the dormant spacecraft, like the faint, erratic heartbeat of a patient in a coma or the worn-out, distressed cries of a tired, lost child. The elated European engineers, who built the spacecraft, knew that *SOHO* was alive and immediately began regaining control of it.

It was a long, slow recovery. The onboard batteries had to be recharged, and the inner workings had to be warmed up after an enforced period of deep freeze. Some of the propellant in its tanks had been frozen solid, and the pipes that carry fuel to the craft's jet thrusters also had to be thawed out. Fortunately, the fuel, named hydrazine, does not expand when it freezes, so the fuel pipes did not crack open as sometimes happens when a building's water pipes freeze during the loss of heat in a severe winter storm. Altogether, it took three months from the initial loss of radio contact to full recovery.

As luck would have it, *SOHO*'s tribulations were not yet over, since its gyroscopes acted up just a few months after it resumed observations. The satellite had to constantly fire its onboard jets to keep it balanced and pointed toward the Sun, and this was rapidly exhausting its fuel supply. Ingenious engineers fixed the problem by instructing the spacecraft to bypass the gyroscopes and use the stars to determine its position, somewhat like ancient mariners who navigated by the stars.

site. These electronically linked sentinels follow the Sun as the Earth rotates. In effect, the Sun never sets on the GONG telescopes.

4.3 LOOKING INSIDE THE SUN

The interior of the Sun is as opaque as a stonewall, and there is no way you can see inside it! But we can illuminate, or sound, the hidden depths of the Sun by recording oscillations of its visible disk, which depend on conditions inside the Sun.

Some of the oscillations are created by sound waves that have moved just beneath the part of the Sun we can see; others arise from sounds that have traveled deep into the Sun's interior. We can therefore use them to open a window into the Sun and look at various levels within it. Moreover, the information can be combined to create a picture of the Sun's large-scale internal structure, somewhat in the manner of an X-ray CAT scan that probes the inside of a human's brain.

Geophysicists similarly construct models of the Earth's interior by recording earthquakes, or seismic waves, that travel to different depths; this type of investigation is called seismology. Most earthquakes occur just beneath the Earth's surface when massive blocks of rock slip and crunch against one another. The reverberations move out and propagate throughout the terrestrial interior, like ripples spreading out from a disturbance on the surface of a pond.

Solar oscillations arise from a persistent, random turbulence in the outer convective regions of the Sun, and similarly shake it to its very center. In fact, astronomers use the name helioseismology to describe such studies of the Sun's interior; it is a hybrid name combining the Greek words *helios* for the "Sun" and *seismos* for "earthquake" or "tremor."

Seismic waves move in all directions through the Earth and their arrival at various places on the Earth's surface is recorded by seismometers. By combining the arrival times of waves that have traveled through the Earth to various points on the surface, seismologists can pinpoint the origin of the waves and trace their motions through the Earth. This enables them to construct a profile of the Earth's interior. In analogy with terrestrial seismology, it is now possible to use observations of the five-minute solar oscillations to isolate sound waves penetrating to a given depth. This has resulted in precise and detailed information about the properties of the solar interior, rivaling our knowledge of the inside of the Earth.

Of course, unraveling the internal constitution of either the Earth or the Sun is not quite as simple as it might seem. One geophysicist has compared seismology to determining how a piano is constructed by listening to one falling down a flight of stairs. The situation is even worse for the Sun, where the observed five-minute oscillations result from millions of different sound waves, with new ones starting up and old ones dying away all the time. One therefore has to measure the wavelength, or frequency, of numerous solar oscillations with incredible precision, and then compare them with theoretical expectations using large computer programs and simplified models of the Sun. Differences between the observed and predicted wavelengths can then be used to test and refine the models.

FIG. 4.5 Radial variations of sound speed Just as scientists can use earthquake measurements to determine conditions under the Earth's surface, measurements of the Sun's oscillations, and the sound waves that produce them, can be used to determine the internal structure of the Sun. This composite image shows the extreme ultraviolet radiation of the solar disk (*orange)* and internal measurements of the speed of sound (*cutaway*). In the red colored layers in the solar interior, sound waves travel faster than predicted by the Standard Solar Model (*yellow*), implying that the temperature is higher than anticipated. Blue corresponds to slower sound waves and temperatures that are colder than expected. The conspicuous red layer, about a third of the way down, shows unexpected high temperatures at the boundary between the turbulent outer region (convective zone) and the more stable region inside it (radiative zone). The disk measurements were made at a wavelength of 30.38 nanometers, emitted by singly-ionized helium, denoted He II, at a temperature of about 60,000 kelvin using the Extreme-ultraviolet Imaging Telescope, abbreviated EIT, aboard the *SOlar and Heliospheric Observatory,* or *SOHO* for short, and the MDI/SOI and VIRGO instruments on *SOHO* made the sound measurements over a period of 12 months beginning in May 1996. (Composite image courtesy of Steele Hill and *SOHO,* a project of international cooperation between ESA and NASA.)

By considering a sequence of waves with longer and longer wavelengths, that penetrate deeper and deeper, it is possible to peel away progressively deeper layers of the Sun and establish the radial profile of the sound speed (Fig. 4.5). A small but definite change in sound speed marks the lower boundary of the convective zone, indicating that it is located at a radius of 71.3 percent, or extends to a depth of 28.7 percent, of the radius of the visible Sun. Moreover, this convective zone depth, measured by the helioseismology technique in 1991, agreed at the time with calculations using the Standard Solar Model of the solar interior.

Nevertheless, there has never been perfect agreement between the measured sound speed and the predictions of the model. Helioseismology measurements of the sound speed just below the convective zone were substantially higher than those of the models. And the accord on the zone's depth was slightly disrupted by measurements, in 2003, of low abundances for the lighter metals in the solar disk.

The Sun appears to contain 30 to 40 percent less carbon, nitrogen, oxygen, neon and argon than previously believed. These elements provide an opacity that impedes the outward flow of radiation, like dirt that blocks light flowing through a window. And when these opacities are used in detailed models of the Sun, they predict slightly different values for the speed of sound than those measured by helioseismology at the same locations.

The solar models constructed with the revised, lower element abundances and corrected opacities yield a depth of the solar convection zone of 27.4 percent, at a radius of 72.6 percent of the Sun's radius, disagreeing with the measured value by helioseismology by a small but significant 1.3 percent. There is also a discord between the helium abundance calculated by the new model and the amount inferred from helioseismology; the new value is lower, at 23 percent by mass, rather than about 25 percent as previously thought. So there is room for improvement in our solar models, perhaps by consideration of magnetic fields, rotation or mixing that were previously omitted in the calculations.

Fortunately, the discrepancies occur at a temperature range that is too cool and distant from the Sun center to significantly affect the helioseismic measurements of the Sun's hotter, deeper temperatures. So the model calculations of the amount of neutrinos emitted by the Sun remain the same.

4.4 BREAKING THE SYMMETRY

To a first approximation, the internal structure of the Sun has spherical symmetry, and the measured properties of the five-minute oscillations depend only on radial variations within the spherical shell in which the sound waves propagate. Upon close inspection, however, there are secondary effects that break this symmetry, such as rotational motions and internal flows. These asymmetric forces can produce a fine structure within frequencies that have been calculated under the assumption that they can be neglected.

The most pronounced symmetry-breaking agent is rotation, which can be expressed as a perturbation to the sound waves in a non-rotating Sun. Waves propagating in the direction of rotation will tend to be carried along by the moving gas, and will move faster than they would in a static medium. A bird or a jet airplane similarly moves faster when traveling with the wind and takes a shorter time to complete a trip. The resonating sound-wave crests moving with the rotation will therefore appear, to a fixed observer, to move faster and their measured periods will be shorter. Waves propagating against the rotation will be slowed down, with longer periods.

Thus, rotation imparts a clear signature to the oscillation periods, lengthening them in one direction and shortening them in the other. These opposite effects make the oscillation periods divide, and such rotational splitting depends on both depth and latitude within the Sun.

Observations of sunspots indicate that the visible solar disk rotates with a period of approximately 25 days at the equator. The solar oscillations have a period of about five minutes, so the rotational splitting is roughly five minutes divided by 25 days, or about one part in seven thousand. The oscillations have to be measured ten or a hundred times more accurately than this to determine subtle variations in the Sun's rotation, or as accurately as one part in a million.

It has been known since 1610, when Galileo Galilei (1564–1642) first used a telescope to observe sunspots, that the solar equator rotates faster than the regions closer to the poles. That is, sunspot observations indicated that the visible solar disk rotates differently at different latitudes, with a faster rate at the equator than the poles and a smooth variation in between (Table 4.1). As on Earth, latitude is the angular distance north or south of the equator. In contrast, the rotation rate of the solid Earth is the same at every latitude, so all points of the globe take the same amount of time to complete a rotation and a day lasts 24 hours everywhere on the planet.

The solar oscillation data indicate that this differential rotation, in which the equator spins faster than the poles, is preserved throughout the convective zone (Fig. 4.6). Within this zone, there is little variation of rotation with depth, and the inside of the Sun does not rotate any faster than the outside. At greater depths, the interior rotation

TABLE 4.1 Differential Rotation of the Sun[a]

Solar Latitude (degrees)	Rotation Period (days)	Rotation Speed (km h^{-1})	Rotation Speed (m s^{-1})	Angular Velocity (nHz)
0 (Equator)	25.67	7097	1971	451
15	25.88	6807	1891	447
30	26.64	5922	1645	434
45	28.26	4544	1262	410
60	30.76	2961	823	376
75	33.40	1416	393	347

[a] Data from the MDI instrument aboard the *SOHO* spacecraft.

no longer mimics that of sunspots, and differential rotation disappears. The internal accord breaks apart just below the base of the convective zone, where the rotation speed becomes uniform from pole to pole. Lower down, within the radiative zone, the rotation rate remains independent of latitude, acting as if the Sun were a solid body. Though gaseous, the radiative interior of the Sun rotates at a nearly uniform rate intermediate between the equatorial and high-latitude rates in the overlying convective zone.

Thus, the rotation velocity changes sharply at the top of the radiative zone, located nearly one-third of the way to the center. There the outer parts of the radiative interior, which rotates at one speed, meet the overlying convective zone, which spins faster in

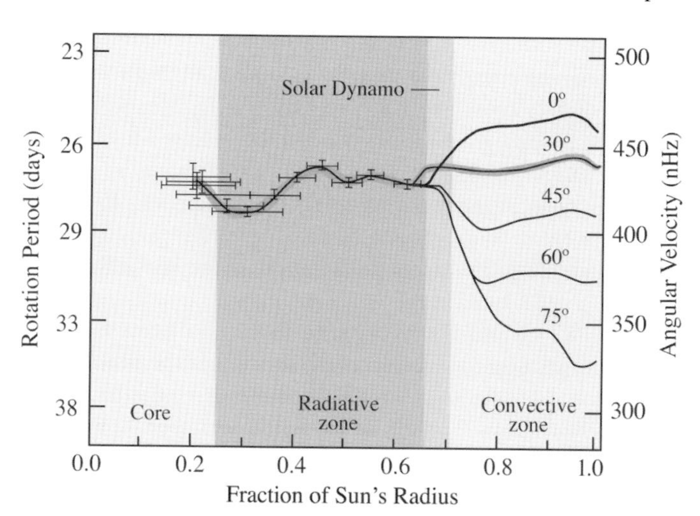

FIG. 4.6 Internal rotation of the Sun The rotation rate inside the Sun, determined from helioseismology. The outer parts of the Sun exhibit differential rotation, with high latitudes rotating more slowly than equatorial ones. This differential rotation persists to the bottom of the convective zone at 28.7 percent of the way down. The rotation period in days is given at the left axis, and the corresponding angular velocity scale is on the right axis in units of nanoHertz, abbreviated nHz, where 1 nHz = 10^{-9}, or a billionth, of a cycle per second. A rotation rate of 320 nHz corresponds to a period of about 36 days (*solar poles*), and a rate of 460 nHz to a period of about 25 days (*solar equator*). The rotation in the outer parts of the Sun, at latitudes of zero (*solar equator*), 30, 45, 60 and 75 degrees, has been inferred from 144 days of data using the Michelson Doppler Imager, abbreviated MDI, aboard the *SOlar and Heliospheric Observatory,* or *SOHO* for short. Just below the convective zone, the rotational speed changes markedly, and shearing motions along this interface may be the dynamo source of the Sun's magnetism. By examining more than five years of low-order acoustic modes, obtained using the GOLF and MDI instruments aboard *SOHO,* the rotation rate has been inferred for the deep solar layers (*error bars*), mainly along the solar equator. There is uniform rotation in the radiative zone, from the base of the convective zone at 0.713 solar radii to about 0.25 solar radii. The acoustic modes (sound waves) do not reach the central part of the energy-generating core. (Courtesy of Alexander G. "Sasha" Kosovichev for the MDI data showing differential rotation in the convective zone, and Sebastien Couvidat, Rafael García and Sylvaine Turck-Chièze for the GOLF/MDI data in the radiative zone. *SOHO* is a project of international cooperation between ESA and NASA.)

its equatorial middle. The transition between these two different regimes takes place in a narrow region, the tacholine, where there is evidently a strong radial rotational shear that most likely plays an important role in the generation of the large-scale solar magnetic field.

The radial rotation profile in the deep solar layers has been measured using more than five years of sound wave observations from the *SOlar and Heliospheric Observatory* (Fig. 4.6). After removing the effects of differential rotation in the convective zone, a uniform rotation is obtained throughout the radiative zone and into the outer parts of the core, down to two tenths of the Sun's radius. To this depth, there is no indication of either a rapidly or slowly rotating center. However, the central parts of the Sun's energy-generating core have not been resolved, so its rotation and structure remain a mystery.

Nevertheless, no asymmetry or oblateness in the shape of the Sun has been detected, as would be expected from exceptionally rapid rotation inside the Sun, and this has important implications for theories of gravity (Focus 4.2).

4.5 INTERNAL FLOWS

White-hot rivers of gas sweep around the Sun's equatorial regions in zonal bands at different speeds, moving just slightly faster or slower than the average rotation at that latitude (Fig. 4.7). The banding is apparently symmetric about the solar equator, with one or two zones of faster rotation and one or two zones of slower rotation in each hemisphere of the Sun. The velocity of the faster zonal flows is about 5 to 10 meters per second higher than gases to either side. This is substantially smaller than the mean velocity of rotation, which is about 2 kilometers per second, so the fast zones glide along in the spinning gas, like a wide, lazy river.

Robert "Bob" Howard (1932–) and Barry LaBonte (1950–) discovered these broad zonal bands in 1980, from Doppler-velocity measurements of the visible solar disk. They named them torsional oscillations, a rather ponderous name. It's clearer if we call them flows, resembling the trade winds in the Earth's atmosphere but on a much larger scale with hotter temperatures.

Instruments aboard the *SOlar and Heliospheric Observatory*, abbreviated *SOHO*, have shown that the internal flows run wide and deep, at about 60,000 kilometers in both directions. After subtracting the time-averaged differential rotation, the *SOHO* scientists found that the zonal flow bands penetrate to a depth of at least 8 percent of the Sun's radius. So they extend through a substantial fraction of the Sun's convective zone, and might involve the entire convective envelope of the Sun.

Ponderous, slow rivers of gas also circulate from the equator toward both poles. This persistent poleward flow, to the north and south, has been dubbed the meridional oscillation. It also extends through a substantial fraction of the convective zone, but scientists have not yet been able to look deep enough to locate the expected return flow toward the equator, which must exist to conserve mass and give the moving material a place to go.

FOCUS 4.2

Confirming Einstein's Theory of Gravity

Instead of always tracing out the same ellipse, the orbit of Mercury pivots around the focus occupied by the Sun. The point of closest approach to the Sun, the perihelion, advances by a small amount, 43 seconds of arc per century, beyond that caused by planetary perturbations. Albert Einstein (1879–1955) invented a new theory of gravity, the *General Theory of Relativity*, which explains this anomalous advance of Mercury's perihelion.

According to this theory, space is distorted and curved in the neighborhood of matter, and this distortion is the cause of gravity. The result is a gravitational effect that departs slightly from Newton's expression, and the planetary orbits are not exactly elliptical. Instead of returning to its starting point to form a closed ellipse after one orbital period, the planet moves slightly ahead in a winding path that can be described as a rotating ellipse.

The observed advance of Mercury's perihelion is in almost exact agreement with Einstein's prediction, but this accord depends on the assumption that the Sun is a nearly perfect sphere. If the interior of the Sun is rotating very fast, it will push the equator out further than the poles, so its shape ought to be somewhat oblate rather than perfectly spherical. After all, even the solid Earth is slightly fatter at the middle because of its rotation, and the effect ought to be more pronounced for a rotating gaseous sphere like the Sun. The size of the oblateness, and the amount that it affects gravity, depend on how fast the interior is rotating.

The gravitational influence of the outward bulge, called a quadrupole moment, will provide an added twist to Mercury's orbital motion, shifting its orbit around the Sun by an additional amount and lessening the agreement with Einstein's theory of gravity. Fortunately, the slow rotation of the outer parts of the Sun, which is inferred from helioseismology, is not enough to produce a substantial asymmetry in its shape, even if the core of the Sun is rapidly rotating. So, we may safely conclude that measurements of Mercury's orbit confirm the predictions of *General Relativity* under the assumption that the Sun is a nearly perfect sphere.

In fact, the small quadrupole moment inferred from the oscillation data, about one ten millionth rather than exactly zero as Einstein assumed, is consistent with a very small difference between radar measurements of Mercury's orbit and Einstein's prediction. So, the Sun does have an extremely small, middle-aged bulge after all.

The entire outer layer of the Sun, to a depth of at least 25,000 kilometers, is slowly but steadily flowing from the equator to the poles with a speed of about 20 meters per second. At this rate, an object would be transported from the equator to a pole in a little more than one year. Of course, the Sun rotates at more than 100 times this rate, completing one revolution at the equator in about 25 days.

Both types of internal flows are linked to the Sun's magnetism. Islands of intense magnetic fields, called sunspots, emerge at the poleward side of the faster zonal bands, marking belts of solar activity. Helioseismologists speculate that sunspots might originate at the boundaries of zones moving at different speeds, where the shearing force and turbulence might twist the magnetic fields and intensify magnetic activity, like two nearby speed boats moving at different velocities and churning the water between them.

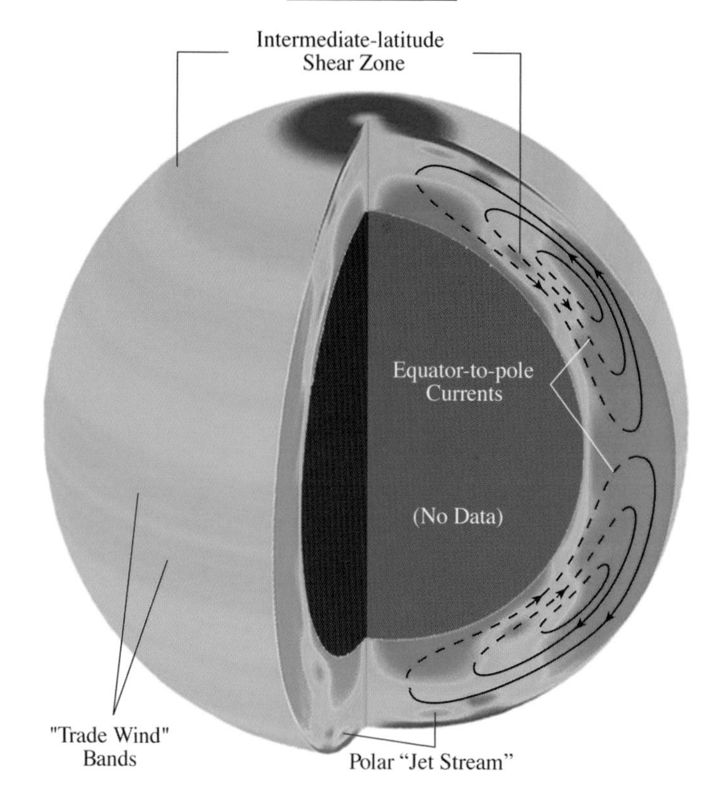

FIG. 4.7 Interior flows Global helioseismology of internal flows in the Sun with rotation removed. Red corresponds to faster-than-average flows, yellow to slower than average, and blue to slower yet. On the left side, deeply rooted zonal flows *(yellow bands),* analogous to the Earth's trade winds, travel slightly faster than their surroundings *(blue regions).* The streamlines in the right-hand cutaway reveal a slow meriodional flow toward the solar poles from the equator; the return flow below it is inferred. This image is the result of computations using one year of continuous observation, from May 1996 to May 1997, with the Michelson Doppler Imager, abbreviated MDI, instrument aboard the *SOlar and Heliospheric Observatory,* or *SOHO* for short. (Courtesy of Philip H. Scherrer and the *SOHO* SOI/MDI consortium. *SOHO* is a project of international cooperation between ESA and NASA.)

The magnetic activity belts and the zonal flow bands migrate together, moving slowly toward the solar equator during the Sun's 11-year cycle of magnetic activity. The meriodional flows also vary with the solar cycle, and these flows may transport leftover magnetic flux to the polar regions, resulting in the reversal of the Sun's global, dipolar magnetic field.

This therefore brings us to a discussion of the magnetic fields that pervade the solar atmosphere.

Chapter Five

A Magnetic Star

GRAINSTACK AT SUNSET NEAR GIVERNY In this painting, the French artist Claude Monet (1840–1926) captures the faint, reddened sunlight and long shadows at the end of the day. In both the Grainstack and Rouen Cathedral paintings he portrayed the same object from dawn to dusk, and from summer to winter, to describe the subtle effects of changing light and shadow. (Courtesy of Museum of Fine Arts, Boston, Juliana Cheney Edwards Collection.)

5.1 MAGNETIC FIELDS IN THE VISIBLE PHOTOSPHERE

The sunlight we see coming from the bright disk of the Sun originates in the photosphere, a word derived from the Greek *photos* for "light" and *sphere* for its spherical shape. The photosphere is a thin tenuous layer, only a few hundred kilometers thick, with a temperature of 5,780 kelvin and a pressure of one ten-thousandth (0.0001) of the pressure of the Earth's atmosphere at sea level. You are looking at the photosphere when you watch the Sun rise in the morning, and continue on its daily journey across the sky.

Although the photosphere is the only part of the Sun we can see, an extended, invisible solar atmosphere lies above it. We look right through this overlying, rarefied gas, just as we see through the Earth's air. The photosphere therefore does not mark the surface of the Sun, for it has no surface that divides the inside from the outside.

Observing the Sun is like looking into the distance on a foggy day. At a certain distance, the total amount of fog you are looking through mounts up to make an opaque barrier. The fog then becomes so thick and dense that radiation can penetrate no further, and we can only see that far. When looking into the solar atmosphere, you can similarly see through only so much gas. For visible sunlight, this opaque layer is the photosphere, the level of the Sun from which we get our light and heat.

The sharp edge of the photosphere is an illusion, caused by unusual ions that make the gas as opaque as a brick wall. In a rare collision, a hydrogen atom in the photosphere can briefly capture a free electron, temporarily becoming an ion with a negative charge. These negative hydrogen ions absorb radiation coming from the solar interior, re-emit visible sunlight, and account for the Sun's sharp edge. At higher levels in the solar atmosphere, there are no hydrogen ions, so the light can escape and we can see right through the overlying gas.

To most of us, the photosphere looks like a perfect, white-hot globe, round, smooth and without a blemish. It's like taking a quick, sideways glimpse of a beautiful woman or handsome man; a sustained, close-up look usually reveals some imperfection. Detailed scrutiny indicates that the Sun is pitted with dark spots called sunspots.

Large sunspots can be seen with the unaided eye through fog or haze, or sometimes at sunrise or sunset, when the Sun's usual brightness is heavily dimmed. But you normally cannot look directly at the Sun without severely damaging your eyes, and most sunspots are too small to be readily visible by naked eye observations, requiring a telescope to be detected. Ancient Chinese records nevertheless indicate that large sunspots have been observed with the unaided eye for nearly 2,000 years.

Sunspots became a lot easier to see around 1610, when not less than four men turned the newly invented telescope toward the Sun, independently and nearly simultaneously confirming the existence of sunspots. One of them, the Italian scientist Galileo Galilei (1564–1642), carried out a detailed scrutiny, announcing that sunspots are embedded in the solar atmosphere, and not in front of it. He also used their apparent motion to show that the Sun is spinning in space, and turning around once every 27 days, as viewed from the moving Earth. Because our planet orbits the Sun in the same direction that the Sun rotates, the rotation rate observed from Earth is about 2 days longer than the Sun's intrinsic rate of spin of about 25 days at the solar equator.

Galileo noticed that sunspots change in size and shape, and that they eventually fade and disappear from view. The sunspots appear from inside the Sun, often remaining visible for only a few days. Some last just a few hours; others persist for weeks and even months. As sunspots form, they can also coalesce, and move past or even through each other.

The dark, ephemeral spots on the apparently serene Sun therefore indicate that it is an imperfect place of constant turmoil and change, contradicting the Greek philosopher Aristotle's (384–322 BC) philosophy of cosmic perfection and immutability.

In Galileo's time it was also discovered that sunspots near the equator rotate more rapidly than those nearer to the poles; this means that different parts of the Sun's visible disk rotate about its axis at different speeds, a phenomenon now known as differential rotation. This effect was thoroughly studied two centuries later by Richard C. Carrington (1826–1875), a wealthy English amateur, from his private observatory at Redhill. He showed in 1863 that the Sun's apparent period of rotation increases systematically with latitude from 27 days at the equator to about 30 days halfway toward the poles, or at a latitude of 30 degrees.

Because they are relatively cool, sunspots appear dark in contrast with their bright surroundings. A sunspot might have a temperature of 3,500 kelvin, for example, instead of the 5,780 kelvin of adjacent regions. Astronomers have even detected the spectral signatures of certain molecules in sunspots, while the surrounding material is so hot that molecules are broken apart. However, appearances can be deceiving, for a sunspot is

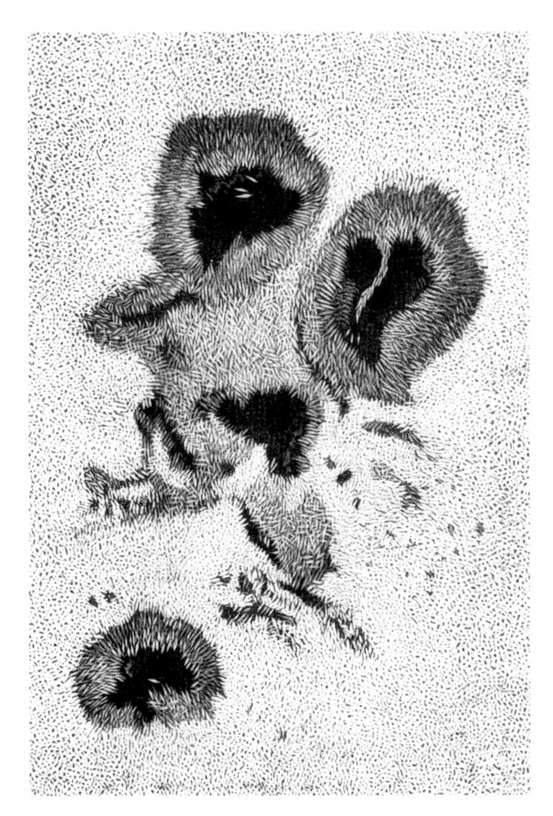

FIG. 5.1 Sunspots This drawing of sunspots was made more than a century ago, in June 1861 by the Scottish engineer and astronomer James Nasmyth (1808–1890). (Adapted from *Le Ciel,* Librairie Hachette: Paris, 1877.)

almost 10 times hotter than the temperature of boiling water, at 373 kelvin, and although sunspots appear dark in comparison to the nearby hotter gas, they still radiate light.

Telescopic observations of large sunspots in normal white light indicate a dark center, the umbra, surrounded by a less dark penumbra, both standing in stark contrast to the rather bland and uniform background (Figs. 5.1, 5.2). Although it looks small in comparison to the solar disk, a large sunspot umbra can surpass the Earth in size.

Modern telescopes have revealed the detailed features of sunspots (Fig. 5.3). A simple sunspot has a dark center, called the umbra, surrounded by a lighter penumbra. The umbra area is a constant fraction (0.17 ± 0.03) of the total area of the spot. A fully developed sunspot has a typical penumbral diameter of between 20,000 and 60,000 kilometers,

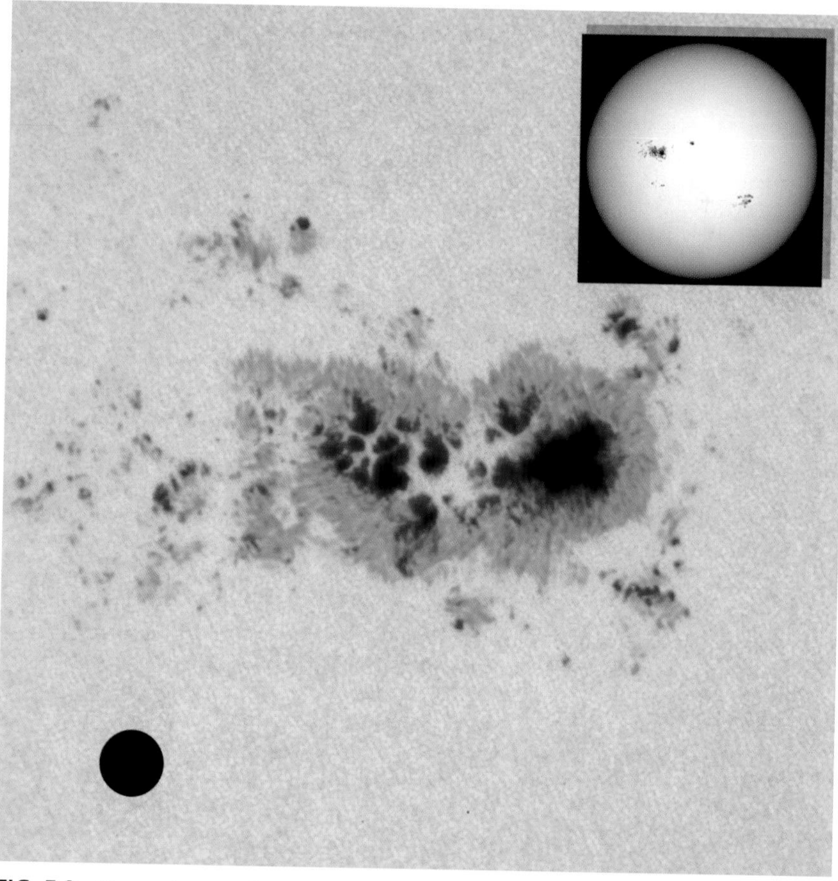

FIG. 5.2 Sunspot group Intense magnetic fields emerge from the interior of the Sun through the photosphere, producing groups of sunspots. The sunspots appear dark because they are slightly cooler than the surrounding photosphere gas. This composite image shows the visible solar disk in white light, or all the colors combined (*upper right*) and an enlarged white-light image of the largest sunspot group (*middle*), which is about 12 times larger than the Earth whose size is denoted by the black spot (*lower left*). (Courtesy of *SOHO*, ESA and NASA.)

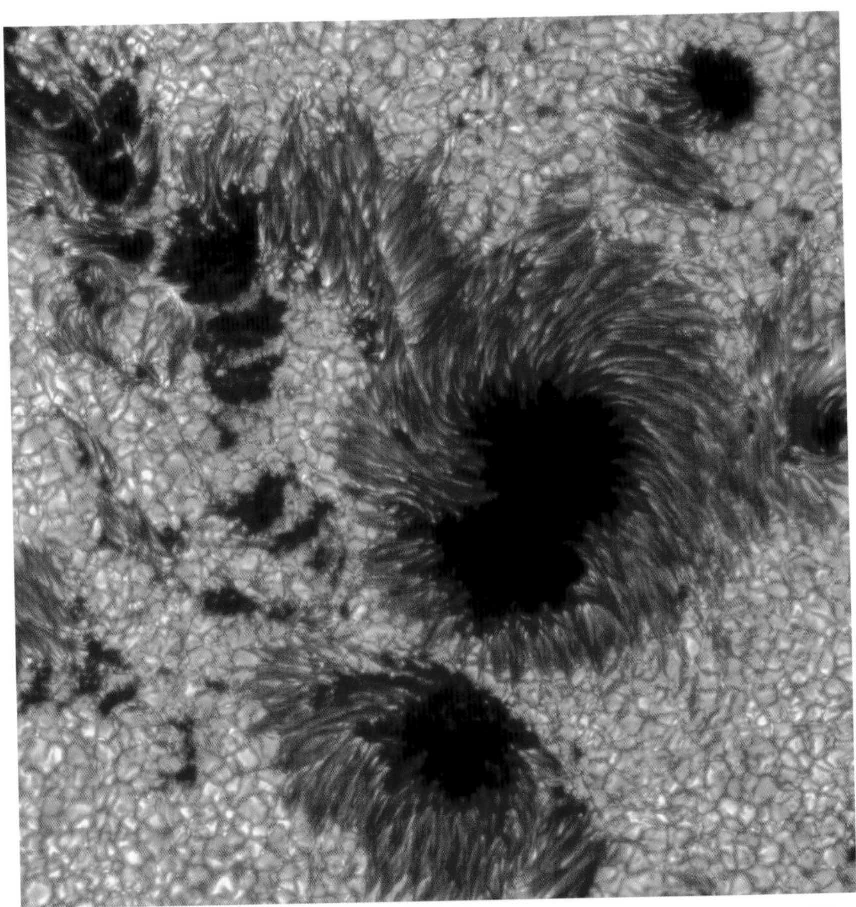

FIG. 5.3 Detailed image of sunspots The Swedish Solar Telescope, abbreviated SST, on the Canary Island of La Palma observes the solar photosphere with unprecedented detail as small as 0.1 seconds of arc, or just 75 kilometers on the Sun, which is about one thousand times better than the angular resolution of our unaided eyes. This image was obtained using a 1-meter adaptive mirror that can change shape 1,000 times a second to compensate for atmospheric blurring and obtain the sharpest-ever pictures of the Sun. Flowing tendrils of ionized gas, or plasma, encircle the dark sunspots; these penumbral filaments contain thin dark cores. Outside the penumbra, the solar granulation is visible. (Courtesy of SST, Göran Scharmer, Institute for Solar Physics, Royal Swedish Academy of Sciences.)

which can be compared with the Earth's mean diameter of 12,700 kilometers. The penumbra consists of radial bright and dark filaments that arch and splay.

In 1908, the American astronomer George Ellery Hale (1868–1938) showed that sunspots are regions of intense magnetism, thousands of times stronger than the Earth's magnetic field. As Hale (Fig. 5.4) suggested, a subtle division and polarization of an atom's spectral lines can measure solar magnetism. This magnetic transformation has been named the Zeeman effect, after the Dutch physicist, Pieter Zeeman (1865–1943), who first noticed it in the terrestrial laboratory in 1896. Hendrik Lorentz (1853–1928)

FIG. 5.4 George Ellery Hale The American astronomer George Ellery Hale (1868–1938), shown here, and the French astronomer Henri Deslandres (1853–1948) independently invented the spectroheliograph in 1891. It is used to image the Sun in the light of one particular wavelength only. Hale subsequently measured the Zeeman effect on the Sun, using laboratory comparisons to establish the existence of strong magnetic fields in sunspots. In later work, Hale and his colleagues found that the majority of sunspots occur in pairs of opposite magnetic polarity, and that the preceding and following spots have opposite polarity that also reverses sign in the northern and southern hemispheres and from one 11-year activity cycle to the next. Hale developed a solar observatory at Mt. Wilson, California in the early 1900s, and even installed a solar telescope in his home in Pasadena together with a large relief of Apollo, the Sun God. (Courtesy of the Archives, California Institute of Technology.)

predicted the effect, and the two Dutch physicists shared the 1907 Nobel Prize in Physics for their investigations of the influence of magnetism on radiation.

When an atom is placed in a magnetic field, it acts like a tiny compass, adjusting the energy levels of its electrons. If the atomic compass is aligned in the direction of the magnetic field, the electron's energy increases; if it is aligned in the opposite direction, the energy decreases. Since each energy change coincides with a change in the wavelength emitted by that electron, a spectral line emitted at a single wavelength by a randomly oriented collection of atoms becomes a group of three lines of slightly different wavelengths in the presence of a magnetic field (Fig. 5.5). The size of an atom's internal adjustments, and the extent of its spectral division, increases with the strength of the magnetic field.

Furthermore, the light at each of these divided wavelengths has a preferred orientation, or circular polarization, that depends on the direction, or polarity, of the magnetic field. Lines that are split by a magnetic field that is directed out along the line of sight have right-hand circular polarization, those pointing in the opposite, inward direction have left-hand circular polarization. So, one can measure the size of the spec-

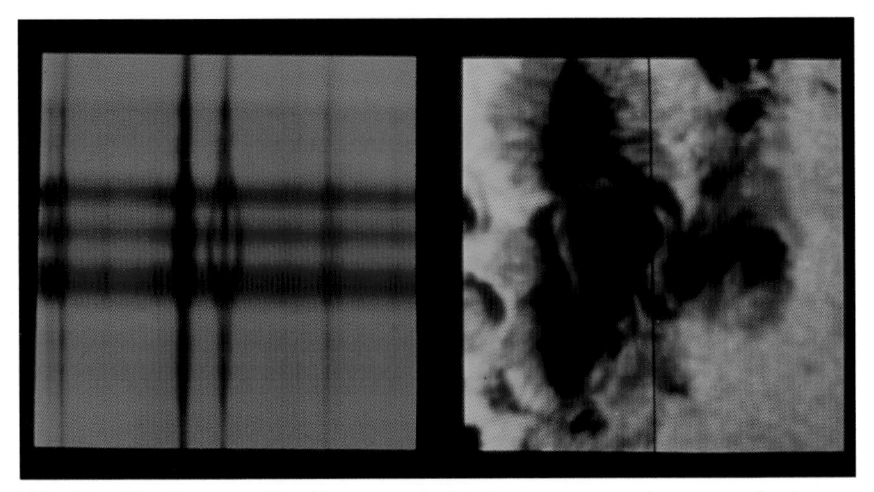

FIG. 5.5 The Zeeman effect The magnetic field in a sunspot can be measured using the Zeeman effect. In a sunspot (*right*) the spectral lines that are normally at a single wavelength become split into two or three components (*left*), depending on the orientation of the field with respect to the line of sight. The vertical line crossing the sunspot denotes the alignment of the observing instrument slit. The separation of the outermost components is proportional to the strength of the magnetic field, in this sunspot about 0.4 tesla, or 4,000 gauss. The components also have a circular polarization, which indicates the direction of the longitudinal magnetic field. (Courtesy of NOAO.)

tral division and use polarized filters to determine both the strength and the direction of the magnetic field using the Zeeman effect.

Hale used the 60-foot tower telescope on Mount Wilson, California to show that spectral lines in the light from sunspots are both divided and polarized, indicating that strong magnetic fields are concentrated within sunspots where the magnetism points into or out from the Sun. By comparing the Zeeman splitting of spectral lines from sunspots with those in laboratory experiments, Hale demonstrated that sunspots have magnetic fields as strong as 0.4 tesla, or 4,000 gauss, extending over areas larger than the Earth itself. The sunspot's magnetism can be thousands of times stronger than the Earth's magnetic field, which orients our compasses; it has a strength of about 0.0003 tesla, or 0.3 gauss, at the equator.

One unit of magnetic field strength is named after the German mathematician Karl Friedrich Gauss (1777–1855), who showed in 1838 that the Earth's dipolar magnetic field must originate within its interior core. Many astronomers still use the gauss, abbreviated G, unit, but the alternate unit of magnetic field strength, the tesla, or T for short, is also used. It is named for the Croatian-born, American electrical engineer Nikola Tesla (1856–1943), a pioneer in the use of alternating-current electricity. For conversion, one tesla = 1T = 10,000 gauss = 10^4 G.

It is powerful magnetic fields that protrude to darken the skin of the Sun, forming dark, Earth-sized sunspots. The intense sunspot magnetism acts as a filter or valve, choking off the heat and energy flowing outward from the solar interior. The strong magnetic fields in sunspots inhibit the convection currents that usually carry hot mate-

FOCUS 5.1
Looking Into and Beneath Sunspots

Scientists were perplexed for decades over what holds sunspots together. Although sunspots can stay organized for several weeks, the outward pressure of their strong magnetic fields ought to make them expand and disperse much more rapidly.

The Chinese astronomer Junwei Zhao (1971–), working with his colleagues Alexander G. "Sasha" Kosovichev (1953–) and Thomas L. Duvall, Jr. (1950–) at Stanford University, has used instruments aboard the *SOlar and Heliospheric Observatory* to trace out the motions of hot, flowing gas in, around and below sunspots. They find that sunspots are held together by powerful converging flows, which force the intense magnetic fields together.

Sound waves move with relatively slow speed within and just below sunspots, as expected since the sunspots are colder than their surroundings. However, the waves speed up at depths of roughly 10,000 kilometers, where the converging flows disappear and the temperatures become roughly uniform.

When observed in the photosphere, sunspots apparently rotate faster than surrounding material, which has less pronounced magnetism. This also suggests that sunspots are relatively shallow, anchored to faster rotating material in the interior, within roots extending about 10,000 kilometers below the photosphere.

rial from deeper layers, so less convective heat bubbles up within them. The intense magnetism thus acts like a refrigerator, keeping sunspots thousands of degrees cooler than the turbulent gas around them. And since it is cooler, a sunspot does not give off as much light as the adjacent material and looks dark in comparison. Indeed, the darkest and coolest parts of sunspots are the sites of the most intense magnetic fields.

At the center of a sunspot, within the umbra, the magnetic field is a few thousand Gauss in strength, and pokes straight out or in from the Sun, like an island of intense magnetism with a fixed direction of north or south polarity. The magnetic field weakens and spreads out in umbrella fashion within the fluted penumbral filaments, where the magnetic field strength is perhaps one thousand Gauss.

Matter flows out from the inner spot penumbra to its surrounding photosphere along the dark penumbral filaments at velocities of a few kilometers per second, in an outflow named after the English astronomer John Evershed (1864–1956) who first observed it in 1908 from the Kodaikanal Observatory in India.

Deeper down, below the Evershed effect, helioseismologists have detected strong converging flows around sunspots and downward directed flows in them. The adjacent streams of gas strengthen and converge towards the sunspots, helping to hold them together (Focus 5.1). Further down, the flows rip through the sunspots, apparently causing them to merge into deeper layers with about the same temperature. This suggests that sunspots are relatively shallow phenomena, extending roughly 10,000 kilometers below the photosphere, which means that sunspots are about as deep as they are broad. It is as if the buoyant, concentrated magnetic fields rose up to the photosphere, gathering together and spreading out like a lotus flower on a lake, but perhaps connected to the depths with slender, thread-like roots.

Nowadays, astronomers use an array of detectors that measure the Zeeman effect at different locations across the photosphere, producing a magnetogram that displays the strength and direction of the solar magnetic fields all across the visible solar disk. Two images are produced, one in each circular polarization, and the difference of these images produces the magnetogram (Fig. 5.6). Strong magnetic fields show up as bright or dark regions, depending on their polarity; weaker ones are less bright or dark.

The magnetograms demonstrate that there is a lot of magnetism in the photosphere outside sunspots. More than 90 percent of these magnetic fields are concentrated into intense magnetic flux tubes that appear, disappear and are renewed in just 40 hours. The individual flux tubes are a few hundred kilometers across and have magnetic field strengths of about 2,000 gauss, comparable to those of the much larger sunspots.

New magnetic flux emerges in the center of large-scale convective cells, known as the supergranulation. Each supergranule cell is about 30,000 kilometers in diameter. Material in the giant cells moves horizontally across the photosphere, sweeping the small magnetic flux tubes from the center to the edges of each supergranule where the magnetism collects and strengthens. The photosphere's magnetic field is thereby concentrated into a magnetic network that traces out the polygonal pattern of the common boundaries of the supergranule cells, like weeds growing in the cracks between mosaics of paving stones.

Thus, the magnetic field in the photosphere is always present in concentrated form, either in sunspots or at the edges of supergranules. The magnetic fields are moved into these concentrations by the motions of material in and below the photosphere. This concentrated magnetism extends up into the overlying solar atmosphere, through the chromosphere and into the corona.

5.2 THE SOLAR CHROMOSPHERE AND ITS MAGNETISM

Just above the photosphere lies a thin gaseous layer called the chromosphere, from *chromos,* the Greek word for "color " and *sphere* for its spherical shape. The chromosphere is so faint that it was first observed during a total eclipse of the Sun. It became visible a few seconds before and after the eclipse, creating a narrow pink, rose or ruby-colored band at the limb of the Sun. The solar limb is the apparent edge of the photosphere disk as viewed from the Earth; during a total solar eclipse the Moon just covers the photosphere, and the edge of the Moon coincides with the solar limb.

When observing the spectrum of visible sunlight during a total solar eclipse, bright emission lines suddenly flash into view at the exact wavelengths of some of the dark absorption lines in the photosphere. This is because you are looking from the side or edge, seeing the energy emitted from the chromosphere without intense photosphere sunlight in the background. The brightest emission line is the hydrogen-alpha line of hydrogen atoms, at 656.3 nanometers, which gives the chromosphere it red color.

During the eclipse of 18 August 1868, the French astronomer Pierre Jules César "P. J. C." Janssen (1824–1907) discovered a bright yellow emission line in the chromosphere at a wavelength of 587.6 nanometers, which did not seem to come from any known element. Soon thereafter, the English astronomer Joseph Norman Lockyer (1836–1920) succeeded in seeing the chromosphere spectral lines outside the solar limb, with a spectroscope and without an eclipse, also giving the unknown substance the name helium, from the word *Helios,* the "God of the Sun" in Greek mythology.

Helium wasn't discovered on Earth until 1895, by heating uranium minerals; during its radioactive decay uranium emits helium nuclei, or alpha particles, that combine with electrons to make helium atoms that get trapped in the rock.

Other prominent emission lines of the chromosphere are the two violet lines of calcium ions. Designated Ca II, the calcium is singly ionized, so the calcium atoms are missing one electron. They emit radiation at wavelengths of 393.4 and 396.8 nanometers, and are often called the calcium K and H lines after Fraunhofer's designation of the corresponding absorption lines in the underlying photosphere.

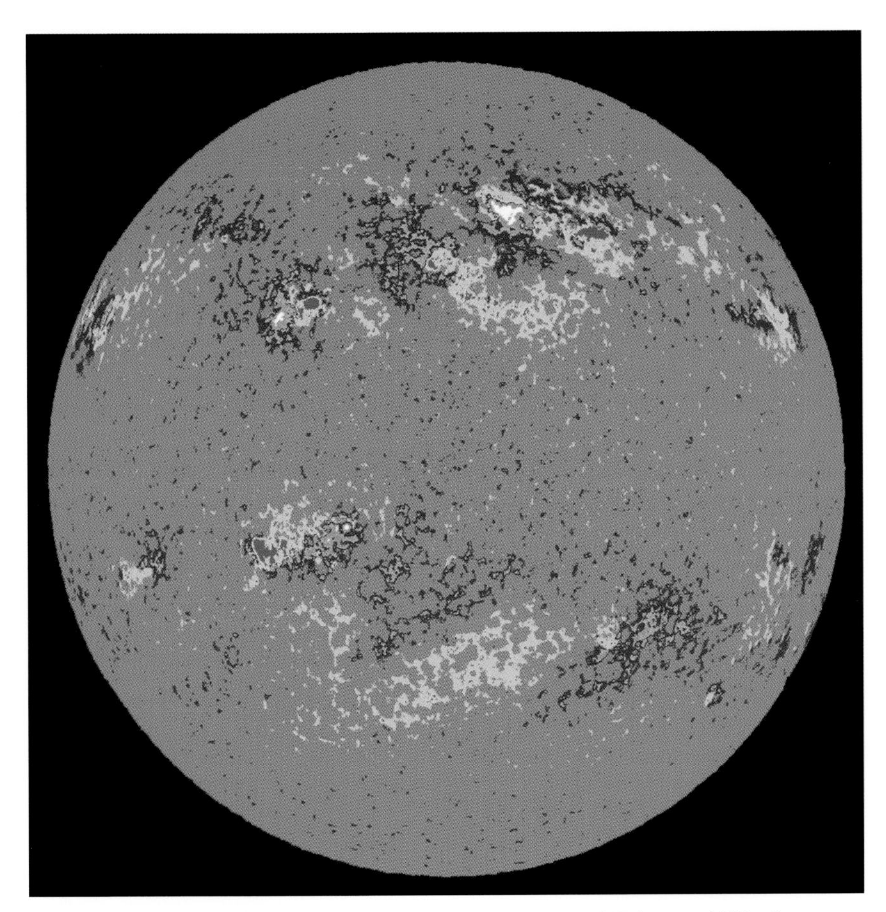

FIG. 5.6 Magnetogram This magnetogram was taken on 12 February 1989, close to the maximum in the Sun's 11-year cycle of magnetic activity. Yellow represents positive or north polarity pointing out of the Sun, with red the strongest fields which are around sunspots; blue is negative or south polarity that points into the Sun, with green the strongest. In the northern hemisphere (*top half*) positive fields lead, in the southern hemisphere (*bottom half*) the polarities are exactly reversed and the negative fields lead. (Courtesy of William C. Livingston, NSO and NOAO.)

In 1891 the French astronomer Henri Deslandres (1853–1948) at Meudon and the American astronomer George Ellery Hale (1868–1938) at Mount Wilson independently invented an entirely new way of observing the chromosphere. Instead of looking at all of the Sun's colors together, they devised an instrument, called the spectroheliograph, meaning Sun spectrum recorder. It creates an image of the Sun in just one color or wavelength, without the blinding glare of all the other visible wavelengths. In a spectroheliograph, the sunlight falls on a vertical slit, and light coming through the slit is spread out in wavelength by a diffraction grating (Fig. 5.7). Light at the wavelength of one of the bright emission lines is then directed through a second slit. The two slits are moved together, with the first slit scanning the Sun from side to side, and an image of the Sun is obtained at the chosen wavelength.

By tuning in the red emission of hydrogen or a violet line of calcium, the spectroheliograph can be used to isolate the light of the chromosphere and produce photographs

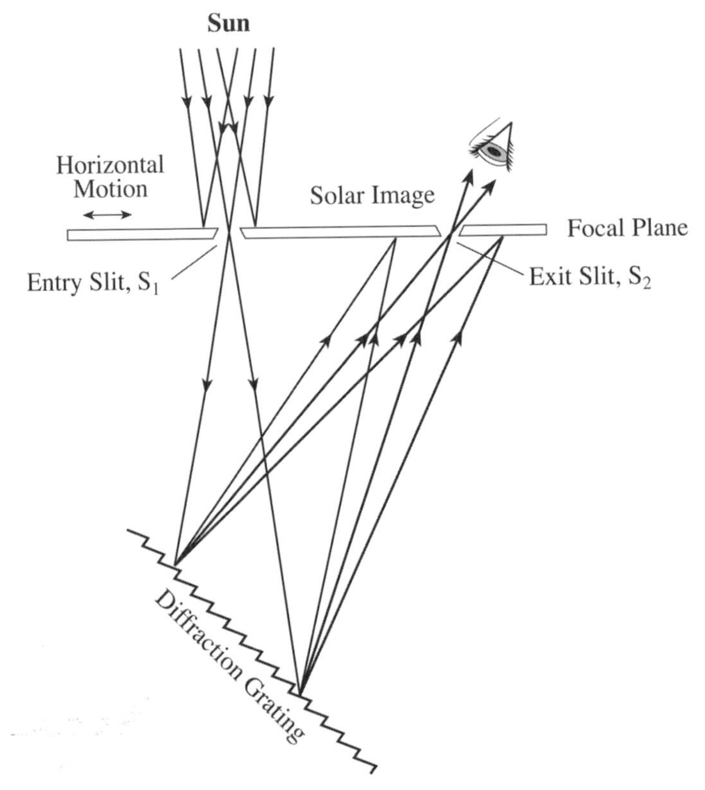

FIG. 5.7 Spectroheliograph A small section of the Sun's image at the focal plane of a telescope is selected with a narrow entry slit, S_1, and this light passes to a diffraction grating, producing a spectrum. A second slit, S_2, at the focal plane selects a specific wavelength from the spectrum. If the plate containing the two slits is moved horizontally, then the entrance slit passes adjacent strips of the solar image. The light leaving the moving exit slit then builds up an image of the Sun at a specific wavelength.

or digital images of it without the blinding glare of all the rest of visible sunlight. In this way, the chromosphere can be observed across the entire disk whenever the Sun is in the sky, rather than just at the edge during a brief, infrequent eclipse.

Sunspots extend from the photosphere into the chromosphere, creating dark regions in hydrogen-alpha photographs (Fig. 5.8). Bright regions, called *plage* from the French word for "beach", glow in hydrogen light; they are often located near sunspots in places with intense magnetism. The plages are chromospheric phenomena detected in monochromatic hydrogen-alpha light; they are associated with, and often confused with, bright patches in the photosphere, called *faculae* – Latin for "little torches," seen near the solar limb in white light. Long, dark filaments also curl across the hydrogen-alpha Sun. They are huge regions of dense, cool gas supported by powerful magnetic

FIG. 5.8 The Sun in hydrogen alpha This global image of the Sun was taken in the light of hydrogen atoms, emitting at the alpha transition that occurs at a particular red wavelength of 656.3 nanometers. It shows small, dark, magnetic sunspots, long, dark, snaking filaments, and bright plage. (Courtesy of the Baikal Astrophysical Observatory, Academy of Sciences, Russia.)

forces. Indeed, the Sun's magnetism dominates the hydrogen-alpha world and gives rise to its startling inhomogeneity.

Although the chromosphere is often described as a thin layer of gas, about 10,000 kilometers thick, it consists of a jagged, dynamic, ever-changing set of little vertical spikes. They were described as early as 1877 by the Italian astronomer and priest Pietro Angelo Secchi (1818–1878), and named spicules by the American astronomer Walter Orr Roberts (1915–1990) in 1945. When you look at the edge, or limb, of the Sun in hydrogen-alpha light, hundreds of thousands of the evanescent, flame-like spicules are observed dancing in the chromosphere at any given moment.

The short-lived spicules rise and fall like chopping waves on the sea or a prairie fire of burning, wind-blown grass (Fig. 5.9). The needle-shaped spicules are about 2 kilometers in width, and shoot up to heights of up to 15,000 kilometers in 5 minutes, moving at speeds of up to 25 kilometers per second. The spicules then fall back down again, but new spicules continually arise as old ones fade away.

For more than a century, no one knew for certain just what causes the upward moving spicules, but the mystery now seems to have been solved. There were two clues

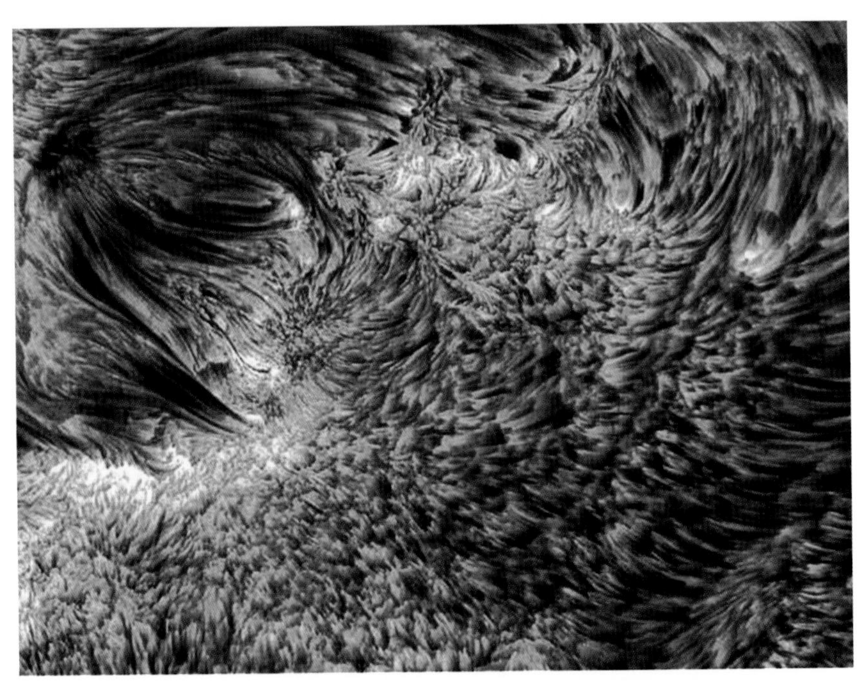

FIG. 5.9 Spicules Thousands of dark, long, thin spicules, or little spikes, dot this high-resolution image of a solar active region, taken with the Swedish Solar Telescope, abbreviated SST, on the Canary Island of La Palma. Layered, needle-shaped spicules (*right side*), each about a kilometer wide, shoot out to more than 15,000 kilometers. The narrow jets of gas are moving out of the solar chromosphere in magnetic channels, or flux tubes, at supersonic speeds of up to 25 kilometers per second. Time-sequenced images have shown that these spicules rise and fall in about five minutes, driven by sound waves beneath them. (Courtesy of SST, Royal Swedish Academy of Sciences, and LMSAL.)

to the solution. First, the spicule lifetimes are comparable to the five-minute period of the photosphere oscillations, and second, the spicules consist largely of ionized material that will follow the direction of magnetic field lines. Some of the sound waves that push the photosphere in and out also move into the chromosphere, powering shocks that drive upward flows along magnetic flux tubes, forming the spicules.

A completely different view of the chromosphere is obtained when it is pictured in the calcium H or K emission lines. Bright regions of calcium light correspond to places where there are strong magnetic fields, both above sunspots and all over the Sun in a network of magnetism (Fig. 5.10). Regions of intense magnetism are probably

FIG. 5.10 Calcium magnetic network This global spectroheliogram of the Sun was taken in the light of the singly ionized calcium, abbreviated Ca II, at the core of the violet K line with a wavelength of 393.4 nanometers. The emission outlines the chromosphere calcium network where magnetic fields are concentrated, like the edges of the tiles in a mosaic; it lies just above the magnetic network of concentrated magnetism in the photosphere. The brightest extended regions are called plages; they are dense places in the chromosphere found above sunspots or other active areas of the photosphere in regions of enhanced magnetic field. (Courtesy of NSO, NOAO, and NSF.)

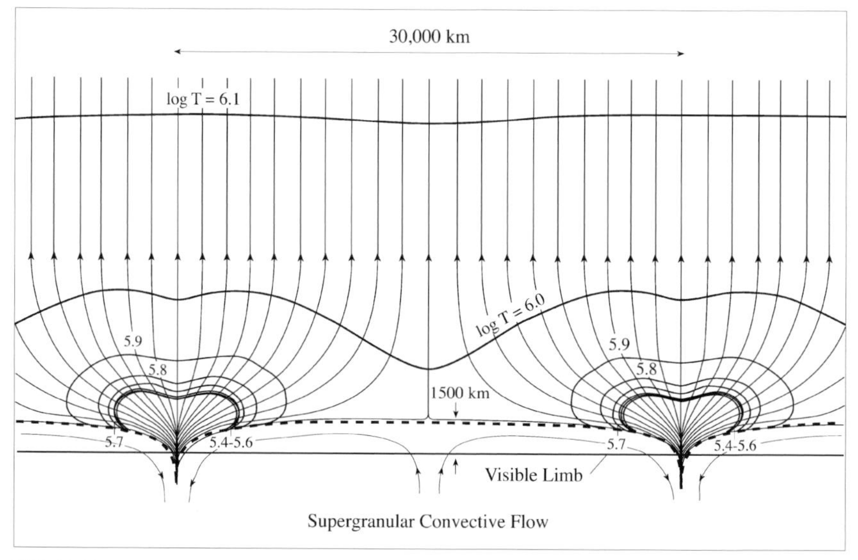

FIG. 5.11 **Magnetic canopy** A two-dimensional, radial cross section of the magnetic network model of the solar transition region. The motion of supergranular convective cells (*bottom*) concentrates magnetic fields at their boundaries in the photosphere. The magnetic fields (*arrowed lines*) are pushed together and amplified up to 0.1 tesla at the cell edges. Heating in the chromosphere above this magnetic network produces bright calcium emission (see Fig. 5.10). The concentrated magnetic fields expand and flare out with height in the overlying corona, producing the magnetic canopy. Temperature, T, contours between log T = 6.1 (*corona*) and log T = 5.4 (*upper transition region*) are marked. [Adapted from Alan H. Gabriel, *Philosophical Transactions of the Royal Society (London)* **A281**, 339–352 (1976).]

associated with heating of the chromosphere, resulting in the bright calcium emission lines that coincide with the supergranulation cell boundaries and the photosphere's magnetic network.

The flux tubes in the photosphere's magnetic network exhibit an expansion with height, opening up into the overlying solar chromosphere and forming a magnetic canopy (Fig. 5.11). The magnetic canopy is a layer of magnetic field which is directed parallel to the solar surface and located in the low chromosphere, like a canopy in a rain forest; the tree-trunks correspond to the magnetic flux tubes in the photosphere that rise in the vertical direction and spread out like branch foliage in the chromosphere.

5.3 BIPOLAR SUNSPOTS, MAGNETIC LOOPS AND ACTIVE REGIONS

Like people, sunspots often group together, often in opposite bipolar pairs. One sunspot of each pair has positive, north, magnetic polarity, or outward-directed magnetism, and its partner has the opposite negative, south, magnetic polarity that is directed inward. Groups of bipolar sunspots are usually oriented roughly parallel to the Sun's equator, in the east-west direction of solar rotation.

The opposite magnetic poles are joined together, like Siamese twins, by long, thin magnetic loops that run between them, rising in arches like bridges that connect the bipolar magnetic islands. The magnetic loops can be visualized in terms of the magnetic lines of force, or magnetic field lines, that align compass needles on Earth. The lines of force emerge nearly radially from the sunspot with the positive north polarity and loop through the overlying solar atmosphere before re-entering the photosphere in the spot with negative south polarity, like the lines of force running between the north and south poles of the Earth or a bar magnet. It's as if a powerful magnet, aligned roughly in the east-west direction, was buried deep beneath each sunspot pair.

The magnetized loops that arch above bipolar sunspots can be seen in images taken at the solar edge or limb (Fig. 5.12). They then appear bright in contrast with the dark background, sometimes extending tens of thousands of kilometers above the solar limb.

The highly magnetized realm in, around and above bipolar sunspot pairs or groups is a disturbed area called an active region. Neighboring sunspots of opposite polarity are joined by magnetic loops that rise into the overlying atmosphere, so an active region consists mainly of sunspots and the magnetic loops that connect them. Energized material is concentrated and enhanced within the solar active regions, where magnetic loops shape, mold and constrain the hot, electrified gas.

Each magnetic loop acts as a barrier to electrically charged particles. They move back and forth along the magnetic field lines but cannot move across them. The hot material is therefore slaved to the magnetism, and becomes trapped within the closed magnetic loops where it emits bright extreme ultraviolet and X-ray radiation (Fig. 5.13).

The ubiquitous magnetic loops are never still, but instead continually alter their shape. They can rise up out of the solar interior, and submerge back within it in hours or days. Sometimes you can see the hot material trying to escape its magnetic cage, in a high-velocity upwelling, or surge (Fig. 5.14).

Active regions begin their life when magnetic loops rise up from inside the Sun. The magnetic structure of an active region then gradually changes in appearance as new magnetic loops surface within it and its sunspots move and shift about. This results in continued alterations of the form and intensity of their visible and invisible radiation. Eventually, the ephemeral active regions simply disappear. Over the course of weeks to months, their magnetic loops break apart, disintegrate, or move back inside the Sun where they came from.

Thus, active regions are never still, but instead continually alter their magnetic shape. They are the seat of profound change and unrest on the Sun! The stressed magnetic fields build up magnetic energy that is waiting to be released, and the ongoing magnetic interaction can trigger sudden and catastrophic explosions such as powerful solar flares. Indeed, the continually evolving magnetic structure and intense radiation, as well as the eruptive solar flares, give active regions their name. The whole range of activity varies with the 11-year solar cycle of magnetic activity, which we now focus attention on.

FIG. 5.12 Magnetic loops made visible An electrified, million-degree gas, known as plasma, is channeled by magnetic fields into bright, thin loops. The magnetized loops stretch up to 500,000 kilometers from the visible solar disk, spanning up to 40 times the diameter of planet Earth. The magnetic loops are seen in extreme ultraviolet radiation of eight and nine times ionized iron, denoted Fe IX and Fe X, at a wavelength of 17.1 nanometers and a temperature of about 1.0 million kelvin. The hot plasma is heated at the bases of loops near the place where their legs emerge from and return to the photosphere. Bright loops with a broad range of lengths all have a fine, thread-like substructure with widths as small as the telescope resolution of 1 second of arc, or 725 kilometers at the Sun. This image was taken with the *Transition Region And Coronal Explorer,* abbreviated *TRACE,* spacecraft. (Courtesy of the *TRACE* consortium, LMSAL and NASA; *TRACE* is a mission of the Stanford-Lockheed Institute for Space Research, a joint program of the Lockheed-Martin Solar and Astrophysics Laboratory, or LMSAL for short, and Stanford's Solar Observatories Group.)

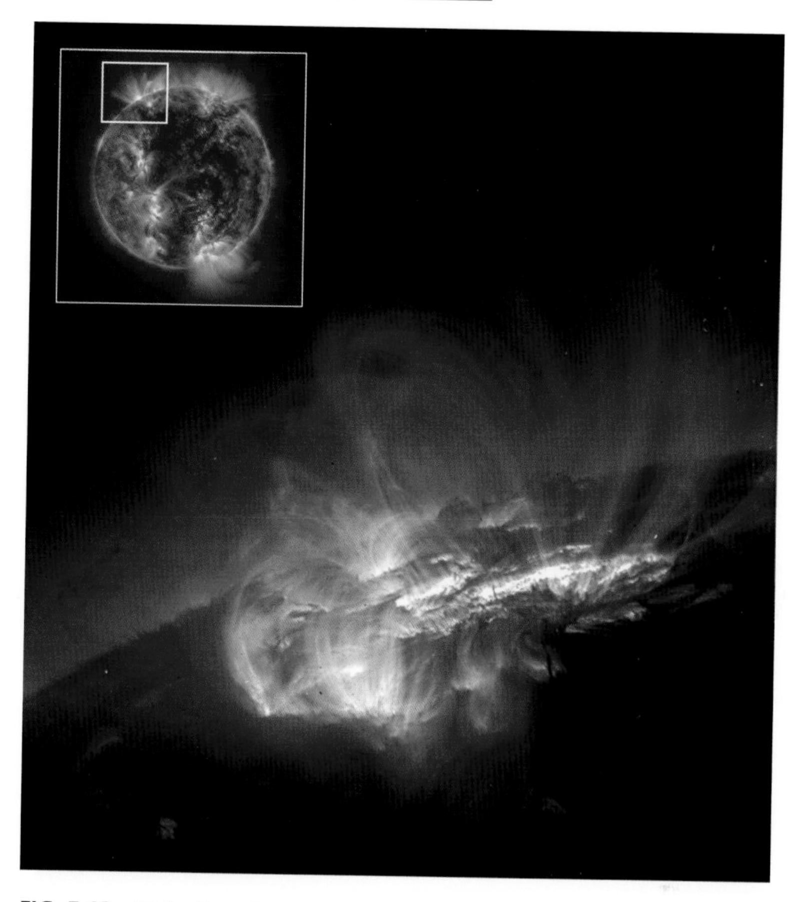

FIG. 5.13 Ubiquitous loops The looping structures of the solar magnetic field are seen in great detail with instruments aboard the *Transition Region And Coronal Explorer,* abbreviated *TRACE,* spacecraft. This *TRACE* image (*center*) was taken at the extreme ultraviolet wavelength of 17.1 nanometers, emitted by eight and nine times ionized iron, denoted Fe IX and Fe X, at a temperature of about 1.0 million kelvin. A full-disk image (*upper left*) shows the ubiquitous loops across the Sun; it was taken on the same day at the same wavelength with the Extreme-ultraviolet Imaging Telescope, abbreviated EIT, aboard the *SOlar and Heliospheric Observatory,* or *SOHO* for short. (Courtesy of the *TRACE* consortium, the *SOHO* EIT consortium, and NASA. *SOHO* is a project of international cooperation between ESA and NASA and *TRACE* is a mission of the Stanford-Lockheed Institute for Space Research, a joint program of the Lockheed-Martin Solar and Astrophysics Laboratory, or LMSAL for short, and Stanford's Solar Observatories Group.)

FIG. 5.14 Surge A sudden high-velocity upwelling tries to break free of the magnetic and gravitational constraints of the Sun. The million-degree gas can generally move only along the curved magnetic loops, but it is sometimes accelerated to such a high speed that the gas is propelled in a straight line. Nevertheless, despite the large initial speed, most of the material is pulled back down by the Sun's tremendous gravity, and the surge cools and darkens. This image was taken at the extreme ultraviolet wavelength of 17.1 nanometers, emitted by eight and nine times ionized iron, denoted Fe IX and Fe X, at a temperature of about 1.0 million kelvin, using the *Transition Region And Coronal Explorer,* abbreviated *TRACE,* spacecraft. (Courtesy of Charles Kankelborg, the *TRACE* consortium and NASA; *TRACE* is a mission of the Stanford-Lockheed Institute for Space Research, a joint program of the Lockheed-Martin Solar and Astrophysics Laboratory, or LMSAL for short, and Stanford's Solar Observatories Group.)

5.4 CYCLES OF MAGNETIC ACTIVITY

Samuel Heinrich Schwabe (1789–1875), a pharmacist and amateur astronomer of Dessau, Germany, diligently and meticulously observed the Sun day after day, and year after year, with his small 5-centimeter (2-inch) telescope. Schwabe was looking for the shadow of a hypothetical planet Vulcan that was predicted to revolve around the Sun inside the orbit of Mercury. His search included counting the number of sunspots, which might be confused with Vulcan's shadow. Schwabe never found the planet, which does not exist, but he did unexpectedly discover that the total number of sunspots visible on the Sun varies periodically, from a maximum to a minimum and back to a maximum, in about 10 years.

After 17 years of observations, Schwabe reported in 1843 that:

> [The total number] of sunspots has a period of about 10 years. The future will tell whether this period persists, whether the minimum activity of the Sun in producing spots lasts one or two years and whether this phenomenon takes longer to build up or longer to decline.[22]

Upon presenting a gold medal to Schwabe, the president of England's Royal Astronomical Society summed up the magnitude of his feat:

> For thirty years never has the Sun exhibited his disk above the horizon of Dessau without being confronted by Schwabe's imperturbable telescope.... This is, I believe, an instance of devoted persistence unsurpassed in the annals of astronomy. The energy of one man has revealed a phenomenon that had eluded even the suspicion of astronomers for 200 years![23]

The sunspot cycle certainly does persist. Other astronomers have now compiled systematic records of the periodic variation in sunspot numbers for more than one hundred years (Fig. 5.15). They sometimes record the total area of the visible Sun covered by the spots, which rises and falls with the number of sunspots and is thought to be more physically significant.

Detailed observations of sunspots have been obtained at England's Royal Greenwich Observatory since 1874, including the numbers, sizes and positions of sunspots. The maximum number of sunspots, at the peak of the cycle, can reach 200 per month, while fewer than a dozen spots are seen near the cycle minimum, five or six years later. The data also show that the positions of sunspots and their associated active regions vary systematically during the sunspot cycle (Fig. 5.15).

With the passage of time during the 11-year activity cycle, new sunspots appear closer and closer to the solar equator, while streams of remnant magnetic flux spread poleward (north or south). In the early part of the cycle, when solar activity rises to its maximum, active regions break out in two belts at about 30 degrees latitude, one north and one south of the solar equator. As on the Earth, latitude is the angular distance north or south of the equator.

When the cycle progresses toward sunspot minimum, the active regions at mid-latitudes fade away, and new ones surface in belts that are closer and closer to the equator. The drifting active-region belts, one in each hemisphere, describe a slow, 11-year churning motion that originates deep inside the Sun and sweeps down across the photosphere toward its equator, until the sunspots come together and disappear at sunspot minimum.

Then, out of the destruction, the cycle renews itself once more, and active regions emerge again at mid-latitudes about one-third of the way toward the poles. But no sunspots are ever found at high latitudes near the polar regions of the Sun.

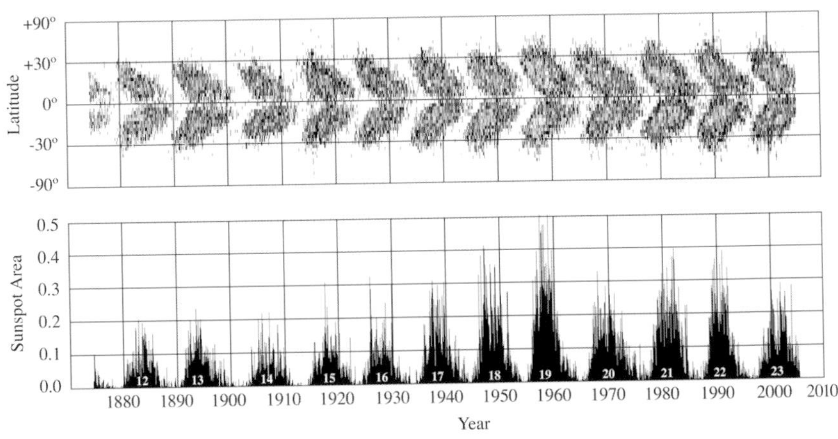

FIG. 5.15 Sunspot cycle The location of sunspots (*upper panel*) and their total area (*bottom panel*) have varied in an 11-year cycle for the past 130 years, but this magnetic activity cycle varies both in cycle length and maximum amplitude. As shown in the upper panel, the sunspots form at about 30 degrees latitude at the beginning of the cycle, within two bands of active latitudes (one in the north and one in the south) that migrate to near the Sun's equator at the end of the cycle. Such an illustration is sometimes called a "butterfly diagram" because of its resemblance to the wings of a butterfly. The upper panel also shows how the cycles overlap with spots from a new cycle appearing at high latitudes while spots from the old cycle are still present in the equatorial regions. The total area covered by sunspots (*bottom panel*), given as a percent of the visible hemisphere of the Sun, follows a similar 11-year cycle; during each cycle the total area often rises quickly from a minimum to a maximum and then drops back to a minimum at a slower rate. There are large variations in total sunspot area, and in solar activity, during each cycle, and from cycle to cycle. (Courtesy of David Hathaway, NASA/MSFC.)

This systematic, 11-year drift of sunspots toward the solar equator was initially noticed in 1858 by the English astronomer Richard C. Carrington (1826–1875), during his studies of differential solar rotation, and then described by the German astronomer Gustav Spörer (1822–1895). It is graphically represented in a plot of sunspot latitude as a function of time, first drawn by E. Walter Maunder (1851–1928) in 1922 and brought up to date in Fig. 5.15. Such an illustration is sometimes called a "butterfly diagram" because of its resemblance to the wings of a butterfly.

As old spots linger near the equator, new ones break out at mid-latitudes, but the magnetic polarities of the new spots are reversed with north becoming south and *vice versa* – as if the Sun had turned itself inside out. During one 11-year cycle, the magnetic polarity of all the leading (westernmost) spots in the northern solar hemisphere is the same, and is opposite to that of leading spots in the southern hemisphere. The magnetic polarity of the leading spots reverses in each hemisphere at the beginning of the next 11-year cycle.

Remnants of old sunspots move to the polar regions to replenish the weaker dipolar (two poles) magnetic field of the Sun. It also reverses at the beginning of each cycle, so the magnetic north pole switches to a magnetic south pole and *vice versa*. Then, for the next 11 years, in the new cycle, all the polarities will be exchanged, including those of all the sunspots and that of the general polar magnetic field. Thus, the full magnetic cycle takes an average of 22 years.

Active regions contain twinned bipolar sunspots that are aligned roughly parallel to the equator, and describe a global pattern of magnetic polarity. All of the sunspot pairs in each active-region belt have the same orientation and polarity alignment, with an exactly opposite arrangement in the two hemispheres. According to Hale's law of polarity, the leading, or westernmost spots (leading in the sense rotation) of any sunspot group in the northern belt of active regions have the same magnetic polarity, while the following (easternmost) spots have the opposite magnetic polarity. In the southern hemisphere, the leading and trailing sunspots of any sunspot group also exhibit opposite polarities, but the magnetic direction of the bipolar sunspots in the southern belt is the reverse of that in the northern one.

Thus, if the leading spot in the northern hemisphere has one magnetic polarity, the leading spot in the southern hemisphere will have the opposite polarity. It's as if men and women always walked down the street in couples, with the men preceding the women on one side of the street and the women leading the men on the other side. Moreover, the couples exactly reverse their orientation at sunspot minimum every 11 years, as if participating in a dance that has been carefully choreographed. The bipolar sunspot magnets then flip and turn around, so the leading spots in each hemisphere have opposite magnetic polarities during successive 11-year cycles. The leading spots resume their original magnetic polarity in approximately 22 years after reversing orientation twice.

Magnetograms indicate that there is still plenty of magnetism at the minimum of the solar cycle of magnetic activity, when there are no large sunspots present. The magnetism then comes up in a large number of very small regions spread all over the Sun (Fig. 5.16). The magnetic field averaged over vast areas of the Sun at activity minimum is only a few ten thousandths of a Tesla, or a few Gauss, but the averaging process conceals a host of fields of small size and large strength. When the telescope resolution is increased, finer and finer magnetic fields are found, with higher and higher field strengths.

The magnetic fields are never smoothly distributed across the photosphere, either during activity minimum or maximum. Instead, they are highly concentrated, inhomogeneous, and everywhere clumped together into intense bundles that cover only a few percent of the photosphere's surface area.

5.5 INTERNAL DYNAMO

Where do the Sun's magnetic fields come from and how are they made? We know that the motion of electric charges can produce magnetic fields, and that changing magnetic fields produce electric currents. And the Sun's interior is totally electrically charged, consisting of electrons and protons. In deep layers, these charged particles are so hot that they conduct electric current as well as copper does at room temperature. The hot circulating gases generate electrical currents that create magnetic fields; these fields in turn sustain the generation of electricity, just as in a power-plant dynamo.

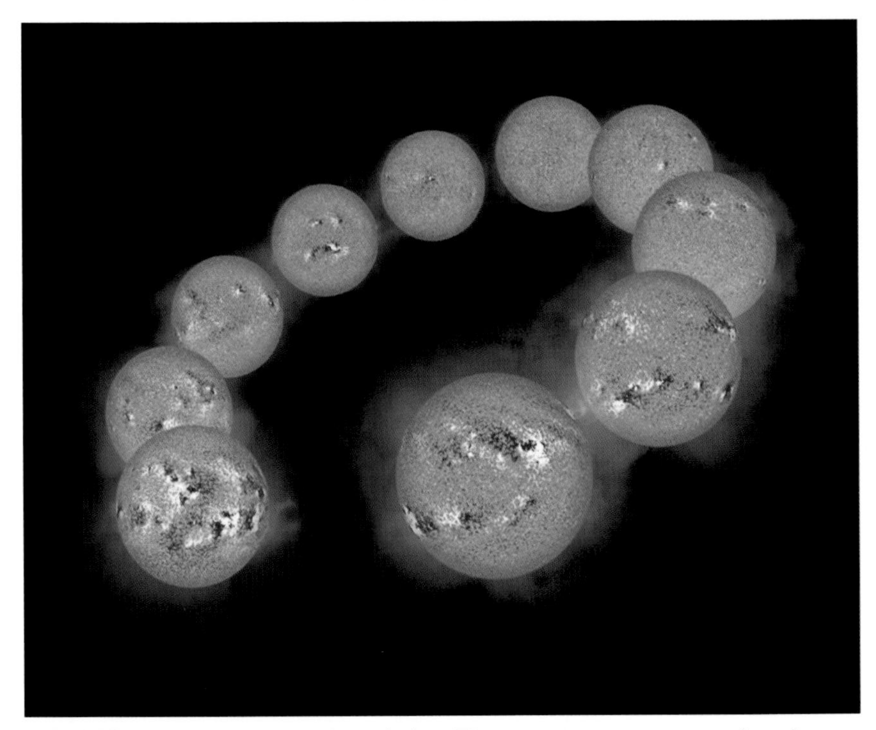

FIG. 5.16 Solar cycle magnetic variations These magnetograms portray the polarity and distribution of the magnetism in the solar photosphere. They were made with the Vacuum Tower Telescope of the National Solar Observatory at Kitt Peak from 8 January 1992, at a maximum in the sunspot cycle (*lower left*) to 25 July 1999, well into the next maximum (*lower right*). Each magnetogram shows opposite polarities as darker and brighter than average tint. When the Sun is most active, the number of sunspots is at a maximum, with large bipolar sunspots that are oriented in the east-west (*left-right*) direction within two parallel bands. At times of low activity (*top middle*), there are no large sunspots and tiny magnetic fields of different magnetic polarity can be observed all over the photosphere. The haze around the images is the inner solar corona. (Courtesy of Karel J. Schrijver, NSO, NOAO and NSF.)

The Earth's magnetic field is supposed to be generated by such a dynamo, operating on a much smaller scale within its molten core.

The magnetic fields of the Sun are entrained and "frozen" into the conducting gas whose particles carry the magnetism with them. As they move along with the gas, the embedded magnetic fields are deformed, folded, stretched, twisted and amplified. The mechanical energy of the motion of the charged gas particles is thereby converted into the energy of magnetic fields. This dynamo mechanism does not explain how the magnetic fields originated, but rather how they are amplified and maintained. The process of field amplification is nevertheless cumulative, so a dynamo can generate an intense magnetic field from an initially weak one.

The solar dynamo is now thought to operate in a thin layer, called the tacholine, located at the interface between the deep interior, which rotates with one speed, and the overlying layer that spins faster in the equatorial middle. Sound waves also speed up more than expected in this shear layer, indicating that turbulence and mixing associated with a dynamo are most likely present. Below the tacholine the Sun rotates like a solid object, with too little variation in spin to drive a solar dynamo. And above this boundary layer, the rotation rates at different latitudes diverge over broad areas that are not focused enough to play much of a role in the dynamo.

Global rotation, differential rotation, and convective motion are supposed to interact strongly at the tacholine, producing dynamo action. In this deep-seated region, the dipolar magnetic field, which runs inside the Sun between its north and south pole, is apparently wound up and stretched into an azimuthal, or toroidal, coil that rises radially through the overlying convective zone to produce sunspots and active regions with their 11-year cyclic behavior.

In technical terms, a global dipole magnetic field is supposed to be sheared and stretched into a toroidal field by differential rotation inside the Sun, and dipolar sunspot pairs are supposed to be produced by a lifting and twisting process related to the rising toroidal magnetic fields. The global dipole and toroidal magnetic field components are alternately destroyed and recreated in a cycle that lasts a total of 22 years. Deep meridional flow, or north-south circulation, is supposed to help regenerate the dipole field, also accounting for its reversal, and explain why sunspots do not form at high latitudes.

Nevertheless, despite all the mathematical complexity, or perhaps because of it, there is no solar dynamo model that explains the different aspects of the Sun's magnetic activity cycle in detail. And no one has yet observed the magnetic fields deep inside the Sun.

Solar astronomers therefore often extrapolate from a conceptually simple model devised in 1961 by the American astronomer Horace W. Babcock (1912–2003). His theory begins at sunspot minimum with a global, dipolar magnetic field that runs inside the Sun from south to north, or from pole to pole. Uneven, or differential, rotation shears the electrically conducting gases of the interior, so the entrained magnetic fields get stretched out and squeezed together. The magnetism is coiled, bunched and amplified as it is wrapped around the inside of the solar globe, eventually becoming strong enough to rise to the surface and break through it in active-region belts with their bipolar sunspot pairs (Fig. 5.17), like a stitch of yarn pulled from a woolen sweater. The surrounding gas buoys up the concentrated magnetism, just as a piece of wood is subject to buoyant forces when it is immersed in water.

As Babcock expressed it:

> Shallow submerged lines of force of an initial, axisymmetric dipolar field are drawn out in longitude by the differential rotation.... Twisting of the irregular flux strands by the faster shallow layers in low latitudes forms "ropes" with local concentrations that are brought to the surface by magnetic buoyancy to produce *bipolar magnetic regions* with associated sunspots and related activity.[24]

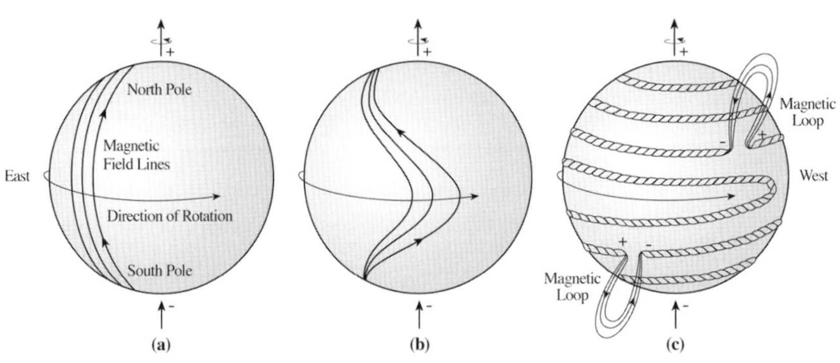

FIG. 5.17 Winding up the field A model for generating the changing location, orientation and polarity of the sunspot magnetic fields. At the beginning of the 11-year cycle of magnetic activity, when the number of sunspots is at a minimum, the magnetic field is the dipolar or poloidal field seen at the poles of the Sun (*left*). The internal magnetic fields then run just below the photosphere from the south to north poles. As time proceeds, the highly conductive, rotating material inside the Sun carries the magnetic field along and winds it up. Because the equatorial regions rotate at a faster rate than the polar ones, the internal magnetic fields become stretched out and wrapped around the Sun's center, becoming deformed into a partly toroidal field (*middle* and *right*). The fields are then concentrated and twisted together like ropes. With increasing strength, the submerged magnetism becomes buoyant, rises and penetrates the visible solar disk, or photosphere, creating magnetic loops and bipolar sunspots that are formed in two belts, one each in the northern and southern hemisphere (*right*). [Adapted from Horace W. Babcock, *Astrophysical Journal* **133**, 572–587 (1961)]

The initial dipole, or poloidal, magnetic field is twisted into a submerged toroidal, or ring-shaped, field running parallel to the equator, or east to west. Apparently, the dynamo generates two toroidal magnetic fields, one in the northern hemisphere and one in the southern hemisphere, but oppositely directed, which bubble up at mid- to low-latitudes to spawn the two belts of active regions, symmetrically placed on each side of the equator. Thus, according to Babcock's scenario, we may view the solar cycle as an engine in which differential rotation drives an oscillation between poloidal and toroidal geometries.

As the 11-year cycle progresses, the internal magnetic field is wound tighter and tighter by the shearing action of differential rotation, and the two belts of new active regions slowly migrate toward the solar equator. Because the active regions emerge, on the average, with their leading ends slightly twisted toward the equator, the leading sunspots in the two hemispheres tend to merge and cancel out, or neutralize, each other at the equator. This leaves a surplus of following-polarity magnetism in each hemisphere, north and south, which is eventually carried poleward at sunspot minimum.

Diffusion and poleward flows apparently sweep remnants of former active regions into streams, each dominated by a single magnetic polarity, that slowly wind their way from the low- and mid-latitude active-region belts to the Sun's poles. By sunspot minimum, when the active regions have largely disintegrated, submerged or annihilated each other, the continued poleward transport of their debris may form a global dipole, like the phoenix rising from its ashes. Because the Sun's polar field is created from the following polarity of decaying active regions, they reverse the overall dipole polarity at sunspot minimum, so the north and south pole switch magnetic direction or polarity. The Earth's dipole magnetic field also reverses itself, but at much longer intervals of about a million years. When the Sun's magnetic flip is taken into account, we see that it takes two activity cycles, or about 22 years, for the overall magnetic polarity to get back where it started.

By the time that sunspot minimum occurs, most of the magnetic flux that emerged in former active regions has been obliterated. It's as if the internal magnetism has been wound so tight that it snaps, like an over-wound watch spring, and no more sunspots can form. Relatively small amounts of magnetism remain, as leftover flux that originated in active regions that are no longer there; it is this flux that has been gradually dispersed over a much wider range of latitudes to form the Sun's global dipole. The internal magnetism has then readjusted to a poloidal form, and the magnetic cycle begins again.

Thus, the dynamo theory seems to explain all of the repetitive aspects of the Sun's magnetism, including the periodic variation in the number of sunspots, their cyclic migration toward the equator (the butterfly diagram), the roughly east-west orientation, location and polarity (Hale's law of polarity) of bipolar sunspot pairs, and the periodic reversal of the overall global dipole. However, many details of the theory are uncertain or incomplete, and so far no dynamo model has succeeded in accounting for all of the magnetic observations.

Chapter Six

An Unseen World of Perpetual Change

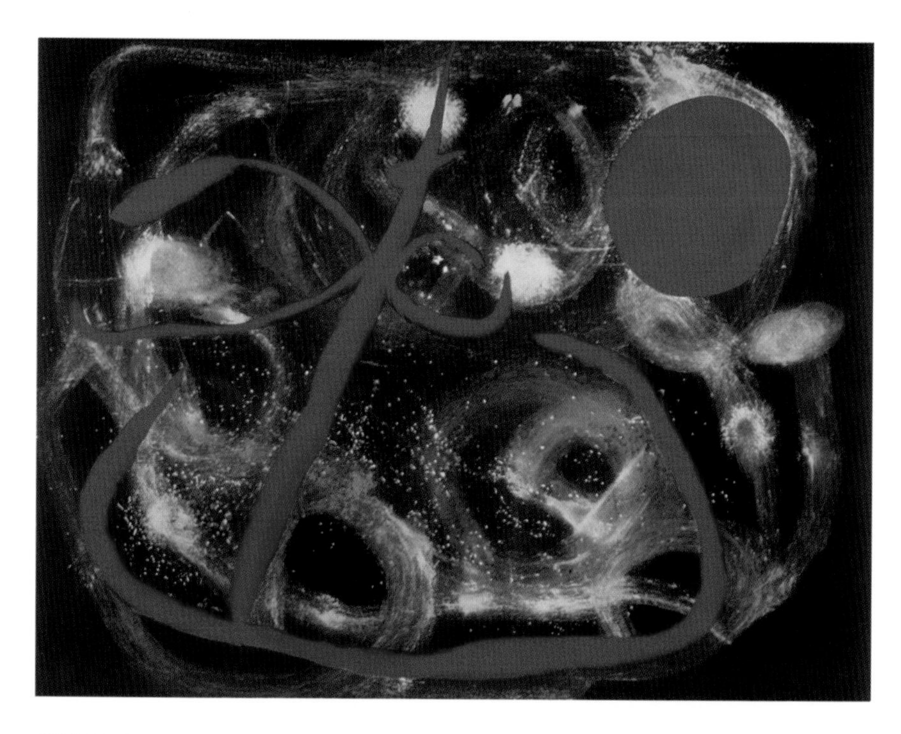

JOY OF A LITTLE GIRL BEFORE THE SUN In this painting by the Spanish artist Joan Miró (1893–1983) a figure, representing a little girl, seems to be dancing with the fire-red Sun and reaching out to the vastness of space. Some white paint has been spread upon the black background, making it appear thin and blue like a spiral nebula against the black night. (Private Collection.)

6.1 THE SUN'S VISIBLE EDGE IS AN ILLUSION

The entire Sun is just a giant incandescent, gaseous ball that seems to extend forever. The gas is compressed at its center, becoming progressively more tenuous further out. And being entirely gaseous, the Sun has no solid surface and no permanent visible features. The specification of a "surface" that divides the inside of the Sun from the outside is therefore largely a matter of choice, depending on the wavelength that provides the required perspective.

At the wavelengths we see with our eye, the Sun is a bright, round disk technically known as the photosphere. It is the lowest, densest level of the solar atmosphere, the layer that forms the Sun you can watch moving across the sky each day.

The gases enveloping the photosphere are so rarefied that we can look right through them, so you can't see anything in front of the photosphere. The overlying material is invisible except during solar eclipses or with special instruments. Indeed, we use the term atmosphere for this tenuous outer part of the Sun because it is relatively transparent at visible wavelengths.

When we look at the visible disk of the Sun, its edge looks sharp, but that is because our eyes can't resolve the details. The edges of clouds in the Earth's atmosphere look sharp for the same reason. We can't see features narrower than about one minute of arc, or 1/30th of the diameter of the Sun, which means that the unaided eye cannot distinguish anything on the Sun smaller than 43,500 kilometers across or smaller than four times the size of the Earth.

So the sharp visible edge of the Sun is an illusion. We can't resolve its details, and there is an extensive, unseen atmosphere that surrounds it. This solar atmosphere, located above the photosphere, is an energized realm of violent change, extreme temperatures and powerful eruptions that can strongly affect the Earth's environment.

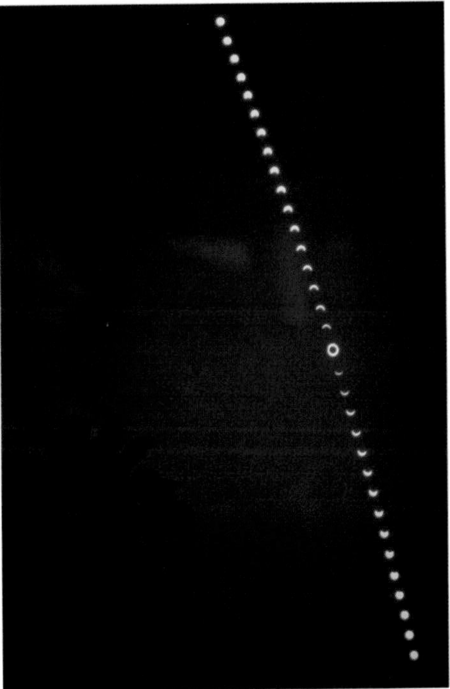

FIG. 6.1 Total eclipse of the Sun A multiple-exposure photograph of a total eclipse of the Sun. The circular form eclipsing the Sun is the Moon. Because the Moon and the Sun have nearly the same angular extent, the Moon blocks out most of the Sun's light during a total solar eclipse. Akira Fujii took this photograph on 16 February 1980.

The solar gases that reside just above the photosphere are momentarily seen during a total eclipse of the Sun, when the bright solar disk is blocked out, or eclipsed, by our own Moon (Fig. 6.1). It is encircled by a narrow band of light known as the *corona,* from the Latin word for "crown." The corona is visible at any total eclipse of the Sun, and has been recorded for more than a thousand years.

But the corona is only seen for a brief interval during an eclipse, for no more than 8 minutes. And it doesn't happen very often. The Moon only passes directly between the Sun and the Earth once every year and a half, on average, and even then, it's only visible from a narrow track along the Earth's surface, often in remote locations (Table 6.1). At any given point on our planet, you and your ancestors would have to wait about four centuries to witness a total eclipse of the Sun.

But seeing a total solar eclipse is certainly worth the trip. The world grows dark, as day turns to night, birds go home to roost, fish rise to the ocean's surface to feed, and bright stars become visible in the sky. It's as if the Sun was taken away, abandoning the Earth. Hence the word *eclipse,* which comes from a Greek word meaning "abandonment" or "to leave."

TABLE 6.1 Total eclipses of the Sun from 2005 to 2024

Date	Maximum Duration (minutes)	Path of Totality
8 April 2005	0.70	Pacific Ocean (annular-total)
29 March 2006	4.12	Eastern Brazil, Atlantic Ocean, Ghana, Togo, Dahomey, Nigeria, Niger, Chad, Libya, Egypt, Turkey, Russia.
01 August 2008	2.47	Greenland, Arctic Ocean, Russia, Mongolia, China
22 July 2009	6.65	India, China, Pacific Ocean
11 July 2010	5.33	South Pacific Ocean, Chile, Argentina
13 November 2012	4.05	Northern Australia, South Pacific Ocean
3 November 2013	1.67	Atlantic Ocean, equatorial Africa (annular-total)
20 March 2015	2.78	North Atlantic and Arctic Oceans
09 March 2016	4.17	Indonesia, Pacific Ocean
21 August 2017	2.67	Pacific Ocean, United States (Oregon, Idaho, Wyoming, Nebraska, Missouri, Kentucky, Tennessee, and South Carolina), Atlantic Ocean
2 July 2019	4.55	South Pacific Ocean, Chile, Argentina
14 December 2020	2.17	Pacific Ocean, Chile, Argentina, South Atlantic Ocean, South Africa
4 December 2021	1.92	Antarctica
20 April 2023	1.27	South Indian Ocean, New Guinea, Pacific Ocean
8 April 2024	4.47	Pacific Ocean, Mexico, United States (Texas, Oklahoma, Arkansas, Missouri, Illinois, Indiana, Ohio, Pennsylvania, New York, Vermont, New Hampshire, Maine), southeastern Canada, Atlantic Ocean

During such a solar eclipse, the Sun's normally invisible atmosphere becomes as bright as the full Moon, a shimmering halo of pearl-white light that extends outward from the lunar silhouette against the blackened sky (Fig. 6.2). A representative description of the spectacular crown of light, observed during the eclipse of 1842, was provided by Francis Bailey (1774–1844), a stockbroker and enthusiastic amateur astronomer in England:

> I was astounded by a tremendous burst of applause from the streets below and at the same moment was electrified at the sight of one of the most brilliant and splendid phenomena that can be imagined. For at that instant the dark body of the Moon was suddenly surrounded with a corona, a kind of bright glory. . . . I had anticipated a luminous circle round the Moon during the time of total obscurity . . . , but the most remarkable circumstance attending this phenomenon was the appearance of three large protuberances apparently emanating from the circumference of the Moon, but evidently forming a portion of the corona.[25]

The protuberances that Bailey observed were arches of incandescent gas, which loop up into the solar corona and are held there by strong magnetic fields. Astronomers now call the looping features *prominences,* the French word for "protuberances."

The reddened prominences can rise up to a few hundred thousand kilometers (Fig. 6.3), large enough to stretch from the Earth to the Moon, hanging there for weeks and months at a time. Powerful magnetic forces hold these quiescent prominences together and suspend them against the force of the Sun's gravity for so long.

FIG. 6.2 Eclipse corona The million-degree solar atmosphere, known as the corona, is seen around the black disk of the Moon, photographed in white light from atop Mauna Kea, Hawaii during the solar eclipse on 11 July 1991. The million-degree, electrically charged gas is concentrated in numerous fine rays as well as larger helmet streamers. (Courtesy of the High Altitude Observatory, National Center for Atmospheric Research.)

FIG. 6.3 Quiescent prominence Dense, cool gas, seen in the red light of hydrogen alpha at the rim of the Sun, outlines magnetic arches that are silhouetted against the dark background. The prominence material, appearing as a flaming curtain up to 65,000 kilometers above the photosphere, is probably injected into the base of the magnetic loops in the chromosphere. (Courtesy of the Big Bear Solar Observatory.)

Other active, eruptive prominences, smaller but still large enough to girdle the Earth, might last only a few minutes or hours before their magnetism becomes unhinged, and the prominence breaks out and away from the Sun.

During a total solar eclipse, we are seeing patterns of free electrons in the corona made visible because they scatter the light that strikes them, like motes in a sunbeam. And because these electrons are constrained and molded by magnetic fields, the corona's form varies as the Sun's variable magnetism changes and shifts its shape. The electrons are densely concentrated within magnetized loops close to the Sun, creating bubble-like, or arch-like, structures known as helmet streamers (Fig. 6.4), which are peaked like old-fashioned, spiked helmets once used in Europe.

A low-lying prominence is often found near the bottom of a helmet streamer. A cavity, visible as a region of reduced brightness, separates the prominence and the arcades of magnetized loops that seemingly support the arched helmet streamer.

In the outer part of the corona, far from the Sun, the streamers become narrower and surmounted or prolonged by long, straight, tenuous stalks that extend far into interplanetary space. The long, graceful stalks can extend at least ten million kilometers, or 14 solar radii, into interplanetary space, as if some invisible agents were pulling or pushing them out like stretched salt-water taffy.

FIG. 6.4 Coronal structures near sunspot minimum The Sun's corona becomes visible to the unaided eye during a total solar eclipse, such as this one observed from Oranjestad, Aruba on 26 February 1998, close to a minimum in the Sun's 11-year cycle of magnetic activity. Several individual photographs, made with different exposure times, were combined and processed electronically in a computer to produce this composite image, which shows the solar corona approximately as it appears to the human eye during totality. Note the fine rays and helmet streamers that extend far from the Sun along the equatorial regions (*left* and *right*), and the polar rays (*top* and *bottom*) that suggest a global, dipolar magnetic field. (Courtesy of Fred Espenak.)

Helmet streamers are rooted within magnetic loops that sometimes straddle active regions and connect regions of opposite magnetic polarity. Streamers also often rise above long-lived, quiescent prominences that are commonly embedded in the closed-magnetic loops at the bottom of streamers. These bright, curved, low-lying magnetic loops constrain the densest coronal material close to the photosphere, within one or two solar radii.

The shape of the corona is molded by magnetic forces that vary with the number of sunspots and the amount of solar activity. At sunspot maximum, when magnetic activity is strongest, the streamers are distributed all around the solar limb and presumably all over the Sun (Fig. 6.5). When the number of sunspots is low, the relatively weak magnetic activity is largely confined to the Sun's equatorial regions, where the sunspots and streamers are localized. Thus, the eclipse observations indicate that the overall shape of the corona changes in synchronism with the 11-year solar activity cycle; near maximum the coronal structures are stretched out in all directions outside the equatorial plane, and near minimum the corona is considerably flattened toward the equatorial regions.

The width and radial extension of the streamers is also related to the solar activity cycle. At the time of maximum activity, streamers are narrower and shorter, near minimum; they are wider and more extended.

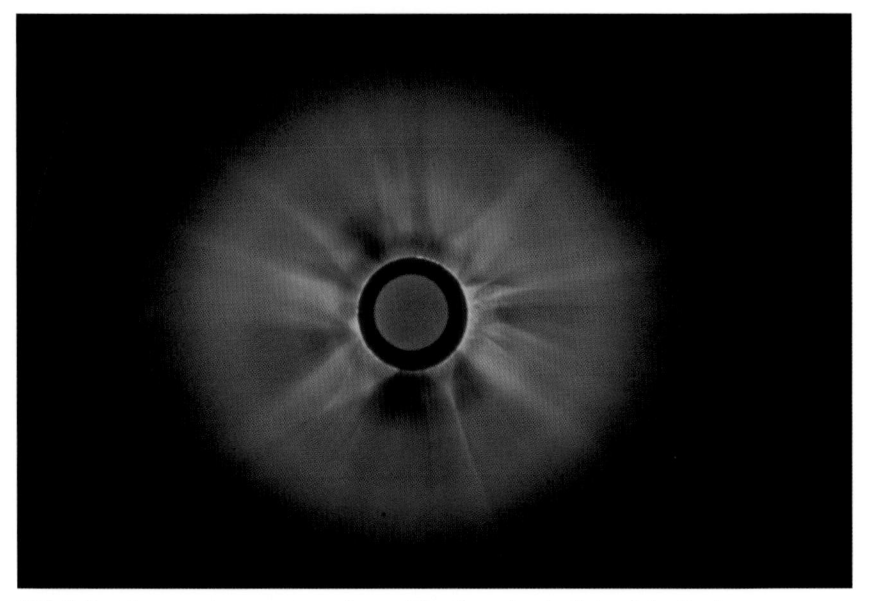

FIG. 6.5 Corona at sunspot maximum The dim, ghostly light of the Sun's outer atmosphere, the corona, during a total solar eclipse on 16 February 1980, observed from Yellapur, India. This was a time near sunspot maximum, when the sunspots are most numerous. The bright helmet streamers were then distributed about the entire solar limb, resembling the petals on a flower. The stalks of the streamers stretch out 4 million kilometers, or about six solar radii, from the center of the Sun in this photograph. It was taken through a radially graded filter to compensate for the sharp decrease in the electron density and coronal brightness with distance from the Sun. (Courtesy of Johannes Durst and Antoine Zelenka, Swiss Federal Observatory.)

The corona can be routinely observed in broad daylight using a special telescope called the coronagraph that has a small occulting disk to mask the Sun's face and block out the photosphere's light (Fig. 6.6). The first coronagraph was invented in 1930 by the French astronomer, Bernard Lyot (1897–1952), and soon installed by him at the Pic du Midi observatory in the Pyrenees. As Lyot realized, such observations are limited by the bright sky to high-altitude sites where the thin, dust-free air scatters less sunlight. The higher and cleaner the air, the darker the sky, and the better we can detect the faint corona around the miniature moon in the coronagraph.

The best coronagraph images with the finest detail are obtained from high-flying satellites where almost no air is left and where the daytime sky is truly and starkly black. Initial discoveries were made from a coronagraph aboard NASA's *Seventh Orbiting Solar Observatory,* abbreviated *OSO 7,* which circled the Earth from 1971 to 1974. The clear, nearly continuous, edge-on views of the corona from NASA's *Skylab* satellite in 1973–74 resulted in the full realization of long-lived holes in the corona, as well as the giant Sun-sized, expanding bubbles dubbed coronal transients or Coronal Mass Ejections, abbreviated CMEs.

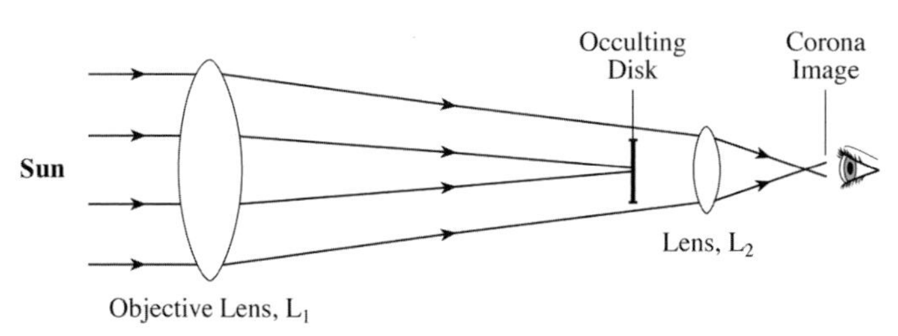

FIG. 6.6 Coronagraph Sunlight enters from the left, and is focused by an objective lens, L_1, on an occulting disk, blocking the intense glare of the photosphere. Light from the corona, which is outside the photosphere, bypasses this occulting disk and is focused by a second lens, L_2, forming an image of the corona. Other optical devices are placed along the light path to divert and remove excess light. The French astronomer Bernard Lyot (1897–1952) constructed the first coronagraph in 1930, and the best coronagraphs are now employed in satellites orbiting above the Earth's obscuring atmosphere.

The Solwind coronagraph aboard the U.S. Air Force's *P78–1* satellite continued investigations of the CMEs from 1979 to 1985, as did the coronagraph aboard NASA's *Solar Maximum Mission* in 1980 and from 1984 to 1989. More might have been learned if the Air Force had not decided to shoot down the *P78–1* satellite during a test of an anti-satellite missile in 1985.

The joint NASA-ESA *SOlar and Heliospheric Observatory* took up the space coronagraph investigations of the Sun from 1996 to 2005, with a remarkable series of discoveries. Nevertheless, all coronagraphs provide only a limited, edge-on view of the corona, just a flat projection against the sky.

6.2 THE MILLION-DEGREE CORONA

During the solar eclipse of 7 August 1869, the American astronomers Charles A. Young (1834–1908) and William Harkness (1849–1900) independently discovered that the spectrum of the solar corona includes a bright, green emission line at a wavelength of 530.3 nanometers. This spectral feature appears as a line when the Sun's radiation intensity is displayed as a function of wavelength. The specific wavelength of other lines had been used to fingerprint the atom or ion from which they originated, but the green coronal emission line could not at first be identified with any known element. So astronomers initially concluded that the corona contains a previously unknown substance, which they named coronium.

Belief in the new element coronium lingered for many years, until it became obvious that there was no place for it in the atomic periodic table, and it must therefore be, not an unknown element, but a known element in an unusual state. The solution to the coronium puzzle nevertheless eluded astronomers for seventy years. During this interval, eclipse observations revealed at least ten coronal emission lines, none of which had been observed to come from terrestrial substances.

Then in 1941 the Swedish astronomer Bengt Edlén (1906–1993) and the German astronomer Walter Grotrian (1890–1954) showed that the coronal emission lines are emitted by ordinary elements, such as iron, calcium and nickel, but from atoms deprived of 10 to 15 electrons. The conspicuous, bright green emission line is, for example, due to iron atoms that have been stripped of half of their 26 electrons, denoted Fe XIV.

Edlén realized that the highly ionized atoms could only be missing so many electrons if the coronal gas was unexpectedly hot, with a temperature of about a million degrees. At this high temperature, many electrons are set free from atoms and move off at high speeds, leaving the ions behind. Only the heavy atoms, such as iron, are able to hold on to any of their orbiting electrons, and even they can only keep some in their grasp.

The searing million-degree temperature in the corona might have also been inferred from the great extent of the corona, the large widths of the coronal emission lines, and the Sun's radio emission (Focus 6.1). Astronomers were probably reluctant to accept all of this evidence because the emission lines might be due to unusual substances, and also because they did not expect heat to flow from the cool photosphere into the surrounding hotter corona.

The reason that the coronal emission lines took so long to identify was that they arise during high-temperature, low-density conditions that are not normally observed on Earth, and are hence designated as "forbidden." According to quantum theory, electrons that are still attached to an ion can be rearranged into certain long-lived orbits when the ion is excited, and the electrons emit the forbidden lines when they eventually move out of these excited orbits. However, even in the best vacuum on Earth, frequent collisions knock the electrons out of these orbits before they have a chance to emit the forbidden lines.

FOCUS 6.1

Taking the Temperature of the Corona

The Sun's enormous gravity would hold a relatively cool gas close to the photosphere, just as the Earth's gravity binds its atmosphere into a thin shell. But this is inconsistent with the larger extent of the corona seen during a total solar eclipse. A temperature of a million degrees is required to keep the corona extended; the motion of the heated gas supports it against the Sun's gravitational pull. At cooler temperatures, the atoms would have to be much lighter than hydrogen to move fast enough and extend the corona out so far, but hydrogen is the lightest element there is.

The detailed character of spectral lines can be used to infer the temperature and motion of the solar gas. The hotter the gas, the faster the motions, and the greater the wavelength shifts from the Doppler effect. These displacements widen the spectral lines, and can be used as a thermometer to measure the corona's temperature. In 1941, for example, the Swedish astronomer Bengt Edlén (1900–1993) noticed that the observed widths of the emission lines in the corona indicate a temperature of about two million degrees.

In 1946, the Russian astrophysicist Vitalii L. Ginzburg (1916–) and the Australian radio astronomers David F. Martyn (1906–1970) and Joseph L. Pawsey (1908–1962) independently used observations of the Sun's meter-wavelength radio radiation to confirm the existence of a million-degree solar corona by measuring the temperature of the radio emission.

The free electrons in the corona, which are not attached to anything, scatter sunlight from the photosphere and illuminate the corona. The electrons bend small amounts of the sunlight into our line of vision, just as tiny dust particles illuminate a sunbeam in your room and air molecules scatter sunlight to make the sky blue. But the coronal electrons are so rarefied that most of the light from the photosphere passes right through the low-density corona. That is the reason the dim corona is a million times fainter than the photosphere at visible wavelengths.

There are about a billion electrons and protons per cubic centimeter (10^9 cm^{-3}) at the base of the corona, which sounds like a lot, but the low corona is one hundred times more tenuous than the chromosphere (10^{11} cm^{-3}) and an additional million times more rarefied than the photosphere (10^{17} cm^{-3}). And even the photosphere is far more rarefied than our air.

The average drop in density from the photosphere to the bottom of the corona is matched by an overall increase in temperature from about 5,780 kelvin in the photosphere to almost 10,000 degrees in the chromosphere and a million degrees at the base of the corona (Fig. 6.7). A very thin transition region, less than 100 kilometers thick,

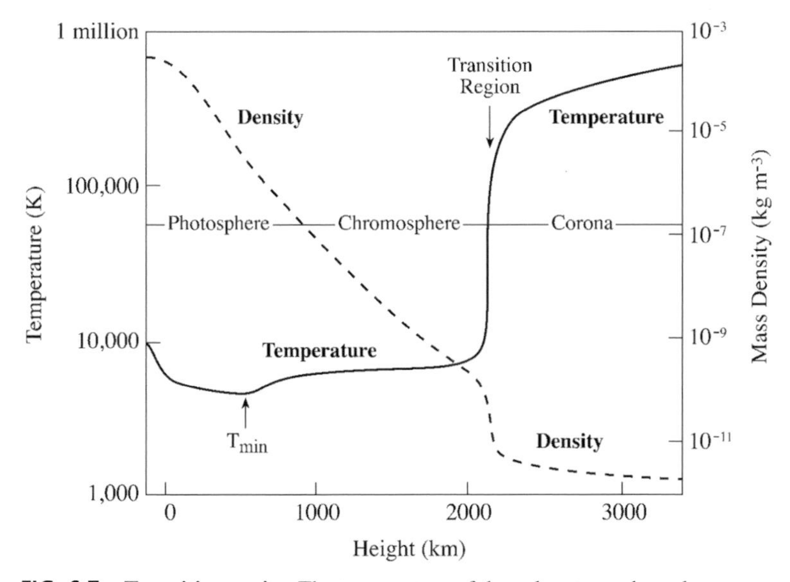

FIG. 6.7 **Transition region** The temperature of the solar atmosphere decreases from values near 6,000 kelvin at the visible photosphere to a minimum value of roughly 4,400 kelvin at the base of the chromosphere, about 500 kilometers higher in the atmosphere. The temperature increases with height, slowly at first, then extremely rapidly in the narrow transition region between the chromosphere and corona, from about 10,000 kelvin to about a million kelvin. Thousands of needle-like spicules, each lasting about five minutes, are detected shooting up from the chromosphere into the transition region, so this static, layered model of the transition region does not completely portray its dynamic, ever-changing aspect. The height is in kilometers, abbreviated km, the temperature in degrees kelvin, denoted K, and the mass density in kilograms per cubic meter, abbreviated kg m^{-3}. (Courtesy of Eugene Avrett, Smithsonian Astrophysical Observatory.)

lies between the chromosphere and corona. Both the density and temperature change abruptly in the transition region; the density decreases as the temperature increases in such a way to keep the gas pressure spatially constant in this region of transition.

The corona then slowly cools with increasing distance from the Sun, slightly decreasing in temperature to about 100,000 kelvin at the Earth's orbit. The corona also thins out as it expands into the increasing volume of space, reaching a density of only about 5 electrons and 5 protons per cubic centimeter at the Earth's orbit.

Although the electrified coronal particles move about at great speed, there are so few of them that the total energy in the corona is quite low. Only about a millionth of the Sun's total energy output is required to heat the corona. And even though the free electrons are extremely hot, they are so scarce and widely separated that an astronaut or a satellite will not burn up when immersed in the rarefied corona just outside the Earth. But if the corona were as dense as the center of the Sun, at a temperature of a million degrees it would contain enough energy to vaporize our planet.

6.3 CORONAL HEATING

More than half a century ago, astronomers knew that the corona is a very hot, rarefied gas. So they knew what the corona is, but not why it exists. The problem, which has not yet been completely solved, is to explain why the corona is so hot.

The visible solar disk, the photosphere, is closer to the Sun's center than the million-degree corona, but the photosphere is several hundred times cooler, and this comes as a big surprise. The essential paradox is that energy should not flow from the cooler photosphere to the hotter corona anymore than water should flow uphill. When you sit far away from a fire, for example, it warms you less.

The temperature of the corona is just not supposed to be so much higher than that of the atmosphere immediately below it. It violates common sense, as well as the second law of thermodynamics, which holds that heat cannot be continuously transferred from a cooler body to a warmer one without doing work. This unexpected aspect of the corona has baffled scientists for decades, and they are still trying to explain where all the heat is coming from.

We know that visible sunlight cannot resolve the heating paradox. Light from the photosphere does not go into the corona; it goes through the corona. There is so little material in the corona that it is transparent to almost all of the photosphere's radiation. Sunlight therefore passes right through the corona without depositing substantial quantities of energy in it, traveling out to warm the Earth and to also keep the photosphere cool.

So, radiation cannot resolve the heating paradox, and we must look for alternate sources of energy. Possible mechanisms involve either the kinetic energy of moving material or the magnetic energy stored in magnetic fields. Indeed, the photosphere seethes with continual motion, and magnetism threads its way through the entire solar atmosphere. Unlike radiation, both of these forms of energy can flow from cold to hot regions.

For several decades, sound waves provided a widely accepted explanation for heating the low-density, million-degree corona. In 1948–49, for example, the German astronomer Ludwig Biermann (1907–1986), the French astronomer Evry Schatzman (1920–) and the German-born American astronomer Martin Schwarzschild (1912–1997) independently proposed that the high coronal temperature could be maintained by acoustical noise produced by solar convection. The up and down motion of the piston-like convection cells, called granules, will generate a thundering sound in the overlying atmosphere, in much the same way that a throbbing high-fidelity speaker drives sound waves in the air. The sound, or acoustic, waves should accelerate and strengthen as they travel outward through the increasingly rarefied solar atmosphere, until supersonic shocks occur that resemble sonic booms of jet aircraft. It was thought that these shocks would dissipate their energy rapidly, and perhaps generate enough heat to account for the high-temperature corona.

Although the majority of the sound waves are reflected back and remain trapped in the Sun, a small percentage of them manage to slip through the photosphere, dissipating their energy rapidly within the chromosphere and generating large amounts of heat. And the low chromosphere does indeed seem to be heated by sound waves that are generated in the convective zone and dissipated by shocks in the chromosphere.

The Sun's chromosphere is composed of hundreds of thousands of tiny spicules, energized and thrust outward by the sound-driven shocks. This method of chromosphere heating is generally consistent with the fact that other stars with outer convection zones have chromospheres; while for stars without outer convection zones no chromosphere is detected.

However, since the *Skylab* observations of ultraviolet spectral lines in the 1970s, it has become apparent that there is very little acoustic energy left over by the time the shock waves reach the upper chromosphere, and sound waves apparently cannot reach the corona. The steep temperature and density gradient in the transition region would reflect sound waves, keeping them from propagating into the corona.

Magnetic fields probably play a pivotal role in heating the solar corona. The hottest and densest material in the low corona is located where the magnetic field is strongest, and it is intense magnetism that molds the corona, producing its highly structured, inhomogeneous shape. The other key ingredient for coronal heating is change. The dynamic corona is magnetically linked to, and driven by, the underlying photosphere and convective zone whose turbulent motions can push the magnetic fields around and shuffle them about. Observations indicate that the corona is always changing on all observed spatial and temporal scales, seething and writhing in tune with the Suns magnetism.

So, recurring themes in explaining the million-degree temperatures of the solar corona are highly structured magnetism and turbulent change. The ultimate source of the energy that heats the corona is convective motions in the solar interior, and that energy is somehow channeled by magnetic fields from the cool photosphere to the hot corona. And the solution to the corona's heating crisis is related to just how energy is transferred to, stored in, and released by the magnetic fields that dominate the corona.

When a magnetic field is disturbed, a tension acts to pull it back, generating magnetic waves that can propagate upward and dissipate energy in the corona. These waves are called Alfvén waves after the Swedish theoretician Hannes Alfvén (1908–1995) who first described them mathematically. He pioneered the study of the interaction of hot gas, or plasma, and magnetic fields, in a discipline called magneto-hydrodynamics, and was awarded the Nobel Prize in Physics in 1970 for this work.

Instruments on a number of spacecraft have detected Alfvén waves from the low corona to more distant regions of interplanetary space. However, once you get energy into an Alfvén wave, it is hard to dampen the wave and extract energy from it. Like radiation, the Alfvén waves seem to propagate right through the low corona without being noticeably absorbed or dissipated there, and without depositing enough energy into the coronal gas to heat it up to the observed temperatures.

Nevertheless, magnetic waves are still viable candidates for some coronal heating, and other energy-carrying waves such as sound (acoustic), magneto-acoustic (magnetic-sound) and shock waves, may play a role. Hot, oscillating magnetic loops have, for example, been observed swaying in the corona under the influence of periodic waves, sometimes for periods of five or ten minutes characteristic of internal sound waves. These back and forth motions have been attributed to magneto-acoustic waves.

Magnetic loops can heat the corona by coming together and releasing stored magnetic energy when they make contact in the corona. Motions down inside the convective zone twist and stretch the overlying magnetic fields, slowly building up their energy. When these magnetic fields are pressed together in the corona, they can merge and join at the place where they touch, releasing their pent-up energy to heat the gas. The magnetic fields reform or reconnect in new magnetic orientations, so this method of coronal heating is termed magnetic reconnection.

Instruments aboard the *SOlar and Heliospheric Observatory,* abbreviated *SOHO,* have shown that tens of thousands of small magnetic loops are constantly being generated, rising up and out of the photosphere, interacting, fragmenting and disappearing within hours or days (Fig. 6.8). The numerous magnetic loops in this "magnetic carpet" are always being replaced, in a sort of self-cleaning action initiated from below. Motions in the photosphere and underlying turbulent convection push the carpet loops around, and when the magnetic fields of adjacent loops meet, they can break and reconnect with each other into simpler magnetic configurations, releasing enough energy to heat the corona. Since the magnetic carpet is continuously replenished every 40 hours, forming new magnetic connections all the time and all over the Sun, it can provide a continual source of coronal heating.

According to another scenario, the steady heating of the hot solar corona is produced by coronal magnetic interactions that produce frequent, numerous, small-scale explosions, called nanoflares or microflares, which occur at seemingly random locations. The coronal magnetic fields can braid, twist and writhe in a perpetual dance, energized from below, rising, falling, intertwining and coupling together to cumulatively produce numerous small explosions that heat the high-temperature corona.

So there is no lack of possible mechanisms for explaining how the corona becomes so hot and stays heated. And that's a good thing, for the corona would collapse in minutes if energy weren't constantly dumped into it to maintain its high temperature.

But exactly where is the corona heated? It becomes hottest at locations where the magnetic fields are strongest. And most of this heating, at least in magnetic loops, is often confined to the lower parts of the corona, near the chromosphere and transition region.

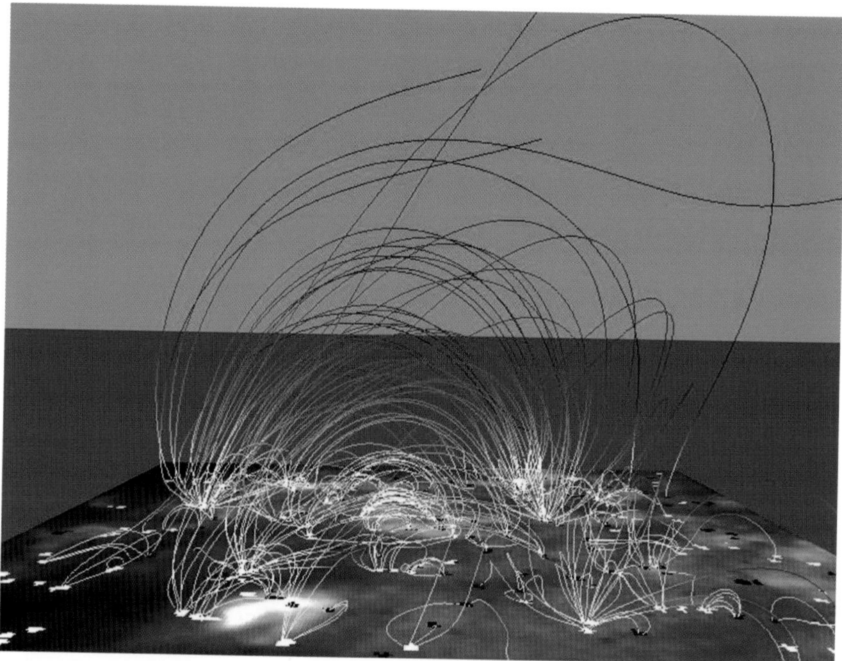

FIG. 6.8 Magnetic carpet Magnetic loops of all sizes rise up into the solar corona from regions of opposite magnetic polarity (*black* and *white*) in the photosphere (*green*), forming a veritable carpet of magnetism in the low corona. Energy released when oppositely directed magnetic fields meet in the corona, to reconnect and form new magnetic configurations, is one likely cause for making the solar corona so hot. (Courtesy of the *SOHO* EIT and MDI consortia. *SOHO* is a project of international cooperation between ESA and NASA.)

And this brings us to the ubiquitous coronal loops that structure the solar atmosphere and the coronal holes that provide an open gateway to interplanetary space.

6.4 CLOSED CORONAL LOOPS AND OPEN CORONAL HOLES

Because of its high, million-degree temperature, the corona emits most of its energy, and its most intense radiation, as X-rays. They can be used to image the hot corona all across the Sun's face with high spatial and temporal resolution. This is because the Sun's visible photosphere, being so much cooler, produces negligible amounts of X-ray radiation, and appears dark under the million-degree corona. In contrast, the faint visible light of coronal emission lines is only detected at the edge of the Sun during a total solar eclipse, and cannot be seen in the intense glare of sunlight outside eclipse.

Since the Sun's X-ray radiation is totally absorbed in the Earth's atmosphere, it must be observed with telescopes lofted into space by rockets or in satellites. Herbert Friedman (1916–2000) and his colleagues at the Naval Research Laboratory obtained the first X-ray pictures of the Sun in 1960, during a brief 5-minute rocket flight. These crude, early images were replaced with high-resolution X-ray photographs taken and returned to Earth by astronauts from NASA's *Skylab* thirteen years later, during a 9-month period in 1973–74. Most recently, the *Yohkoh,* or "sunbeam" spacecraft, launched on 30 August 1991 by the Japanese Institute of Space and Astronautical Science, provided millions of X-ray images of the Sun for a ten year period, until 2001. *Yohkoh's* images at invisible X-ray wavelengths are almost as sharp and clear as pictures made in visible wavelengths from the ground (Fig. 6.9).

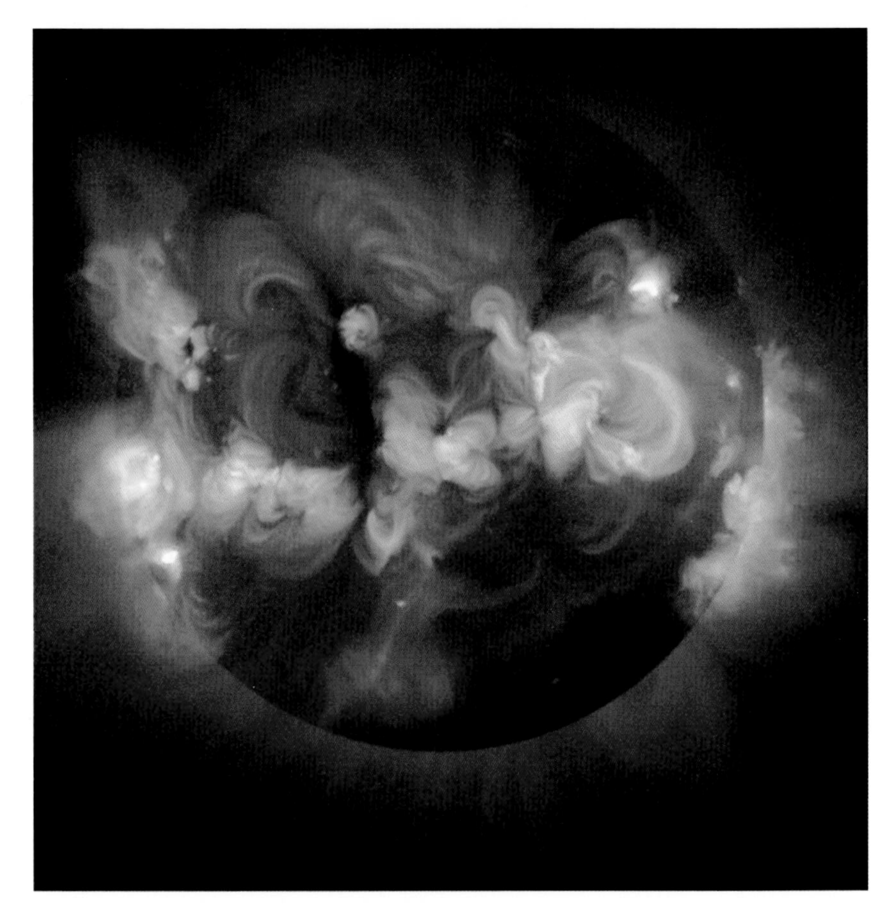

FIG. 6.9 The Sun in X-rays The bright glow seen in this X-ray image of the Sun is produced by ionized gases at a temperature of a few million kelvin. It shows magnetic coronal loops that thread the corona and hold the hot gases in place. The brightest features are called active regions and correspond to the sites of the most intense magnetic field strength. This image of the Sun's corona was recorded by the Soft X-ray Telescope (SXT) aboard the Japanese *Yohkoh* satellite on 1 February 1992, near a maximum of the 11-year cycle of solar magnetic activity. Subsequent SXT images, taken about five years later near activity minimum, show a remarkable dimming of the corona when the active regions associated with sunspots have almost disappeared, and the Sun's magnetic field has changed from a complex structure to a simpler configuration. (Courtesy of Gregory L. Slater, Gary A, Linford, and Lawrence Shing, NASA, ISAS, the Lockheed-Martin Solar and Astrophysics Laboratory, the National Astronomical Observatory of Japan, and the University of Tokyo.)

The X-ray telescopes aboard *Skylab* and *Yohkoh* have shown that the corona is stitched together by bright, thin magnetized loops, which are in a constant state of agitation. They provide the woven fabric of the entire corona. Each closed loop rises into the corona from a region of positive magnetic polarity in the underlying solar photosphere and then turns back into a region of negative polarity in the photosphere. Material is concentrated to higher densities and temperatures within these loops, so they emit X-rays more intensely than their surroundings. The intense X-ray emission thus outlines the magnetic shape and structure of the Sun's outer atmosphere.

Although gas pressure is greater than magnetic pressure in the photosphere and convective zone, where the hot gases confine the magnetic fields and push them about, it is the other way around in the corona, where the ionized gas is confined by magnetic fields that permeate the solar atmosphere, determining the shape and form of the corona. But the coronal magnetic fields emerge from underneath the photosphere, where they are rooted, and they are continually displaced and replaced by convective motions just below the photosphere.

As a result, the dynamic, ever-changing corona has no permanent features, but is instead always in a continued state of metamorphosis, continuously varying in brightness and structure on all detectable spatial and temporal scales, with effects felt throughout the solar atmosphere. Like most of the rest of the Universe, the Sun's outer atmosphere is never still, and there is no such thing as a quiet, inactive corona.

The Extreme-ultraviolet Imaging Telescope, abbreviated EIT, aboard the *SOlar and Heliospheric Observatory,* or *SOHO* for short, has kept the transition region and inner corona under careful watch for more than a decade, since *SOHO* was launched on 2 December 1995. It has taken full-disk images at three lines of ionized iron, Fe IX, Fe XII and Fe XV, and one line of ionized helium, He II; these are the permitted lines emitted by ionized atoms, sensitive to temperatures from 60 thousand to 2.1 million kelvin (Table 6.2), and not the forbidden lines detected at visible wavelengths. While EIT produced images of the entire solar disk, the *Transition Region And Coronal Explorer,* abbreviated *TRACE,* was used to zero in on specific regions, obtaining extreme ultraviolet images with high spatial and temporal resolution. After its launch,

TABLE 6.2 Some prominent solar emission lines in the transition region and low corona[a]

Wavelength (nm)	Emitting ion	Formation temperature (K)
12.16	Hydrogen, Lyman alpha	
17.11	Iron, Fe IX and Fe X	1 000 000
19.51	Iron, Fe XII	1 400 000
28.42	Iron, Fe XV	2 100 000
30.38	Helium, He II	60 000
154.8	Carbon, C IV	110 000

[a] Subtract one from the Roman numeral to get the number of missing electrons. The wavelengths are in nanometers, abbreviated nm, where 1 nm = 10^{-9} meters. Astronomers sometimes use the Ångström unit of wavelength, abbreviated Å, where 1 Å = 10^{-10} meters = 10 nm.

on 1 April 1998, *TRACE* routinely observed the Lyman-alpha line of hydrogen, a line of ionized carbon, C IV and the three iron lines used with *SOHO*.

Because a given stage of ionization occurs within a narrow range of temperature, the different spectral lines can be used to tune in coronal loops at particular temperatures (Fig. 6.10). They have revealed that there are cool and hot loops, each with its own unique temperature and location. All of these magnetic loops, either in the transi-

FIG. 6.10 Loops at different temperatures A composite, color-coded image of the Sun, combining three different extreme ultraviolet wavelengths, reveals solar features unique to each. The hot glowing gas is constrained by magnetic structures, usually looping from the solar interior into the overlying solar atmosphere. The different wavelengths are sensitive to gas at various temperatures. Red shows the hottest material, with a temperature of about 2.5 million kelvin, emitted at a wavelength of 28.4 nanometers from 14 times ionized iron, designated Fe XV. Blue corresponds to radiation at 17.1 nanometers emitted from 8 and 9 times ionized iron, denoted Fe IX and Fe X, at about 1.0 million kelvin, and yellow indicates emission at a wavelength of 19.5 nanometers from 11 times ionized iron, denoted Fe XII, at 1.5 million kelvin. The simultaneous images were obtained with the Extreme-ultraviolet Imaging Telescope, or EIT for short, aboard the *SOlar and Heliospheric Observatory,* abbreviated *SOHO*. (Courtesy of the *SOHO* EIT consortium. *SOHO* is a project of international cooperation between ESA and NASA.)

tion region or the low corona, are extremely dynamic or time variable, emerging continuously from the photosphere and changing on timescales from hours to days. So the apparently steady corona that is frozen into a single image or photograph is an illusion.

In contrast to the dense, bright areas, the corona also contains less dense regions called coronal holes. These so-called holes have so little material in them that they are not very luminous and are difficult to detect. They therefore appear as large dark areas on X-ray or extreme ultraviolet images, seemingly devoid of radiation. Coronal holes cannot be distinguished in visible light on the solar disk, and neither can coronal loops, but both the holes and loops have been recorded at the solar limb in visible light using coronagraphs since the 1950s.

At times of low solar activity, near the minimum of the Sun's 11-year magnetic activity cycle, coronal holes cover the north and south polar caps of the Sun. During more active periods, closer to the cycle maximum, coronal holes can appear at all solar latitudes, even at or near the solar equator.

Unlike the arched and closed magnetism of coronal loops, the coronal holes have relatively weak and open magnetic fields that stretch out radially into interplanetary space and do not return directly to another point on the Sun. These fields eventually turn around, but they are distended to great distances from the Sun. The normally constraining magnetic forces therefore relax and open up in the coronal holes to allow an unencumbered outward flow of electrically charged particles into interplanetary space (Fig. 6.11). Solar particles pour out of a coronal hole along the open magnetic field lines, apparently unabated, keeping the hole's density low.

Coronal holes appear like a dark, empty void, as if there were a hole in the corona, but the rarefied coronal holes are not completely empty. Moreover, the coronal holes are neither constant nor permanent; they appear, evolve and die away in periods ranging from a few weeks to several months. So coronal holes continuously change in content, shape and form, like everything else on the Sun.

6.5 THE ETERNAL SOLAR WIND

Just half a century ago, most people visualized our planet as a solitary sphere traveling in a cold, dark vacuum around the Sun. But we now know that the wide-open spaces in our Solar System are not empty. They are filled with tiny pieces of the Sun, in an ever-flowing solar wind.

The notion that something is always being expelled from the Sun first arose from observations of comet tails. Comets appear unexpectedly almost anywhere in the sky, moving in every possible direction, but with tails that always point away from the Sun. A comet therefore travels headfirst when approaching the Sun and tail-first when departing from it. Ancient Chinese astronomers concluded that the Sun must have a *chi,* or "life force", that blows the comet tails away. And in the early 1600s, the German astronomer Johannes Kepler (1571–1630) proposed that the pressure of sunlight pushes the comet tails away from the Sun.

Modern scientists noticed that a comet could have two tails. One is a yellow tail of dust and dirt, which can litter the comet's curved path. The dust is pushed away from the Sun by the pressure of sunlight. The other tail is electric blue, shining in the light of ionized particles. The ions in comet tails always stream along straight paths away from

FIG. 6.11 Magnetic fields near and far In the low solar corona, strong magnetic fields are tied to the Sun at both ends, trapping hot, dense electrified gas within magnetized loops. Far from the Sun, the magnetic fields are too weak to constrain the outward pressure of the hot gas, and the loops are vastly extended, allowing electrically charged particles to escape, forming the solar wind and carrying magnetic fields away. (Courtesy of Newton Magazine, the Kyoikusha Company.)

the Sun with velocities many times higher than could be caused by the weak pressure of sunlight.

In the early 1950s, the German astrophysicist Ludwig Biermann (1907–1986) proposed that streams of electrically charged particles, called corpuscular radiation, poured out of the Sun at all times and in all directions to shape the comet ion tails. Summing up his work in 1957, Biermann concluded that:

> The acceleration of the ion tails of comets has been recognized as being due to the interaction between the corpuscular radiation of the Sun and the tail plasma. The observations of comets indicate that there is practically always a sufficient intensity of solar corpuscular radiation to produce an acceleration of the tail ions of at least about twenty times solar gravity.[26]

Thus, the ion tails of comets act like an interplanetary windsock, demonstrating the existence of a continuous, space-filling flow of charged particles from the Sun.

So the Sun is continuously blowing itself away, filling the Solar System with a perpetual flow of electrified matter called the solar wind. And every second the Sun blows away about a million tons, or a billion kilograms, of material that must be replaced from below, but this is a small amount compared with the enormous total mass of the Sun. At the present rate, it would take ten billion years for the Sun to lose only 0.01 percent of its mass by the solar wind, and the Sun will evolve into a giant star long before it blows away completely.

Thus, the space between the planets is not completely empty; it contains an eternal solar wind, a rarefied mixture of protons and electrons that stream out radially in all directions from the Sun. The planets move through this wind as if they were ships at sea, and the wind wraps itself around the Earth. So we live inside the Sun.

The solar gale brushes past the planets and engulfs them, carrying the Sun's corona out to interstellar space. The radial, supersonic outflow thereby creates a huge bubble of plasma, with the Sun at the center and the planets inside, called the heliosphere, from *helios* the "God of the Sun" in Greek mythology (Fig. 6.12).

We also know that the million-degree corona is so hot that it cannot stand still. Indeed, the solar wind consists of an overflow corona, which is too hot to be entirely constrained by the Sun's inward gravitational pull. The hot gas creates an outward pressure that tends to oppose the inward pull of the Sun's gravity; and at great distances, where the solar gravity weakens, the hot protons and electrons overcome the Sun's gravity and accelerate away to supersonic speed, like water overflowing a filled bathtub or a dam.

So, the solar corona is really the visible, inner base of the solar wind, and the solar wind is just the hot corona expanding into cold, vacuous interstellar space. In 1957, geophysicist Sydney Chapman (1888–1970) demonstrated mathematically that the Sun's million-degree corona conducts heat so well that its temperature will stay high far out into space, and that its electrons and protons must extend beyond the Earth's orbit, even if it is gravitationally held to the Sun. In the following year, Eugene Parker (1927–), a young astrophysicist at the University of Chicago, showed that the extended corona will not only extend to the Earth, it will expand and flow out there. So Biermann's continual bombardment by solar corpuscles, which propelled comet ion tails, could be attributed to the outward expansion of Chapman's extended corona.

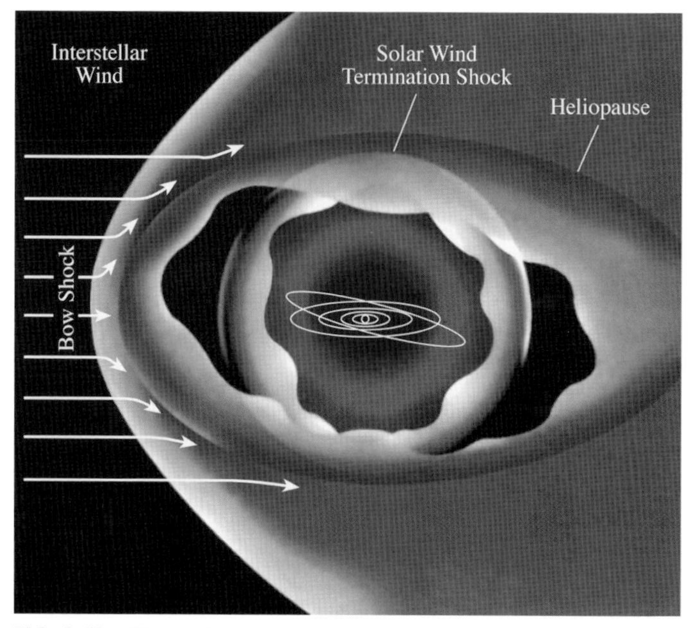

FIG. 6.12 The heliosphere With its solar wind going out in all directions, the Sun blows a huge bubble in space called the heliosphere. The heliopause is the name for the boundary between the heliosphere and the interstellar gas outside the Solar System. Interstellar winds mold the heliosphere into a non-spherical shape, creating a bow shock where they first encounter it. The orbits of the planets are shown near the center of the drawing.

Parker showed from hydrodynamics that the corona must expand rapidly outward because it is extremely hot, and that as the outer corona disperses, gases welling up from below will replenish it. The hydrodynamics shows that the expansion would begin slowly near the Sun, where the solar gravity is the strongest, and then continuously accelerate outward into space, gaining speed with distance and reaching the supersonic velocities needed to account for the acceleration of comet tails. This would create a strong, persistent, solar wind, forever blowing at speeds of hundreds of kilometers per second throughout the Solar System.

Any doubts about the existence of the solar wind were removed by *in situ* (Latin for "in original place", or literally "in the same place") measurements made by instruments on board the Soviet *Lunik 2* spacecraft on the way to the Moon in 1959 and by those aboard NASA's *Mariner II* spacecraft during its trip to Venus in 1962. The solar wind has now been sampled for nearly half a century, and it has never stopped blowing.

Measurements from spacecraft indicate that the solar wind has a fast and slow component. The fast, uniform wind blows at about 750 kilometers per second, and the variable, gusty slow one moves about half as fast. Both winds are supersonic, moving at least ten times faster than the sound speed in the solar wind. The density and temperature of the slow wind are about twice those of the fast one, but both components are much more tenuous, hotter and faster than any wind on Earth.

The solar wind has been diluted to rarefied plasma by the time it reaches the Earth. Near the Earth's orbit, there are about 5 million electrons and 5 million protons per cubic meter of the solar wind. Space probes have also shown that the magnetism entrained in the solar wind has been dragged, stretched, and enormously weakened by the time it reaches the Earth's orbit.

Since the electrified wind material is an excellent conductor of heat, the temperature falls off only gradually with distance from the Sun, reaching about 100,000 kelvin at the Earth's distance. The Sun's wind also rushes on with little reduction in speed, for there is almost nothing out there to slow it down.

The expansion of the solar wind combined with the Sun's rotation determines the magnetic structure of interplanetary space, and thereby establishes the magnetic pathways for energetic particles leaving the Sun. While one end of the solar magnetic field remains firmly rooted in the solar photosphere and below, the other end is extended and stretched out into space by the solar wind. As the wind streams radially outward, the Sun's rotation bends the radial pattern into a spiral shape within the plane of the Sun's equator, coiling the magnetism up (Fig. 6.13).

The spiral magnetic field creates an interplanetary highway that can connect the site of a solar eruption, or flare, to the Earth. Energetic charged particles, that are fewer in total number and much greater in energy than those in the solar wind, are hurled out from the Sun during these brief eruptions, creating powerful gusts in the solar wind. If they occur in just the right place, near the west limb and the solar equator, the energized material will connect to the interplanetary spiral and travel along it to the Earth in about half an hour, threatening astronauts or satellites. The spiral magnetic pattern

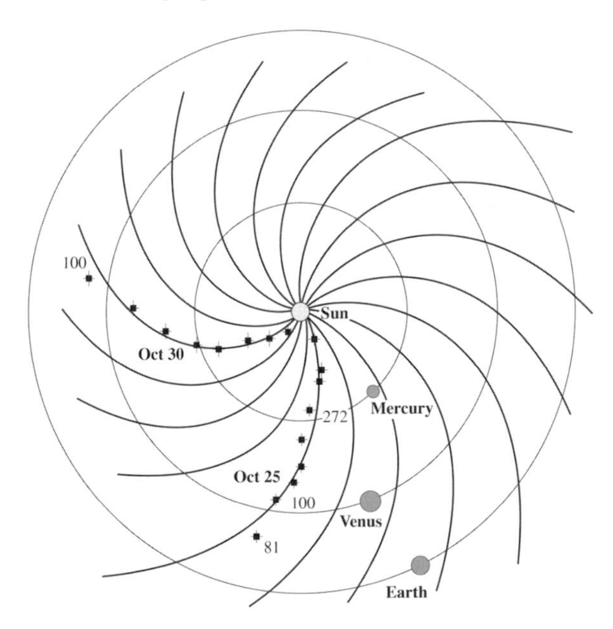

FIG. 6.13 Spiral path of interplanetary electrons The trajectory of flare electrons in interplanetary space as viewed from above the Sun's polar regions using the *Ulysses* spacecraft. As the high-speed electrons move out from the Sun, they excite radiation at successively lower plasma frequencies; the numbers denote the observed frequency in kiloHertz, or kHz. Since the flaring electrons are forced to follow the interplanetary magnetic field, they do not move in a straight line, but instead move along the spiral pattern of the interplanetary magnetic field, shown by the solid curved lines. The squares and crosses indicate *Ulysses* radio measurements of type III radio bursts on 25 and 30 October 1994. The approximate locations of the orbits of Mercury, Venus and the Earth are shown as circles. (Courtesy of Michael J. Reiner. *Ulysses* is a project of international collaboration between ESA and NASA.)

has, in fact, been confirmed by tracking the radio emission of charged particles thrown out during such solar flares, as well as by spacecraft that have sampled the interplanetary magnetism near the Earth.

6.6 *ULYSSES, SOHO* AND THE ORIGIN OF THE SUN'S WINDS

Where do the Sun's winds come from? Since spacecraft generally measure the solar wind streams near the Earth, far from their source on the Sun, scientists have to use other measurements to extrapolate back to the place of their origin. Comparisons of wind-velocity measurements with X-ray images of the Sun indicated, in the 1970s, that whenever a high-speed stream in the solar wind swept by Earth, a coronal hole rotated into alignment with our planet. This naturally suggested that at least some of the high-speed wind is gushing out of coronal holes on the Sun. Observations of the scintillation, or twinkling, of radio signals from remote radio sources indicated that the gusty, slow-speed wind comes from a different place, confined to low latitudes near coronal streamers when the Sun is near the minimum of its activity cycle.

Nevertheless, until recently spacecraft have only been able to directly measure a limited, two-dimensional section, or slice, of the solar wind within the plane in which the Earth orbits the Sun. This plane, called the ecliptic, is tilted only 7 degrees from the plane of the solar equator. Interplanetary spacecraft usually travel within the ecliptic, both because they normally rendezvous with another planet and also because their launch vehicles obtain a natural boost by traveling in the same direction as the Earth's spin and in the plane of the Earth's orbit.

Then in 1994–95, the *Ulysses* spacecraft, a collaborative project of NASA and the European Space Agency, abbreviated ESA, traveled outside the ecliptic plane for the first time to sample the Sun's wind over the full range of solar latitudes, including previously unexplored regions above the Sun's poles.

Ulysses' velocity data indicated that a relatively uniform, fast wind pours out at high latitudes, both near the solar poles and far outside them, and that a capricious, gusty, slow wind emanates from the Sun's equatorial regions.

As the winds blow away, they must be replaced by hot gases welling up from somewhere on the Sun. However, since *Ulysses* never passed closer to the Sun than the Earth does, simultaneous observations with other satellites were required to tell exactly where the winds come from. Fortunately, the *Ulysses* data were obtained near a minimum in the Sun's 11-year activity cycle, with a particularly simple corona characterized by marked symmetry and stability. There was a pronounced coronal hole at and near the Sun's north and south poles, and coronal streamers encircled the solar equator.

Comparisons of *Ulysses*' high-latitude passes with *Yohkoh* soft X-ray images showed that coronal holes were then present at the poles of the Sun, as they usually are during activity minimum. Much, if not all, of the high-speed solar wind therefore seems to blow out from polar coronal holes along open magnetic field lines, whereas the slow wind emanates from the stalks of coronal streamers, above closed magnetic field lines, at least during the minimum in the 11-year cycle of magnetic activity.

But the fast wind is not only found above polar coronal holes during solar activity minimum. At roughly the Earth's distance from the Sun, *Ulysses* found that the fast wind is almost everywhere outside the ecliptic, even at low latitudes near the solar equator and outside the radial projection of the coronal hole edges. The charged particles could be guided to low latitudes by magnetic fields that originate in coronal holes and bend outward toward the equator.

But the polar coronal-hole boundaries might extend radially out from the Sun, instead of diverging at large distances. The fast winds would not then be bent to lower latitudes. They might instead emanate radially from all parts of the inactive Sun, outside coronal holes, streamers or active regions. Or they could sometimes shoot straight out from coronal holes close to the equator.

Comparisons of *Ulysses* data with coronagraph images pinpointed the equatorial streamers as the birthplace of the slow and sporadic wind during the minimum in the 11-year activity cycle. Hot gas is bottled up in the closed coronal loops at the bottom of the helmet streamers. The capricious slow wind can therefore only leak out along the elongated, stretched-out streamer stalks. The part that manages to escape varies in strength as the result of the effort.

Ulysses has now completed a second orbit around the Sun, during a maximum in the 11-year cycle of solar activity, permitting a comparison of solar wind speed at activity minimum and maximum (Fig. 6.14). Near the solar activity minimum, a persistent, fast, tenuous and uniform solar wind was found at high solar latitudes, arising from coronal holes that covered both solar poles during these portions of the solar cycle. During more active parts of the 11-year cycle, fast, low-latitude flows originated from multiple low and mid-latitude coronal holes. Highly variable flows were observed at all latitudes near solar maximum, arising from a mixture of sources including coronal streamers, coronal mass ejections, coronal holes and possibly active regions.

So, with the help of other spacecraft, *Ulysses* has located the general place of origin for both the fast and slow components of the solar wind. Like its namesake, the great explorer in Greek mythology, Ulysses chose

> To venture the uncharted distances;
> to feel life and the new experience
> Of the uninhabited world behind the Sun. . . .
> To follow after knowledge and excellence.[27]

and we would therefore agree with Joachim du Bellay (1522–1560) that:

> Happy he who like Ulysses has made a glorious voyage.[28]

Oppositely directed magnetic fields run in and out of the Sun on each side of the long, narrow stalks of coronal streamers, providing an open channel for the slow flow once it gets out. *SOHO*'s instruments have observed spurt-like blobs of material moving out along the stalks, like water working its way down a clogged pipe in your bathtub or sink. Observations of scintillating radio signals from spacecraft confirm the localized nature of the slow wind at activity minimum, suggesting that it is associated with the narrow stalks of coronal streamers near the solar equator.

A streamer could get so stretched out and constricted that it pinches itself off and snaps at just a few solar radii from Sun center. The lower parts of the streamer would

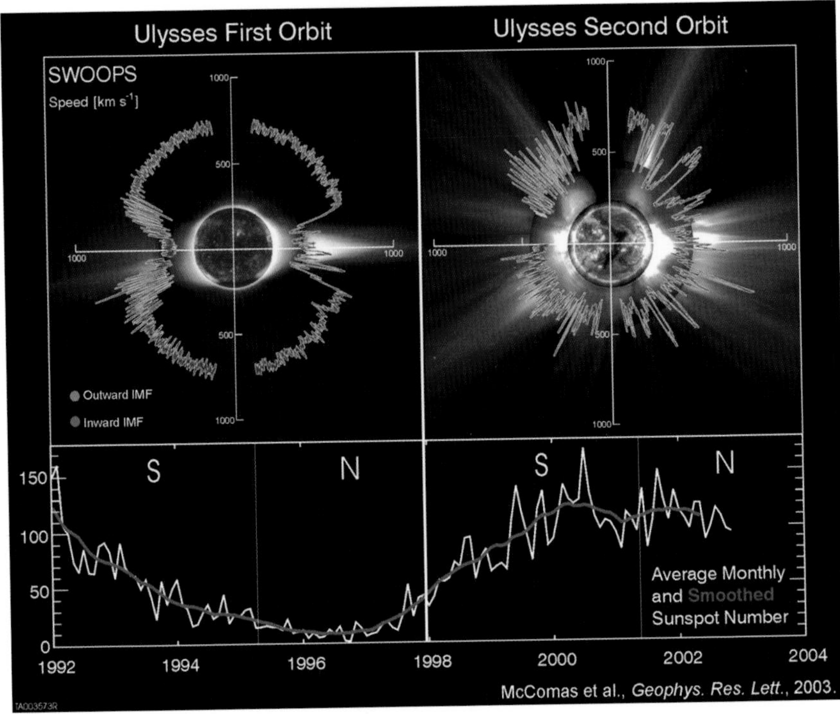

FIG. 6.14 Distributions of solar wind speeds at solar minimum and maximum Plots of the solar wind speed as a function of solar latitude, obtained from two orbits of the *Ulysses* spacecraft (*top panel*). The north and south poles of the Sun are at the top and bottom of each plot, the solar equator is located along the middle, and the velocities are in units of kilometers per second, abbreviated km s^{-1}. The sunspot numbers (*bottom panel*) indicated that the first orbit (*top left*) occurred through the declining phase and minimum of the 11-year solar activity cycle, and that the second orbit (*top right*) spanned a maximum in activity. Ultraviolet images of the solar disk and white-light images of the inner solar corona form a central backdrop for the wind speed data, and indicate the probable sources of the winds. Near solar minimum (*top left*), polar coronal holes, with open magnetic fields, give rise to the fast, low-density wind streams, whereas equatorial streamer regions of closed magnetic field yielded the slow, gusty, dense winds. At solar maximum (*top right*), small, low-latitude coronal holes gave rise to fast winds, and a variety of slow-wind and fast-wind sources resulted in little average latitudinal variation. (Courtesy of David J. McComas and Richard G. Marsden. The *Ulysses* mission is a project of international collaboration between ESA and NASA. The central images are from the Extreme ultraviolet Imaging Telescope and the Large Angle Spectrometric Coronagraph aboard the *Solar and Heliospheric Observatory*, abbreviated *SOHO*, as well as the Muana Loa K-coronameter.)

then close down and collapse, and the outer disconnected segment would be propelled out to form a gust in the slow solar wind.

Coronal loops are found down at the very bottom of streamers, and the expansion of these magnetized loops may provide the energy and mass of the slow component of the solar wind. Sequential soft X-ray images, taken from the *Yohkoh* spacecraft, have shown that magnetic loops expand out into space, perhaps contributing to the slow wind. And *SOHO* instruments have shown that small magnetic loops are nearly continuously emerging into the Sun's equatorial regions at activity minimum, where they reconnect with the Sun's global magnetic field, perhaps becoming the engine that drives the slow solar wind. The small expanding loops also carry material with them, perhaps accounting for the running blobs detected further out in the low-latitude, slow-speed solar wind.

Instruments aboard the *SOlar and Heliospheric Observatory,* abbreviated *SOHO,* have demonstrated that the fast winds accelerate rapidly, like a racehorse breaking away from the starting gate, reaching high speed very low in the corona.

Some of the high-speed outflow from polar coronal holes is apparently concentrated at the boundaries of the magnetic network formed by underlying supergranular convection cells. These edges are places where the magnetic fields are concentrated into inverted magnetic funnels that open up into the overlying corona. The strongest high-speed flows apparently gush out of the crack-like edges of the network.

The fast wind emanating from coronal holes may be accelerated to high speeds at higher altitudes, in the magnetic funnels. Closed magnetic loops may be swept by underlying convection into the funnel regions where they undergo magnetic reconnection with existing open magnetic field lines, releasing energy to power the fast wind in polar coronal holes.

Another *SOHO* instrument has demonstrated that heavier particles in polar coronal holes move faster than light particles in coronal holes (Fig. 6.15). Above two solar radii from

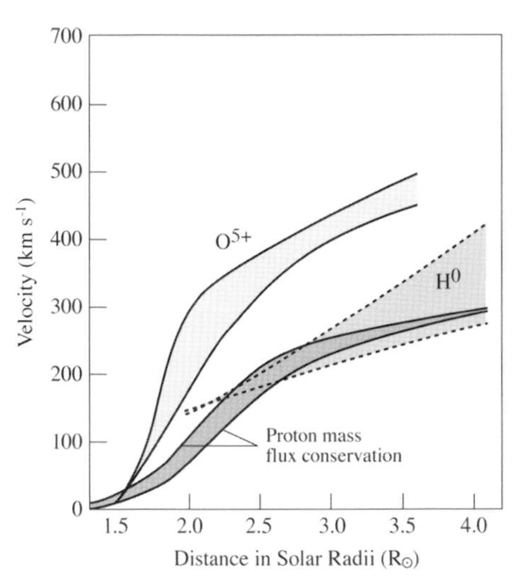

FIG. 6.15 Heavier ions move faster than lighter ones in coronal holes Outflow speeds at different distances over the solar poles for hydrogen atoms, denoted by H^0 or H I, and ionized oxygen, designated O^{5+} or O VI. Here the distances are given in units of the solar radius, denoted $R_\odot$. These data were taken in late 1996 and early 1997 with the UltraViolet Coronagraph Spectrometer, abbreviated UVCS, aboard the *SOlar and Heliospheric Observatory,* or *SOHO* for short. They show that the heavier oxygen ions move out of coronal holes at faster speeds than the lighter hydrogen, and that the oxygen ions attain supersonic velocities within 2.5 solar radii from the Sun center. The dark shaded area denotes the hydrogen outflow speed derived from mass flux conservation; for a time-steady flow, the product of the density, speed and flow-tube area should be constant. (Courtesy of the *SOHO* UVCS consortium. *SOHO* is a project of international collaboration between ESA and NASA.)

FOCUS 6.2
Solar-B

Japan's Institute of Space and Astronautical Science, abbreviated ISAS, is expected to launch its *Solar-B* Mission in 2006, providing years of detailed study of the root causes of the corona, the solar wind, and exploding solar activity. It is an international collaboration building on the highly successful *Solar-A*, renamed *Yohkoh* after launch, involving Japan, the United States and the United Kingdom.

Solar-B will investigate the Sun's changing magnetic fields from the photosphere to the low corona using a coordinated set of optical, extreme ultraviolet and X-ray telescopes. They will provide an improved understanding of the mechanisms of solar magnetic variability and how this variability modulates the Sun's output.

The processes of magnetic reconnection and wave dissipation will be investigated; they are believed to be responsible for the conversion of magnetic energy into coronal heat, ultraviolet and X-ray radiation, and the expanding solar wind. *Solar-B* will additionally improve our understanding of solar eruptions, from flares to coronal mass ejections, which are triggered and powered by some similar magnetic processes.

The ongoing creation and destruction of the Sun's magnetic fields will also be investigated, over time scales from seconds to years, revealing mechanisms for low-level modulation of the Sun's luminosity, as well as the internal dynamo process that sustains the ever-changing solar magnetism.

the Sun's center, oxygen ions have the higher outflow velocity, approaching 500 kilometers per second in the holes, while hydrogen moves at about half this speed. That violates common sense. It would be something like watching people jogging around a racetrack, with heavier adults running much more rapidly than lighter, slimmer youngsters. Something is unexpectedly and preferentially energizing the heavier particles in coronal holes.

Magnetic waves might accelerate the heavier ions more than lighter ones by pumping up gyrations around the open magnetic fields. More massive ions gyrate with lower frequencies where the magnetic waves are most intense, thereby absorbing more magnetic-wave energy and becoming accelerated to higher speeds. *Ulysses* has detected magnetic fluctuations, attributed to Alfvén waves, blowing further out in the winds far above the Sun's poles. These Alfvén waves might provide an extra boost that pushes the polar winds to higher speeds.

In another approach, the speed of the solar wind as it blows past the Earth has been tied to deep roots within the chromosphere. NASA's *Advanced Composition Explorer,* abbreviated *ACE,* spacecraft measured the wind velocity near Earth, while NASA's *Transition Region And Coronal Explorer,* or *TRACE* for short, was used to measure the time sound waves took to travel between two heights in the chromosphere. The comparison indicated that the speed of the solar wind emerging from a given area of the solar corona could be estimated from the thickness of the underlying chromosphere. It is stretched thin and opened wide in coronal holes, with their open magnetic fields and fast, tenuous solar wind, but the chromosphere is compressed below magnetically closed regions associated with the gusty, slow dense solar wind outflow (Fig 6.16).

An improved understanding of the fundamental causes of coronal heating and the Sun's winds are amongst the several objectives of the *Solar-B* mission of the Japanese space agency, scheduled for launch in 2006 (Focus 6.2).

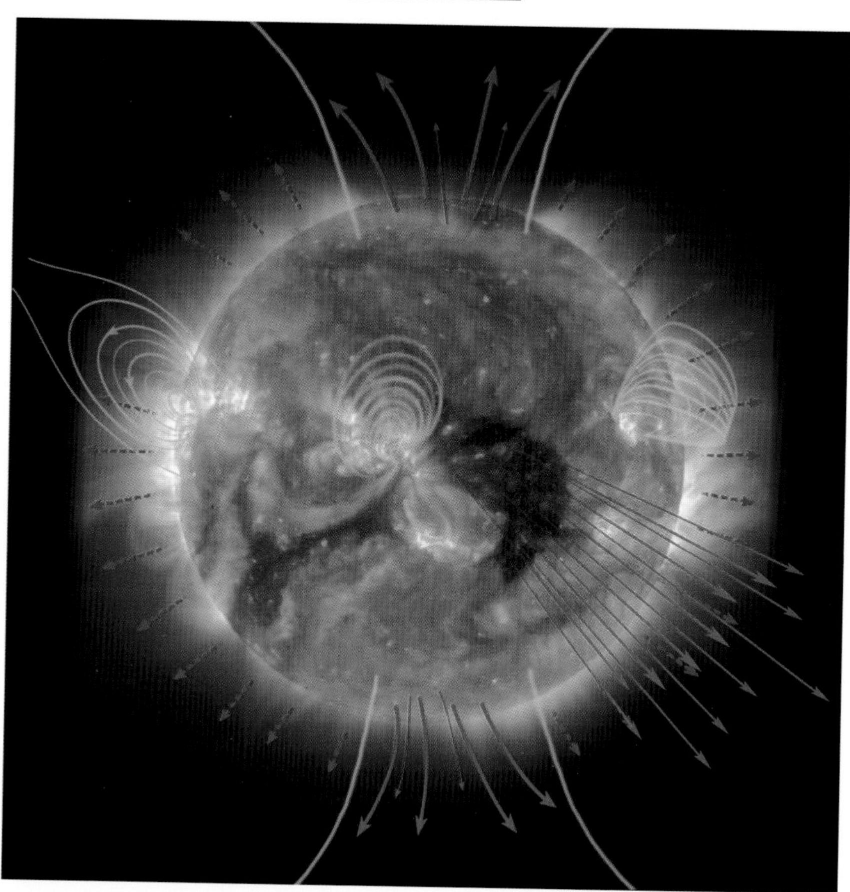

FIG. 6.16 Fast and slow winds, open and closed magnetism The solar atmosphere, or corona, is threaded with magnetic fields (*yellow lines*). Regions with open magnetic fields, known as coronal holes, give rise to fast, low density, solar wind streams (*long, solid red arrows*). In addition to permanent coronal holes at the Sun's poles (*top* and *bottom*), coronal holes can sometimes occur closer to the Sun's equator (*center*). Areas with closed magnetic fields yield the slow, dense wind (*short, dashed red arrows*). Comparisons of *TRACE* images with solar wind *ACE* data indicate that the speed and composition of the solar wind emerging from a given area have deep roots in the chromosphere. There is a shallow dense chromosphere below the strong, closed magnetic regions with a slow, dense, solar-wind outflow; deep, less dense chromosphere is found below the open magnetic regions with fast, tenuous, solar-wind outflow. This image was taken on 11 September 2003 with the Extreme-ultraviolet Imaging Telescope, abbreviated EIT, aboard the *SOlar and Heliospheric Observatory,* or *SOHO* for short. (Courtesy of the *SOHO* EIT consortium. *SOHO* is a project of international collaboration between ESA and NASA.)

But to sum up our current knowledge, winds of different velocities seem to originate in places with different magnetic field configurations, which may be related to phenomena beneath them. The fastest solar winds are traced back to coronal holes with open magnetic fields, those of intermediate speed to the inactive Sun with moderate magnetism, and the slow wind to streamer stalks and underlying active regions of intense closed magnetic fields. And each of these magnetic regions is rooted in the transition region and chromosphere, even into the photosphere and convective flows just beneath it.

6.7 THE DISTANT FRONTIER

All of the planets are immersed in the hot gale that blows from the Sun. It moves past the planets and beyond the most distant comets to the very edge of the Solar System, creating the heliosphere, a vast region centered on the Sun and enclosed by the interstellar medium. Within the heliosphere, physical conditions are dominated, established, maintained, modified and governed by the Sun.

The solar wind becomes increasingly rarefied as it spreads out into space. By the time it has reached the Earth's orbit, there are about 5 million protons and 5 million electrons per cubic meter in the solar wind, which is nearly a perfect vacuum by terrestrial standards. As it moves into a greater volume, the density of the solar wind decreases even further, as the inverse square of the distance from the Sun, and eventually blends with the gas between the stars.

How far does the Sun's influence extend, and where does it all end? Somewhere out there the solar wind becomes too dispersed to continue pushing with sufficient vigor against the interstellar medium, no longer dense or powerful enough to repel the ion-

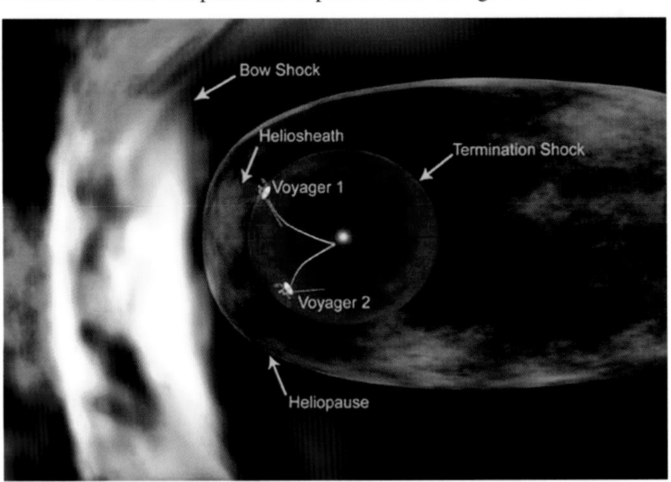

FIG. 6.17 Edge of the Solar System *Voyager 1* and *2* spacecraft, located at a distance of about 90 AU and 70 AU, approach the place where the Solar System ends and interstellar space begins. One AU is the mean distance between the Earth and the Sun, and the edge of the Solar System is located at roughly 100 times this distance. At the termination shock, the supersonic solar wind abruptly slows from an average speed of 400 kilometers per second to less than one quarter that speed. Beyond the termination shock is the heliosheath, a vast region where the turbulent and hot solar wind is compressed as it presses outward against the interstellar wind. The edge of the Solar System is found at the heliopause, where the pressure of the solar wind balances that of the interstellar medium. A bow shock likely forms as the interstellar wind approaches and is deflected around the heliosphere, forcing it into a teardrop-shaped structure with a long, comet-like tail. (Courtesy of JPL and NASA.)

ized matter and magnetic fields coursing between the stars. The radius of this celestial standoff distance, in which the pressure of the solar wind falls to a value comparable to the interstellar pressure, has been estimated at about 100 AU, or one hundred times the mean distance between the Earth and the Sun.

Instruments aboard the twin *Voyager 1* and *2* spacecraft, launched in 1977 and now cruising far beyond the outermost planets, are approaching this edge of the Solar System (Fig. 6.17). In 2005, scientists announced that *Voyager 1* had crossed the termination shock, where the pressure of the interstellar gas slows the outward supersonic flow from the Sun, terminating the solar wind's power. At about 90 AU from the Sun, the *Voyager 1* instruments recorded a sudden increase in the strength of the magnetic field carried by the solar wind, as expected when the solar wind slows down and its particles pile up at the termination shock.

Voyager 1 has therefore crossed into the vast, turbulent heliosheath, the region where the interstellar gas and solar wind start to mix. Both *Voyager* spacecraft are equipped with plutonium power sources expected to last until 2020. So *Voyager 2,* at about 70 AU in 2005, should record the termination shock in the future, and both spacecraft ought to eventually measure the heliopause, at the outer edge of the heliosheath. It is the place where the Solar System ends and interstellar space begins.

The motion of the interstellar gas, with its own wind, compresses the heliosphere on one side, producing a non-spherical shape with an extended tail (Fig. 6.17). A bow shock is formed when the interstellar wind first encounters the heliosphere; just as a bow shock is created when the solar wind strikes the Earth's magnetosphere. And the graceful arc of a bow shock, created by an interstellar wind, has been detected around the young star LL Orionis (Fig. 6.18).

Closer to home, space physicists are concerned about the impact of powerful solar eruptions on the Earth's environment in space.

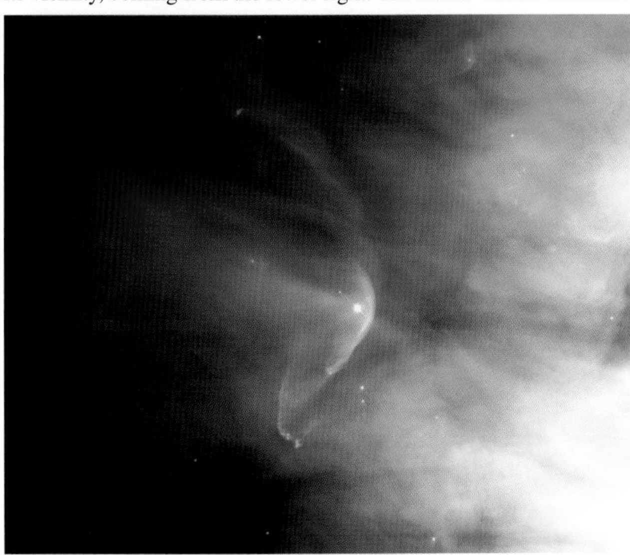

FIG. 6.18 Stellar bow shock A crescent-shaped bow shock is formed when the material in the fast wind from the bright, very young star, LL Ori (*center*) collides with the slow-moving gas in its vicinity, coming from the lower right. The stellar wind is a stream of charged particles moving rapidly outward from the star. It is a less energetic version of the solar wind that flows from the Sun. A second, fainter bow shock can be seen around a star near the upper right-hand corner of this image, taken from the *Hubble Space Telescope.* Both stars are located in the Orion Nebula; an intense star-forming region located about 1,500 light-years from the Earth. (Courtesy of NASA, the Hubble Heritage Team, STScI, and AURA.)

Chapter Seven

The Violent Sun

LONDON, THE HOUSES OF PARLIAMENT: SUN BREAKING THROUGH THE FOG The French painter Claude Monet (1840–1926) wrote that he dreamed of painting the Sun "setting in an enormous ball of fire behind the Parliament", but the Sun's movement and changing appearance in the dreary London weather made this difficult. Instead, Monet used diffuse orange and reddish hues to capture the effect of sunset in a dense, foggy atmosphere, flattening the massive, neo-Gothic building to a dark silhouette of towers and pinnacles. (Courtesy of the Musée d'Orsay, Paris. Photograph: Musées Nationaux, Paris.)

7.1 ENERGETIC SOLAR ACTIVITY

Without warning, the relatively calm solar atmosphere can be torn asunder by sudden, catastrophic outbursts of incredible energy. These transient brightenings, called solar flares, flood the Solar System with intense radiation from X-rays and extreme-ultraviolet radiation to radio waves. The powerful flares are easily observed at these invisible wavelengths, where they can briefly dominate the Sun's output and sometimes outshine all other astronomical sources (Fig. 7.1).

FIG. 7.1 Extreme ultraviolet flare An explosion on the Sun sends material out 150 thousand kilometers from the visible edge of the Sun. The emitting gas shines in the extreme ultraviolet emission from hydrogen atoms at 12.16 nanometers, called the Lyman alpha transition. This image shows that the radiation is emitted from numerous long, thin magnetic filaments. It was taken from the *Transition Region And Coronal Explorer,* abbreviated *TRACE.* (Courtesy of the *TRACE* consortium and NASA; *TRACE* is a mission of the Stanford-Lockheed Institute for Space Research, a joint program of the Lockheed-Martin Solar and Astrophysics Laboratory, or LMSAL for short, and Stanford's Solar Observatories Group.)

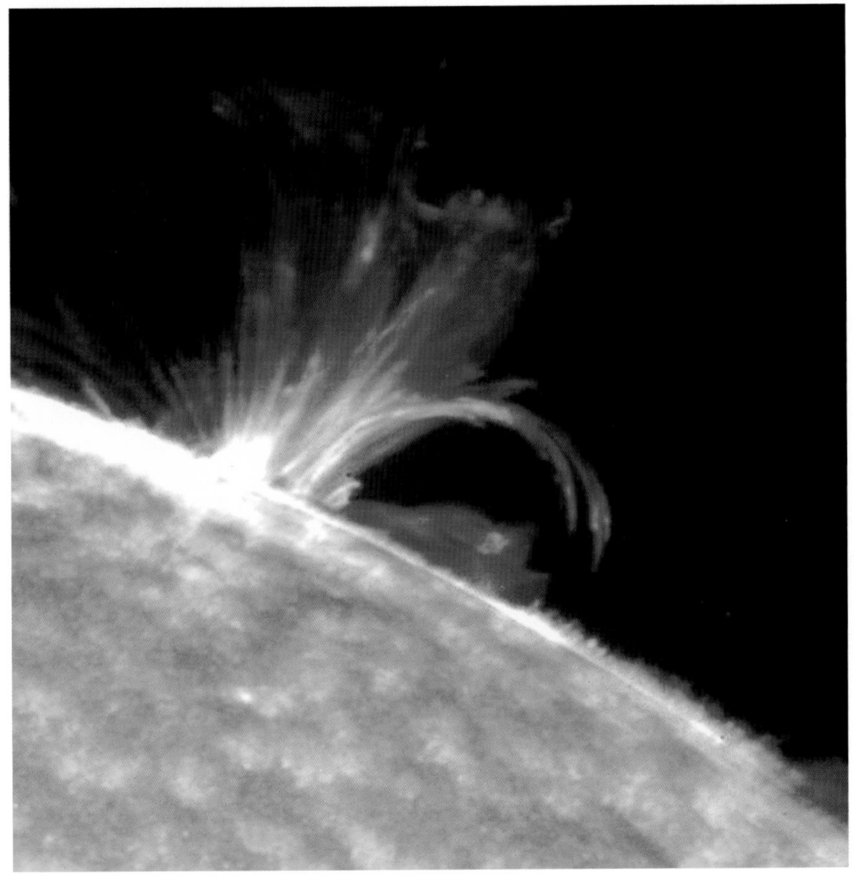

Solar flares are sudden, rapid outbursts of awesome power and violence, on a scale unknown on Earth, the biggest explosions in the Solar System. In minutes, the disturbance spreads along concentrated magnetic fields, releasing energy equivalent to millions of 100-megaton hydrogen bombs exploding at the same time, and raising the temperature of Earth-sized regions to tens of millions of degrees. Solar flares are therefore hotter than the corona. Sometimes they temporarily go out of control and lose equilibrium, becoming hotter than the center of the Sun for a short period of time.

In another type of energetic solar activity, relatively cool, elongated prominences suddenly and unpredictably open up and expel their contents, defying the Sun's enormous gravity (Fig. 7.2). The erupting prominences are associated with coronal mass ejections; giant magnetic bubbles that expand as they propagate outward from the Sun to rapidly rival it in size (Fig. 7.3). These violent eruptions throw billions of tons of material into interplanetary space. Their associated shocks accelerate and propel vast quantities of high-speed particles ahead of them.

The rates of occurrence of solar flares, erupting prominences and coronal mass ejections all vary with the 11-year cycle of magnetic activity, becoming more frequent at sunspot maximum. Truly outstanding flares are infrequent, occurring only a few times a year even at times of maximum solar activity; like rare vintages, they are denoted by their date. Flares of lesser magnitude occur much more frequently; several tens of such events may be observed on a busy day near the peak of the 11-year solar cycle of magnetic activity. Up to five coronal mass ejections can occur daily at the peak of the cycle.

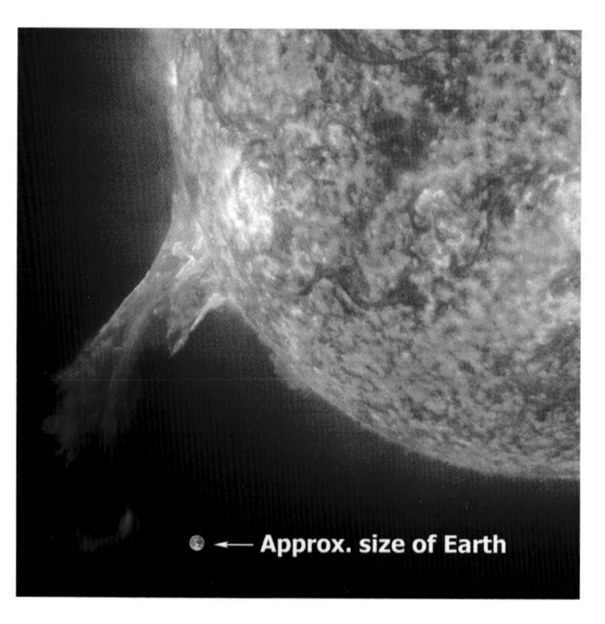

FIG. 7.2 Erupting prominence A gigantic erupting prominence is seen escaping from the Sun in the lower left part of this image, taken at a wavelength of 30.4 nanometers, emitted by singly ionized helium, denoted He II, at a temperature of about 60,000 kelvin. The image also portrays the Sun's upper chromosphere and lower transition region at this temperature, with bright blinkers and even brighter active regions. The Earth inset gives the approximate scale of the image, where the diameter of the Earth is 12,700 kilometers. The image was taken on 1 July 2002 with the Extreme-ultraviolet Imaging Telescope, abbreviated EIT, aboard the *SOlar and Heliospheric Observatory,* or *SOHO* for short. (Courtesy of the *SOHO* EIT consortium. *SOHO* is a project of international collaboration between ESA and NASA.)

All kinds of solar activity therefore seem to be related to the sudden release of stored magnetic energy, but the exact relation between them is unclear. Large coronal mass ejections can occur together with an eruptive prominence, with a solar flare, or without either one of them. And although the frequency of solar eruptions increases with the number of sunspots, their strength does not necessarily increase. The most intense solar flares and coronal mass ejections are spread throughout the solar cycle.

Even though we are 150 million kilometers away, we still notice the disrupting effects of solar eruptions. Intense radiation from powerful solar flares can travel to the Earth in just 499 seconds, or about 8 minutes, altering our atmosphere, disrupting long-distance radio communications, and affecting satellite orbits. Very energetic

FIG. 7.3 Coronal mass ejection A huge coronal mass ejection is seen in this coronagraph image, taken on 5 December 2003 with the Large Angle Spectrometric COronagraph, abbreviated LASCO, on the *SOlar and Heliospheric Observatory,* or *SOHO* for short. The black area corresponds to the occulting disk of the coronagraph that blocks intense sunlight and permits the corona to be seen. An image of the singly ionized helium, denoted He II, emission of the Sun, taken at about the same time, has been appropriately scaled and superimposed at the center of the LASCO image. The full disk helium image was taken at a wavelength of 30.4 nanometers, corresponding to a temperature of about 60,000 kelvin, using the Extreme-ultraviolet Imaging Telescope, or EIT for short, aboard *SOHO.* (Courtesy of the *SOHO* LASCO and EIT consortia. *SOHO* is a project of international cooperation between ESA and NASA.)

particles, accelerated during the flare process, can take an hour or less to reach Earth, where they can endanger unprotected astronauts or destroy satellite electronics. The coronal mass ejections arrive at the Earth one to four days after a major eruption on the Sun, resulting in strong geomagnetic storms with accompanying auroras and the threat of electrical power blackouts. These various terrestrial effects of solar flares and coronal mass ejections are of such vital importance that national centers employ space weather forecasters and continuously monitor the Sun from ground and space to warn of threatening solar activity.

The entire Universe is full of similar cataclysms. The dark night sky, with its steady stellar beacons, may give the impression of stillness and serenity, but it is actually a place of cosmic violence. Indeed, the acceleration of energetic particles, with the associated emission of powerful, invisible radiation, characterizes much of the Universe, from black holes and pulsars to the most distant quasar. Investigations of the Sun's energetic activity provide important insights to understanding violence throughout the Universe, for only the Sun is close enough and bright enough to be studied with sufficient detail.

7.2 SOLAR FLARES

Our perceptions of solar flares have evolved with the development of new methods of looking at them. Despite the powerful cataclysm, most solar flares are not, for example, detected on the visible solar disk. Although emitting awesome amounts of energy, enough to power the United States for decades, a solar flare releases less than one-tenth the total energy radiated by the Sun every second. And since most of this radiation is in visible sunlight, solar flares are only minor perturbations in the combined colors, or white light, of the Sun. The first record of a solar flare, observed in the white light of the photosphere, therefore did not occur until the mid 19th century.

The English astronomer Richard C. Carrington (1826–1875) was in the habit of observing sunspots daily, demonstrating that their positions occur closer and closer to the solar equator during the 11-year activity cycle. And on 1 September 1859, he suddenly saw a "sudden conflagration", consisting of two brief, intense sources of light near some sunspots. According to his account:

> Two patches of intensely bright and white light broke out. . . . I hastily ran to call some one to witness the exhibition with me, and on returning within 60 seconds, was mortified to find that it was already much changed and enfeebled. Very shortly afterwards the last trace was gone.[29]

Carrington observed a relatively rare event, in which a flare's light was enhanced sufficiently over the background sunlight to be visible by contrast.

His friend, Richard Hodgson (1804–1872), also an English amateur astronomer, chanced to be observing the Sun at the same time and confirmed this first account of a white-light flare in the astronomical literature. Such powerful events do not occur very often, for a solar flare has to be exceptionally intense to be seen against the light of the white-hot Sun.

Carrington also noticed that the sunspots were precisely the same after the sudden conflagration as they were before, leading him to conclude that:

> The phenomenon took place at an elevation considerably above the general [visible] surface of the Sun, and, accordingly, altogether above and over the great [sunspot] group in which it was seen projected.[30]

To this day, more than 146 years later, no solar observer has yet recorded a definitive change in the magnetic fields at the photosphere level during a solar flare, and it is generally believed that solar flares originate in the overlying coronal atmosphere.

A new perspective, that demonstrated the frequent occurrence of moderate solar flares, was made possible when they were observed in the red hydrogen-alpha Balmer transition at 656.3 nanometers, originating in the chromosphere. This wasn't possible until the early 20th century, after the invention of a device used to image the Sun in the light of just one wavelength – in 1891 by the French astronomer Henri Deslandres (1853–1948) and by the American astronomer George Ellery Hale (1868–1938). Routine observations of solar flares at this wavelength began in the 1930s, and continue today using automatic cameras and telescopes.

For more than half a century, astronomers throughout the world have carried out this vigilant flare patrol, like hunters waiting for the sudden flash of game birds. These systematic patrol observations have provided a two-dimensional picture of solar flares. They detect one slice through the Sun's atmosphere at the chromosphere level, showing that this part of a solar flare can consist in simplest form as two extended, parallel flare ribbons.

A fundamental understanding of the physical processes responsible for solar flares had to wait until they were detected at invisible wavelengths from both the ground and space, beginning in the 1960s and 1970s and continuing with increased sophistication to the present day. When combined, the visible and invisible observations provide a full three-dimensional view of solar flares.

Although flares usually produce only minor perturbations in the visible white light of the Sun, the flaring radio and X-ray emission is frequently several thousand times more energetic than the Sun's normal radiation at these wavelengths. Relatively small telescopes can therefore be used to detect invisible flares.

Because solar flares have very high temperatures, the bulk of their radiation is emitted at X-ray and ultraviolet wavelengths. This radiation is absorbed in the Earth's atmosphere, so astronomers observe it with telescopes in outer space.

Satellites that are specifically intended to investigate flare emission have now been launched to coincide with three successive maxima in the 11-year cycle of solar activity (Focus 7.1). *Skylab* X-ray photographs taken in 1973–74 showed that the low solar corona is shaped and constrained by the ubiquitous coronal loops, and indicated that solar flares can occur in coronal loops anchored in underlying sunspots. Such X-ray images reveal the presence of a very hot gas with temperatures of about 10 million kelvin, and also outline or trace the magnetic configuration of the coronal material. The soft X-ray flare emission can form looping arcades that bridge the hydrogen-alpha flare ribbons in the underlying chromosphere.

FOCUS 7.1
Observing Solar Flares From Space

Our knowledge of the basic physics of solar flares has increased dramatically as the result of observations from space. They began with primitive instruments aboard sounding rockets and continued with increasingly sophisticated telescopes, to the present day with Earth-orbiting satellites specifically designed to study flares. NASA's *Seventh Orbiting Solar Observatory,* abbreviated *OSO 7,* detected gamma ray lines from solar flares in 1972, and NASA's *Skylab* mission obtained high-resolution images of solar flares at ultraviolet and X-ray wavelengths in 1973–74. It was followed by the *Solar Maximum Mission,* abbreviated *SMM,* launched by NASA in 1980 and the Japanese *Hinotori* spacecraft, meaning "fire-bird" in Japanese, launched in 1981; both detected the high-energy, invisible radiation of solar flares.

Then in 1991, Japan launched the *Yohkoh,* or "sunbeam," spacecraft. It carried a coordinated set of four co-aligned instruments designed to study transient high-energy phenomena in solar flares at soft X-ray, hard X-ray and gamma-ray wavelengths.

For more than a decade, beginning in 1995, the ESA-NASA *SOlar and Heliospheric Observatory,* abbreviated *SOHO,* monitored the Sun for solar flares and other activity at extreme ultraviolet wavelengths, as has NASA's *Transition Region And Coronal Explorer,* or *TRACE* for short, since its launch in April 1998, but with greater angular resolution and enhanced detail. And in February 2002, NASA launched the *Reuven Ramaty High Energy Spectroscopic Imager,* abbreviated *RHESSI,* spacecraft to study gamma rays emitted from solar flares.

The short-lived solar flares unleash their vast power from a relatively small volume within active regions, the magnetized atmosphere in, around and above sunspots. The largest solar flares cover but a few tenths of a percent of the solar disk, and are intimately related to the magnetized coronal loops that link sunspots of opposite magnetic polarity.

7.3 FLARE RADIATION FROM ENERGETIC ACCELERATED PARTICLES

The Solar Particle Accelerator

Intense invisible radiation is not the only thing unleashed during a solar flare. Up to half the total explosion energy is used to accelerate electrons and protons to speeds approaching the velocity of light, about 300 thousand kilometers per second. Solar flares are indeed the most powerful particle accelerators in the Solar System, comparable only to another type of solar eruption, the coronal mass ejection, discussed a bit later. The electrons and protons accelerated during a solar flare are beamed down into the Sun, where the protons initiate nuclear reactions.

The electrons and protons are also ejected from the low corona outward along the spiral-shaped interplanetary magnetic field, traveling at nearly the speed of light. These high-speed particles sometimes travel unimpeded all the way to the Earth where they are detected by instruments aboard spacecraft, and endanger both astronauts and satellites.

Radiation from High-Speed Flare Electrons

The electrons accelerated during solar flares emit invisible radiation across the electromagnetic spectrum, from the shortest X-rays and gamma rays to the longest radio

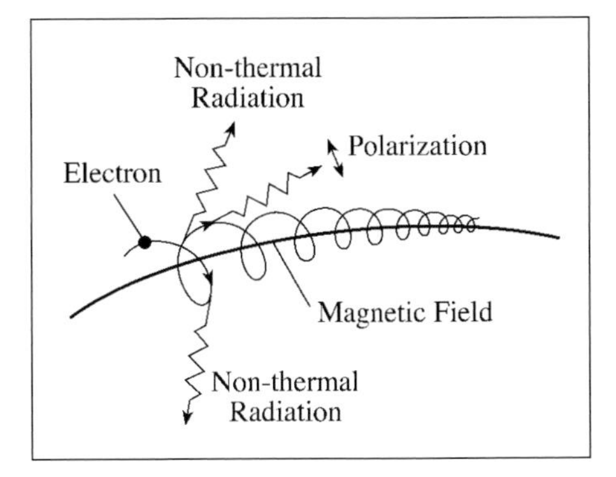

FIG. 7.4 Synchrotron radiation High-speed electrons moving at velocities near that of light emit a narrow beam of synchrotron radiation as they spiral around a magnetic field. This emission is sometimes called non-thermal radiation because the electron speeds are much greater than those of thermal motion at any plausible temperature. The name "synchrotron" refers to the man-made, ring-shaped synchrotron particle accelerator where this type of radiation was first observed; a synchronous mechanism keeps the particles in step with the acceleration as they circulate in the ring.

waves, but the protons are too massive to emit intense radio and X-ray radiation. The radio waves provide information about the location, size, magnetic fields and temperature during a solar flare. They additionally demonstrate that flare electrons have been accelerated to very high speeds. Energetic electrons that have been accelerated during solar flares are also hurled out into interplanetary space, emitting intense radio emission in the process.

As the high-speed electrons spiral down along magnetic channels, they generate intense radio emission; sometimes-called radio bursts to emphasize their brief, energetic and eruptive characteristics. The magnetic fields bend the paths of the electrons into a circle, resulting in the emission of a narrow beam of synchrotron radiation (Fig. 7.4), named for the man-made synchrotron particle accelerator where it was first observed.

Most of the energy radiated during a solar flare is emitted as hard and soft X-rays. They provide detailed information about the flare process, including why and where they occur. The X-ray radiation is named *bremsstrahlung,* the German word for "braking radiation." It is produced when a free electron, unattached to an atom, passes near an ion (Fig. 7.5). Their electrical interaction bends the path of the electrons, which radiate soft X-rays in the process. There is thermal bremsstrahlung, which depends on the random thermal motion of the hot electrons; it is emitted when electrons have been heated during the flare to temperatures of about 10 million kelvin. Electrons are also briefly accelerated to energies much higher than the mean energies of thermal plasma at any plausible temperature, and these very high-speed electrons give rise to non-thermal bremsstrahlung (Fig. 7.6).

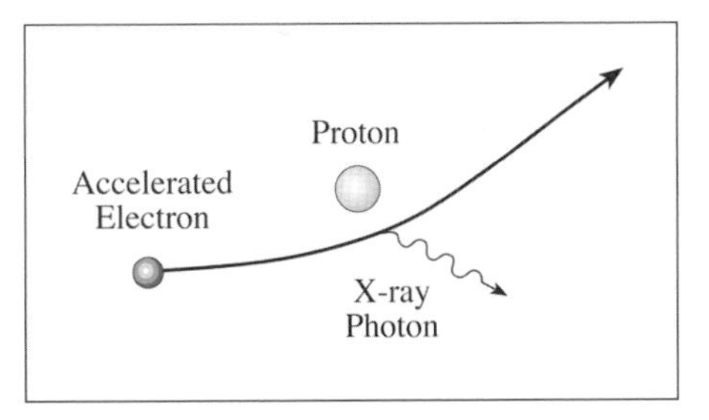

FIG. 7.5 Bremsstrahlung When a hot electron moves rapidly and freely outside an atom, it inevitably moves near a proton in the ambient gas. There is an electrical attraction between the electron and proton because they have equal and opposite charge, and this pulls the electron toward the proton, bending the electron's trajectory and altering its speed. The electron emits electromagnetic radiation in the process. This radiation is known as *bremsstrahlung* from the German word for "braking radiation".

X-ray astronomers distinguish between soft X-rays and hard X-rays, with the softer variety having less energy than the hard type. The energy that X-rays carry is a measure of the energy of the electrons that produce it, often specified in units of kilo-electron volts, abbreviated keV. Soft X-rays have energies between 1 and 10 keV, and hard X-rays lie between 10 and 100 keV. In contrast, gamma rays can have energies greater than one Mev, or a million electron volts. And radiation at different X-ray wavelengths describes different aspects of the flare time profile (Fig. 7.7). So what you see during a solar flare depends on how you look at it.

At the impulsive stage of a solar flare, electrons are accelerated rapidly, in a second or less, to energies that can exceed one MeV. The high-energy electrons emit hard X-rays and gamma rays that mark the flare onset. These energetic, high-speed, non-thermal electrons are believed to be accelerated above the tops of coronal loops, and to radiate energy by non-thermal bremsstrahlung and synchrotron radiation as they are beamed down along the looping magnetic channels into the low corona and chromosphere.

During the decay, or post-impulsive, phase, energy is gradually released on longer time-scales of tens of minutes. Soft X-rays slowly build up in strength and usually reach peak intensity during the post-impulsive decay phase. Initially cool material in the chromosphere is heated by down flowing flare electrons that emit the hard X-rays. The heated gas expands upward into the low-density corona along magnetic loops that shine brightly in soft X-rays.

Major solar flares are often accompanied by the disruption of a filament that was held up by the magnetic loops in the active region. These closed magnetic fields are blown open during the impulsive phase; but the newly opened magnetic field lines reconnect, forming an arcade of post-impulsive flare loops, aligned like the bones in your rib cage or the arched trestle in a rose garden (Fig. 7.8). They are subsequently filled by plasma flowing up from the chromosphere at their footpoints.

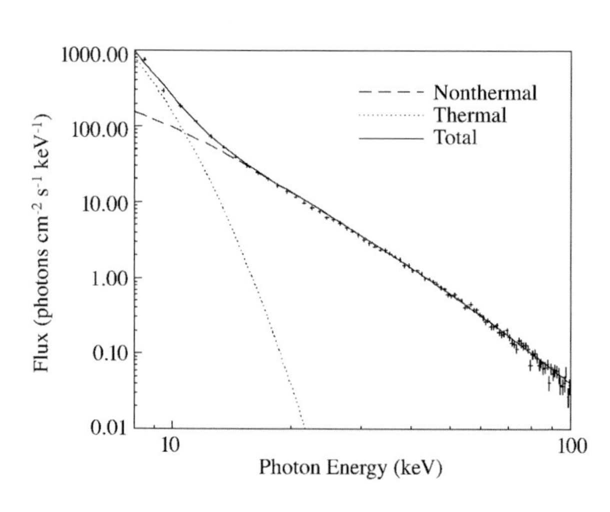

FIG. 7.6 Energy spectrum of flare radiation The spatially integrated energy spectrum of the radiation photons during a 14-second time interval at the peak of the solar flare on 20 February 2002. The observed low-energy spectrum is described by a thermal spectrum at a temperature of 15 million kelvin. The high-energy emission is attributed to non-thermal emission of energetic electrons with a power-law spectral index of 4.4. This data was obtained from the *Ramaty High Energy Solar Spectroscopic Imager,* abbreviated *RHESSI,* a NASA spacecraft. (Courtesy of Brian R. Dennis, NASA.)

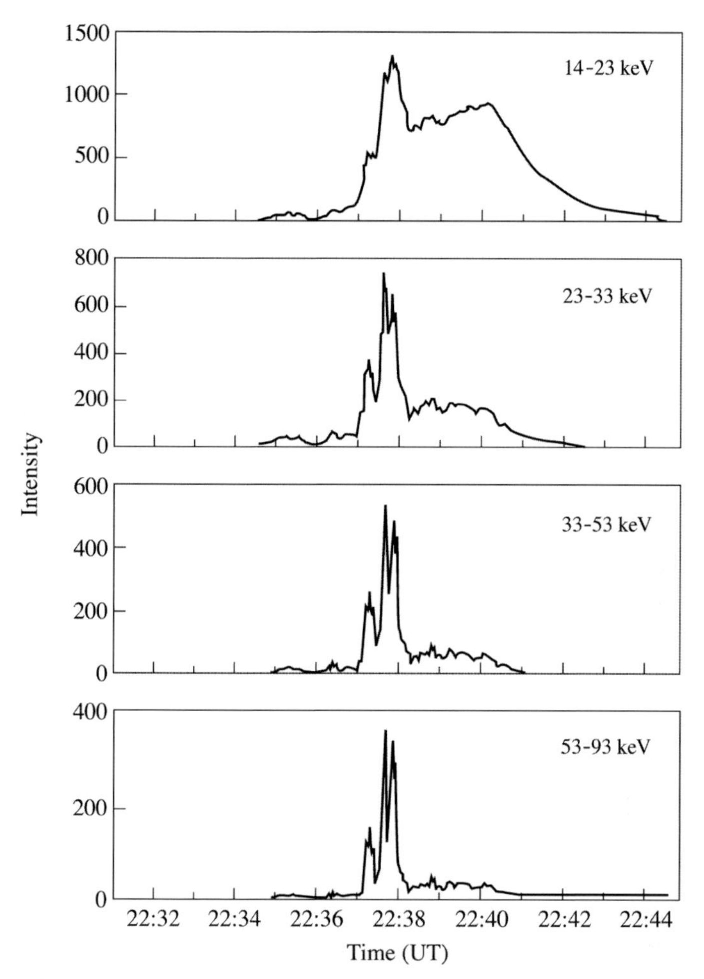

FIG. 7.7 Impulsive and gradual phases of a flare The time profile of a solar flare observed at hard X-ray energies, above 30 keV (*bottom*), is characterized by an impulsive feature that lasts for about one minute. This impulsive phase coincides with the acceleration of high-speed electrons that emit non-thermal bremsstrahlung at hard X-ray wavelengths and non-thermal synchrotron radiation at centimeter radio wavelengths. The less-energetic emission, below 30 keV (*top*), can be composed of two components, an impulsive component followed by a gradual one. The latter component builds up slowly and becomes most intense during the gradual decay phase of solar flares when thermal radiation dominates. At even lower soft X-ray energies (about 10 keV), the gradual phase dominates the flare emission. This data was taken on 15 November 1991 with the Hard X-ray Telescope (HXT) aboard *Yohkoh*. (Courtesy of NASA, ISAS, the Lockheed-Martin Solar and Astrophysics Laboratory, the National Astronomical Observatory of Japan, and the University of Tokyo.)

FIG. 7.8 Flare loops Radiation from an arcade of magnetic loops was produced shortly after a solar flare erupted on 7 July 2002. This image was taken at a wavelength of 17.11 nanometers, emitted by eight and nine times ionized iron, denoted Fe IX and Fe X, at a temperature of about 1.0 million kelvin, using the *Transition Region And Coronal Explorer,* abbreviated *TRACE.* (Courtesy of the *TRACE* consortium and NASA; *TRACE* is a mission of the Stanford-Lockheed Institute for Space Research, a joint program of the Lockheed-Martin Solar and Astrophysics Laboratory, or LMSAL for short, and Stanford's Solar Observatories Group.)

When replenished and full, the post-impulsive flare loops contain plasma heated to temperatures of tens of millions of degrees, even hotter than the center of the Sun at 15.6 million kelvin. They appear to migrate upwards and outwards as successive field lines are reconnected at higher altitudes. The post-flare loops bridge the magnetic neutral line between opposite polarity regions in the photosphere, stitching together and healing the wound inflicted by the solar flare and related events.

The Gamma-Ray Sun

Normally you can't detect gamma rays from the Sun, even from outer space, for their intensity is so low. But when protons and heavier ions are accelerated to high speed during solar flares, and beamed down into the chromosphere, they produce nuclear reactions and generate gamma rays, the most energetic kind of radiation detected from solar flares. The gamma rays have energies of about one MeV, equivalent to a thousand keV, so the gamma rays are ten to one hundred times more energetic than the hard X-rays and soft X-rays detected during solar flares. Like X-rays, the gamma rays are totally absorbed in our atmosphere and must be observed from space.

The high-speed flare protons slam into the dense, lower atmosphere, like a bullet hitting a concrete wall, shattering nuclei in a process called spallation. The nuclear fragments are initially excited, but then relax to their former state by emitting gamma rays. Other abundant nuclei are energized by collision with the flare-accelerated protons, and emit gamma rays to get rid of the excess energy. Nuclear de-excitation gamma-ray lines observed during solar flares include those from excited nuclei of carbon, oxygen, nitrogen, neon, magnesium, silicon and iron; most recently detected from the *Reuven Ramaty High Energy Spectroscopic Imager,* abbreviated *RHESSI.*

During bombardment by flare-accelerated ions, energetic neutrons can be torn out of the nuclei of atoms. Many of these neutrons are eventually captured by ambient, or non-flaring, hydrogen nuclei, the protons, in the photosphere, making deuterons, the nuclei of deuterium atoms, and emitting one of the Sun's strongest gamma-ray lines at 2.223 MeV (Fig. 7.9). The neutrons must slow down and lose some energy by collisions before the protons can capture them, so the gamma-ray line is delayed by a minute or two from the onset of impulsive flare emission.

Neutrons produced by accelerated particle interactions during solar flares can also escape from the Sun, avoiding capture there. Neutrons with energies above 1,000 MeV have even been directly measured in space near Earth, in the 1980s from the *Solar Maximum Mission,* abbreviated *SMM,* and in the 1990s from the *Compton Gamma Ray*

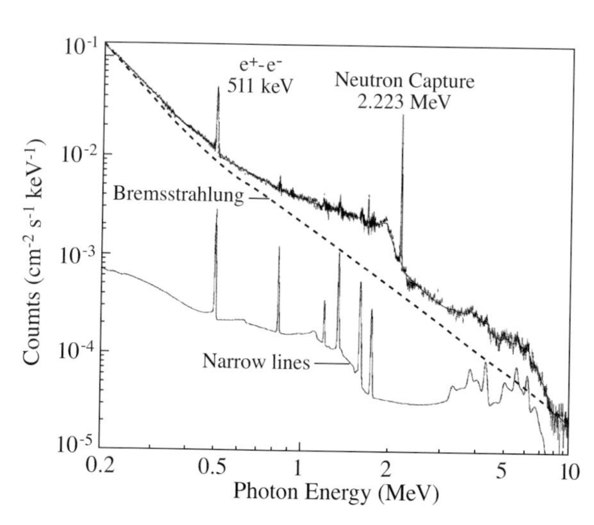

FIG. 7.9 Flare spectral lines at high energy The energy spectrum of radiation from a flare on 28 October 2003 exhibits two prominent spectral lines and numerous less intense, narrow ones. The line with an energy of 511 keV, or 0.511 MeV, is emitted when electrons collide with their anti-matter counterparts, the positrons or "positive electrons"; both particles are destroyed while emitting the high-energy radiation. The neutron-capture line at 2.223 MeV, and several narrow lines from accelerated protons are also seen. Underlying the line features is the bremsstrahlung continuum from accelerated electrons. This data was obtained from the *Ramaty High Energy Solar Spectroscopic Imager,* abbreviated *RHESSI,* a NASA spacecraft. (Courtesy of Brian R. Dennis, NASA.)

Observatory, or *CGRO* for short. In the most energetic flares, the associated neutrons can reach the Earth and produce a signal in ground level neutron monitors.

Another strong gamma-ray line emitted during solar flares is the 0.511 MeV positron annihilation line (Fig. 7.9). Positrons, the anti-matter counterpart of electrons, are released during the decay of radioactive nuclei produced when flare-accelerated protons and heavier nuclei interact with the lower solar atmosphere. The positrons annihilate with the electrons, producing radiation at 0.511 MeV, which is the energy contained in the entire mass of a non-moving electron. Before the positrons can interact with the ambient thermal electrons, they must also slow down by collisions until they have similar velocities. The positron can even combine with an electron to briefly produce a positronium "atom", before self-destructing with the production of the 0.511 MeV spectral features.

The detection of gamma-ray lines due to solar flares is not a recent thing. The American astronomer Edward L. Chupp (1927–) observed the positron annihilation, neutron capture, and carbon and oxygen de-excitation lines in 1972, using an instrument aboard NASA's *Seventh Orbiting Solar Observatory,* abbreviated *OSO 7.* But our understanding of the high-energy processes that produce gamma rays and neutrons has been significantly improved as the result of observations in the 1980s using the Gamma Ray Spectrometer aboard the *SMM* satellite, in the 1990s with the *CGRO,* and after 2002 from *RHESSI* .

7.4 ENERGIZING SOLAR FLARES

Magnetic Energy in the Corona

What energizes solar flares, and how do they accelerate particles to nearly the speed of light? They release more energy than that present in the heat of the entire corona, so they cannot draw their energy from the hot, surrounding gas. But they occur above sunspots in active regions with their powerful magnetic fields, where coronal loops congregate. So flares are most likely powered by magnetic energy that has been built up in the low solar corona in active regions, and then abruptly released, like the sudden flash and crack of a lightning bolt from a storm cloud.

The coronal magnetic fields are rooted deep inside the Sun, where differential rotation and turbulent convection prevail. The internal gases are therefore in constant motion, carrying the magnetic fields with them. So the coronal magnetism becomes sheared, stretched, tangled and twisted by internal churning motions, slowly accumulating stored energy in the process, in much the same way as continually twisting a rubber band increases the tension and stores energy in its kinks and bends.

The energy keeps growing until the magnetized coronal loops are pushed beyond their limits, or an outside force intervenes and destabilizes them. Then in a matter of seconds the stored energy is rapidly and violently unleashed, like the quick snap of a rubber band that has been twisted too tightly. A solar flare erupts accompanied by copious acceleration of charged particles and the emission of intense invisible radiation.

Magnetic Reconnection

But why does a solar flare occur? What triggers the instability and suddenly ignites an explosion from magnetic fields that remain unperturbed for long intervals of time?

It might be triggered when magnetized coronal loops are pressed together, driven by motions beneath them, meeting to touch each other and merge (Fig. 7.10).

Magnetic fields have a direction associated with them, and if oppositely directed magnetic fields are pushed together, they can interact. When the merging magnetic fields are closed coronal loops, they will be broken open to release magnetic energy in the form of flare heating and particle acceleration. But the magnetic fields are not permanently broken, and they simply reconnect back to a closed state. For this reason, the technical name for this merging and coupling is magnetic reconnection.

In 1956, Peter A. Sweet (1921–2005), then at the University of London Observatory, realized that electric current would form between the merging, oppositely directed magnetic fields. The current flows in a flat, two-dimensional plane, shaped like a sheet on a well made bed, and is hence called a current sheet. And since the magnetic direction cancels into neutrality at their meeting place, the term neutral current sheet is also used. The American solar physicist Eugene Parker (1927–) derived the detailed mathematics of magnetic reconnection during solar flares in 1963, and in the same year another American Harry E. Petscheck (1930–2005) showed how stored magnetic energy might be rapidly released during reconnection in the current sheet. Nowadays, the two modes of magnetic reconnection are called Sweet-Parker or slow reconnection and Petchek or fast reconnection.

So, we now think of these powerful outbursts as stemming from the interaction of coronal loops. They are always moving about, like swaying seaweed or wind-blown grass, and existing coronal loops may often be brought into contact by these movements. Magnetic fields coiled up in the solar interior, where the Sun's magnetism is produced, can also bob into the corona to interact with pre-existing coronal loops. In either case, the coalescence leads to the rapid release of magnetic energy through magnetic reconnection.

The explosive instability has been compared to an earthquake, with the moving roots or footpoints of a sheared magnetic loop resembling two tectonic plates. As the plates move in opposite directions along a fault line, they grind against each other and build up stress and energy. When the stress is pushed to the limit, the two plates cannot slide further, and the accumulated energy is released as an earthquake. That part of the fault line then lurches back to its original, equilibrium position, waiting for the next

FIG. 7.10 **Magnetic connection** A pair of oppositely directed, twisted coronal loops come in contact, releasing magnetic energy to power a solar flare. Arrows indicate the direction of the magnetic field lines and the sense of twist. Such magnetic encounters can occur when newly emerging magnetic fields rise through the photosphere to join pre-existing ones in the corona, or when the twisted coronal loops are forced together by underlying motions. [Adapted from Thomas Gold and Fred Hoyle, "Origin of Solar Flares", *Monthly Notices of the Royal Astronomical Society* **120,** 89–105 (1960).]

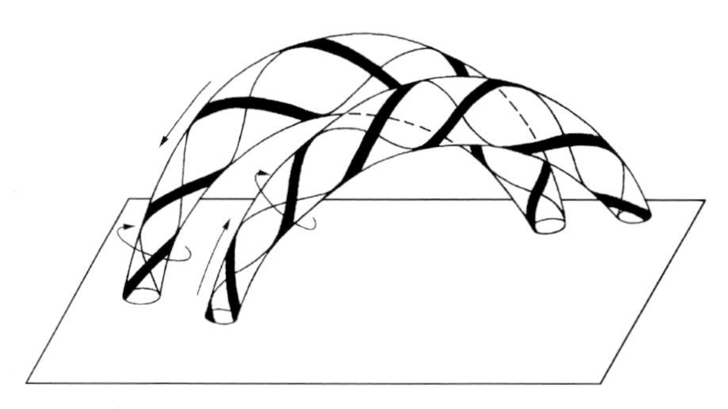

earthquake. In this analogy, the magnetic fields become stressed to the breaking point and similarly regain their composure after an explosive convulsion on the Sun, fusing together and becoming primed for the next outburst.

Observations of Magnetic Reconnection During Solar Flares

The Soft X-ray Telescope, or SXT, aboard the *Yohkoh* spacecraft has revealed the probable location of the magnetic reconnection site during solar flares, showing that the rounded magnetism of a coronal loop can be pulled into a peaked shape at the top during a solar flare. The sharp, cusp-like feature marks the place where oppositely directed field lines stretch out nearly parallel to each other and are brought into close proximity. Here the magnetism comes together, merges and reconnects, releasing the energy needed to power a solar explosion.

Many gradual soft X-ray flares observed from *Yohkoh* show cusp-shaped loop structures suggesting magnetic reconnection, and they are often associated with coronal mass ejections. Moreover, electron time-of-flight measurements with the *Compton Gamma Ray Observatory,* abbreviated *CGRO,* satellite, confirm the existence of a coronal acceleration site for flares observed with both *CGRO* and *Yohkoh.*

Co-aligned *Yohkoh* images also show a compact, impulsive hard X-ray source well above the corresponding soft X-ray loop structure for at least one impulsive solar flare, in addition to the double-footpoint hard X-ray emission. The hard X-rays at the loop footpoints are produced when high-speed, non-thermal electrons collide with dense material in the chromosphere. However, there is no similar dense material out in the tenuous corona, so the hard X-ray radiation near the loop top has to be emitted during the electron acceleration process, most likely representing the site where oppositely directed magnetic fields meet and electrons are accelerated to high energy. These electrons rapidly move down toward the footpoints, explaining the similarity in the time variations of all three hard X-ray sources.

RHESSI has apparently confirmed that large-scale magnetic reconnection in the low corona is the most likely explanation for how solar flares suddenly release so much energy. A compact coronal X-ray source was observed above a series of bright X-ray flare loops during the impulsive phase of an intense solar flare. The loop-top source rose up, moving outward, and a coronal mass ejections was released. The *RHESSI* astronomers have interpreted this in terms of a neutral current sheet that forms above the nested coronal magnetic loops. Other loops, formed by magnetic reconnection in the current sheet, move downward onto the existing ones, which then shine brightly in X-rays, and the other reconnected magnetic fields and plasma flow and twist upward until they are severed, releasing the coronal mass ejection.

As expected, *RHESSI* also observes two bright hard X-ray sources, accelerated during the impulsive phase of the solar flare and generated by non-thermal electrons beamed down from the reconnection site to the two footpoints of magnetic loops or arcades. Instruments aboard *RHESSI* have additionally provided the first gamma-ray images of solar flares, in the intense neutron-capture line. As reported by Robert P. "Bob" Lin (1942–) and his colleagues, the gamma-ray footpoints are unexpectedly displaced from the hard X-ray ones (Fig. 7.11). This suggests a difference in the acceleration of the energetic ions and electrons at the reconnection site, or differences in their transport down to the lower solar atmosphere.

All of these observations, from *Yohkoh* to *RHESSI,* nevertheless provide only indirect evidence for magnetic reconnection in solar flares. No one has ever detected individual magnetic fields coming together on the Sun during a solar flare, and they probably never will. But it has given rise to a widely accepted model for solar flares that will be thoroughly tested by future observations.

A Model of Solar Flares

In summary, a well-developed magnetic theory for solar flares has been developed (Fig. 7.12). Since flares apparently originate in the low corona, and the ubiquitous coronal loops dominate its structure, it is perhaps not surprising that this model involves a single coronal loop. Magnetic reconnection triggers the primary energy release of a flare just above the loop top, where electrons and protons are accelerated. During an

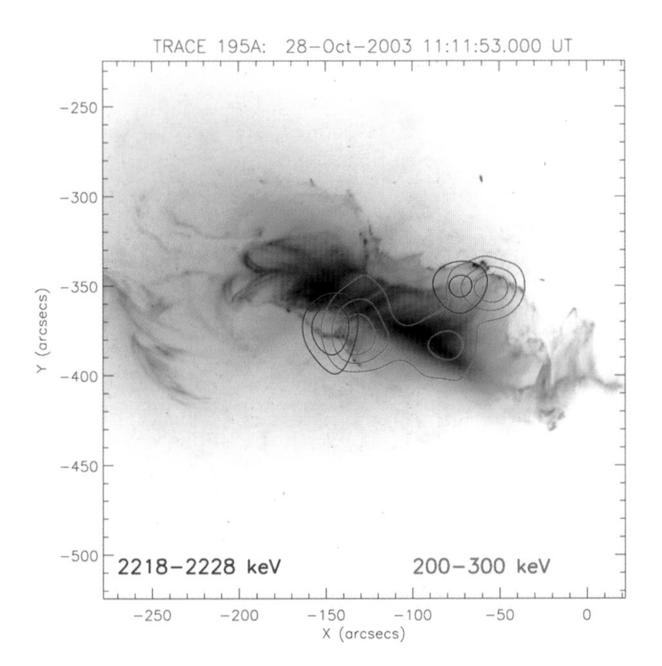

FIG. 7.11 Gamma ray flare image Relative locations of the gamma-ray emission (*blue*), the hard X-ray emission (*red*), and the extreme ultraviolet, or EUV, radiation (*green and black with black being the most intense*) of a solar flare on 28 October 2003. The gamma-ray contours show the location of the interaction between the neutrons, generated by flare-accelerated ions, and the dense cooler material in the solar photosphere, the hard X-ray contours represent the non-thermal bremsstrahlung from electrons accelerated during the flare, and the EUV image shows the location of the thermal emission of a hot dense plasma. Note the offset of the gamma ray and hard X-ray footpoints that appear along the bright (*black*) EUV ribbons, suggesting a difference in acceleration and/or propagation between the accelerated ions and electrons. This image is 300 by 250 seconds of arc across, where 1 second of arc corresponds to 750 kilometers at the Sun. The gamma-ray and hard X-ray data are from the *Ramaty High Energy Solar Spectroscopic Imager,* abbreviated *RHESSI,* and the extreme ultraviolet image was taken from the *Transition Region And Coronal Explorer,* or *TRACE* for short. The *RHESSI* images have an angular resolution of about 35 seconds of arc, and were taken for a 22-minute period during the decay phase of the flare at the 2.223 MeV neutron capture gamma-ray line and for hard X-rays from 200 to 300 keV. The *TRACE* image was taken within the same time interval at the 19.51 nanometer line emitted by eleven times ionized iron, denoted Fe XII, at a temperature of 1.4 million kelvin. (Courtesy of Brian R. Dennis, Gordon J. Hurford, the *TRACE* consortium and NASA.)

impulsive solar flare, the magnetic energy released during magnetic reconnection is converted to charged particle kinetic energy. In less than a second, electrons are accelerated to nearly the speed of light, producing intense radio signals. Protons are likewise accelerated, and both the electrons and protons are hurled down into the Sun and out into space.

The high-speed electrons emit non-thermal, hard X-ray bremmstrahlung and synchrotron radio radiation at or near the loop apex during the impulsive phase of a solar flare. Energy is transported downward by the electrons and protons, which follow the coronal loop's arching magnetism into the denser chromosphere and underlying photosphere, producing nuclear reactions and creating hard X-rays and gamma rays at small, localized areas that mark the loop footpoints. The rarely seen, white-light flares are most likely produced by the downward impact of the non-thermal electrons.

Since the chromosphere has been heated very rapidly by the accelerated particles, it explodes up into the corona to get rid of the excess energy. This may be followed by the more gradual release of energy when the coronal loop relaxes into a more stable configuration during the decay phase of a solar flare.

7.5 ERUPTING PROMINENCES

Looping arches of magnetism hold up, suspend and insulate elongated structures filled with material at a temperature of about 10,000 kelvin, which is hundreds of times cooler and denser than the surrounding million-degree corona. These structures are called prominences when seen in hydrogen-alpha photographs taken at the apparent edge of the Sun, perhaps because they prominently stand out against the dark background,

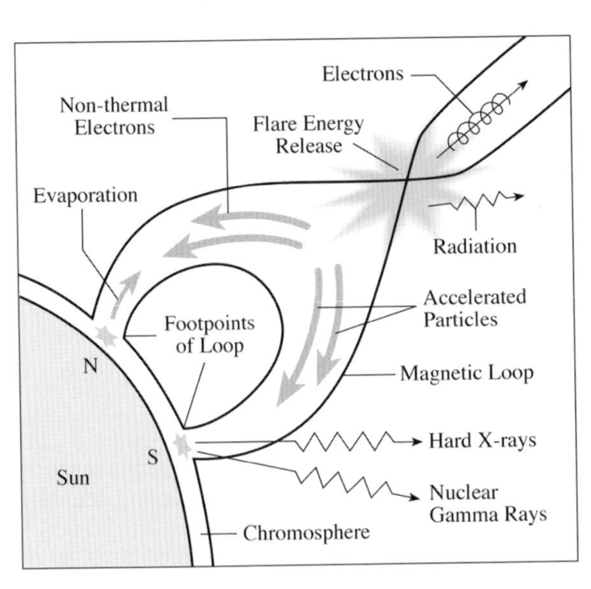

FIG. 7.12 Solar flare model A solar flare is powered by magnetic energy released from a magnetic interaction site above the top of the loop shown schematically here. Electrons are accelerated to high speed, generating a burst of radio energy as well as impulsive loop-top hard X-ray emission. Some of these non-thermal electrons are channeled down the loop and strike the chromosphere at nearly the speed of light, emitting hard X-rays by electron-ion bremsstrahlung at the loop footpoints. When beams of accelerated protons enter the dense, lower atmosphere, they cause nuclear reactions that result in gamma-ray spectral lines and energetic neutrons. Material in the chromosphere is heated very quickly and rises into the loop, accompanied by a slow, gradual increase in soft X-ray radiation. This upwelling of heated material is called chromospheric evaporation.

resembling arched viaducts or bridges, like the Pont Neuf in Paris. They appear as dark, snaking features, called filaments, when projected against the bright chromosphere disk. So, prominence and filament are essentially two words that describe different perspectives of the same thing. Their long sinuous forms trace out a region of magnetic neutrality that separates large areas of opposite magnetic direction, or polarity, in the underlying photosphere.

Coronal loops that are anchored within individual active regions are about ten times smaller than the total length of the largest prominences or filaments. Active regions sometimes appear underneath a prominence, and a coronal streamer is often found above it. In such a helmet streamer, a low-density cavity surrounds the prominence with a dense helmet dome overlying the cavity.

Prominences and filaments can remain suspended and almost motionless above the photosphere for weeks or months, but there comes a time when they cannot bear the strain. Then, without warning, the supporting magnetism becomes unhinged (Fig. 7.13), most likely because it got so twisted out of shape that it snapped. And a surprising thing happens! Instead of falling down under gravity, the stately, self-

FIG. 7.13 Filament lifts off A filament is caught at the moment of erupting from the Sun. The dark matter is relatively cool, around 20,000 kelvin, while the bright material is at a temperature of about a million kelvin. The structure extends 120,000 kilometers from top to bottom. This image was taken from the *Transition Region And Coronal Explorer,* abbreviated *TRACE,* on 19 July 2000 at a wavelength of 17.11 nanometer, emitted by eight and nine times ionized iron, denoted Fe IX and Fe X, at a temperature of about 1.0 million kelvin. (Courtesy of the *TRACE* consortium and NASA; *TRACE* is a mission of the Stanford-Lockheed Institute for Space Research, a joint program of the Lockheed-Martin Solar and Astrophysics Laboratory, or LMSAL for short, and Stanford's Solar Observatories Group.)

contained structures erupt, often rising as though propelled outward through the corona by a loaded spring (Fig. 7.14).

It's as if the lid had been taken off the caged material. Cool gas is then flung outward in slingshot fashion, tearing apart the overlying corona and ejecting large quantities of matter into space.

These eruptive prominences, as they are called when viewed at the Sun's apparent edge, are hurled outward at speeds of several hundreds of kilometers per second, releasing a mass equivalent to that of a small mountain in just a few hours. Such an event is sometimes called a *disparition brusque* for its sudden disappearance (Fig. 7.15). Eruptive prominences can be larger, longer lasting and more massive than solar flares.

Prominences, or filaments, often re-form in the same shape and place after an explosive convulsion. It is as if some minor irritation builds up beyond the limit of tol-

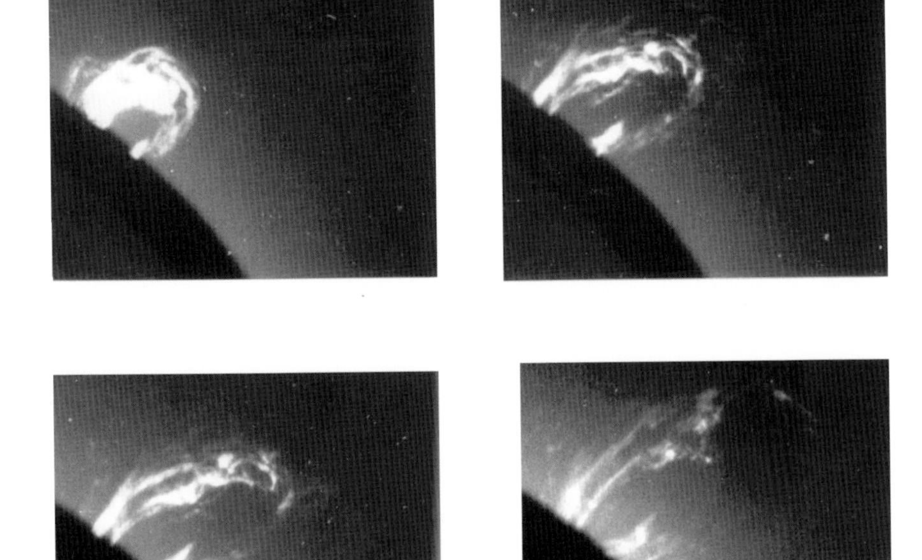

FIG. 7.14 A prominence erupts Rapid, sequential hydrogen-alpha photography catches an erupting prominence, which had not been detected as a filament during previous days. It suddenly rose from an active region and expanded at an apparent velocity of 375 kilometers per second, hurling material far away from the Sun. Here the magnetic loops rise to a maximum visible extent of 360,000 kilometers in just 16 minutes. This sequence of hydrogen-alpha images was taken on the west edge, or limb, of the Sun using the automatic flare patrol heliograph at the Meudon Station of the Observatoire de Paris; the solar disk has been occulted to give a better view of the event. (Courtesy of Madame Marie-Josephe Martres, and observer Michel Bernot, Observatoire de Paris, Meudon, DASOP.)

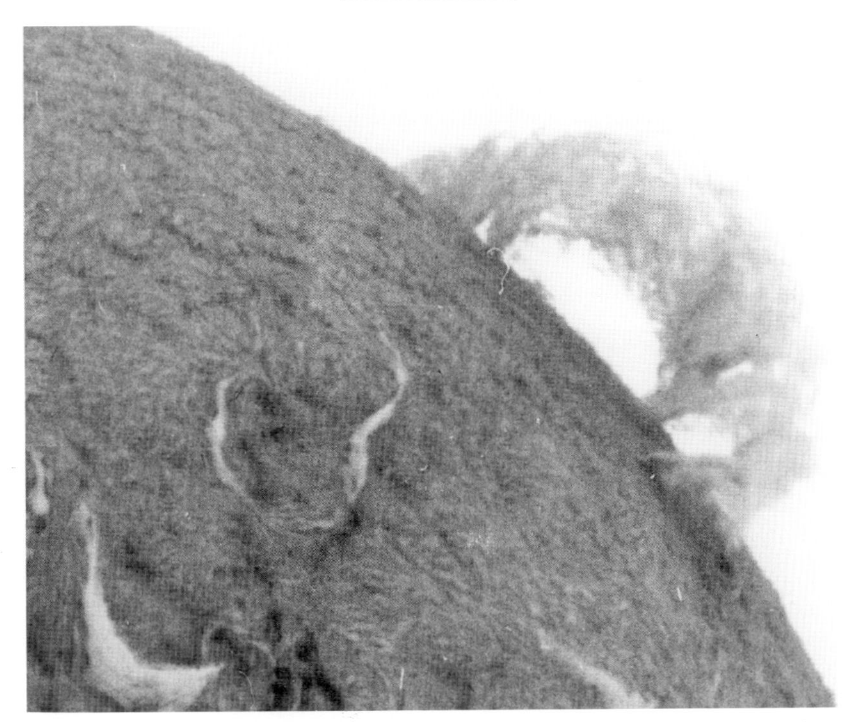

FIG. 7.15 Disparition brusque A prominence shines brightly at the south-east edge of the Sun in the red light of hydrogen alpha, printed here as a negative image for contrast. It had been observed as a dark filament for weeks at a time during several previous rotations of the Sun. Then, in less than 40 hours after this picture was taken, the prominence was no longer visible. It probably rose and disappeared high in the corona within just a few hours. The French astronomers use the term disparition brusque to describe this sudden disappearing act of a large prominence rooted in a quiet region of the Sun. (Courtesy of Madame Marie-Josephe Martres, Observatoire de Paris, Meudon, DASOP.)

erance, and the magnetic structure tosses off the pent-up frustration, like a dog shaking off the rain. Long, dark filaments rise and disappear, replaced by an elongated arcade of bright X-ray loops across the initial filament position, so their magnetic backbone regroups as before beneath the erupting prominence.

Closed magnetic loops apparently support the long filament, like parallel hammocks, at heights that are about ten times the diameter of the Earth. This arcade of closed loops is anchored in the Sun, but is opened up at the top by the rising filament, like removing the top of a jack-in-the-box. The magnetism subsequently reconnects and closes up again beneath the erupting filament or prominence, forming a new arcade of closed loops that shines in X-rays and resembles a giant's rib cage.

The disappearing prominences are strongly correlated with another form of energetic solar activity, the coronal mass ejections, that play an important role in solar-terrestrial interactions.

7.6 CORONAL MASS EJECTIONS

The most spectacular solar eruptions are gigantic magnetic bubbles, called coronal mass ejections, which expand outward from the Sun (Fig.7.16). A typical coronal mass ejection carries about 10 billion tons, or 10 million million kilograms, of coronal material as it lifts off into space, removing about a tenth of the total coronal mass. The outward-moving coronal mass ejections stretch the magnetic field until it snaps, leaving behind only bright rays rooted in the Sun. They can expand to become larger than

FIG. 7.16 Coronal mass ejection A contorted coronal mass ejection is seen in this coronagraph image, taken on 4 January 2002. The white circle denotes the edge of the photosphere, so the length of the ejected material is about twice the size of the visible disk of the Sun. The dark area corresponds to the occulting disk that blocked the intense sunlight from the photosphere, revealing the surrounding faint corona. This image was taken with the Large Angle Spectrometric COronagraph, abbreviated LASCO, on the *SOlar and Heliospheric Observatory*, or *SOHO* for short. (Courtesy of the *SOHO* LASCO consortium and NASA. *SOHO* is a project of international collaboration between ESA and NASA.)

the visible solar disk, streaming outward past the planets and dwarfing everything in their path.

Thousands of mass ejections have been observed during the last few decades with white-light coronagraphs aboard the seventh *Orbiting Solar Observatory, OSO-7* (1972), *Skylab* (ATM, 1973–1974), *P 78-1* (Solwind, 1979–1985), the *Solar Maximum Mission* (C/P, 1980 and 1984–1989), and the *Solar and Heliospheric Observatory,* abbreviated *SOHO* (LASCO, 1996–2005).

Such events work only in one direction, always moving away from the Sun into interplanetary space and never falling back in the reverse direction. They often exhibit a three-part structure – a bright outer loop, followed by a depleted region, or cavity, that rises above an erupted prominence. The leading bright loop, or coronal mass ejection, is a rapidly expanding, bubble-like shell that opens up and lifts off like a huge umbrella in the solar wind, piling the corona up and shoving it out like a snowplow.

The material moves outward, an immense cosmic storm that moves with speeds of several hundred kilometers per second. The energy of this mass motion is comparable to the net radiated energy of a large solar flare, with the destructive potential of a million hurricanes on Earth.

Mass ejections erupt from the Sun as self-contained structures of hot material and magnetic fields, apparently resulting from a rapid, large-scale restructuring of magnetic fields in the low corona. They are most likely triggered by large-scale magnetic reconnection events in the corona, similar to those that ignite solar flares.

Although subatomic particles are accelerated to high energies during solar flares, it is now thought that coronal mass ejections may be the main source of the most energetic solar particles arriving at the Earth. They are responsible for most, if not all, of the largest energetic proton events that bombard the Earth. Strong mass ejections serve as pistons to drive huge shock waves ahead of them, plowing into the slower-moving solar wind, like a car out of control, and producing shock waves millions of kilometers across. Electrified particles energized by the shocks travel outward with it, somewhat like surfers riding the ocean waves.

Coronal mass ejections can energize subatomic particles across large regions of interplanetary space. Those that head toward Earth usually take three or four days to get there. In contrast, the particles accelerated during flares follow well-defined paths described by the interplanetary magnetic spiral, and they can take only minutes to arrive at Earth.

When a coronagraph in a satellite near Earth detects a mass ejection ballooning out from the Sun's apparent edge, the ejection is not headed for Earth. Those directed toward or away from our planet can be observed as a halo around the Sun, detected by a differencing technique. Another method of detecting an Earth-directed coronal mass ejection is the dimming of soft X-ray or extreme-ultraviolet radiation, caused by the removal of low corona material near the center of the visible solar disk.

When coronal mass ejections were discovered in the early 1970s, it was thought that they were an explosive consequence of the bright flares; at the time flares were the most energetic eruptive phenomena known on the Sun. Then a key piece of evidence was provided by the heretofore largely ignored prominences. When astronomers began to make association studies of coronal mass ejections and other forms of solar activity, they found to their surprise that the erupting prominences were best associated with the mass ejections. Since intense flares do not usually accompany erupting prominences, it was concluded that flares were not required to drive mass ejections.

In addition, the physical size of the mass ejections dwarfs that of flares and even the active regions in which flares occur, and flares are much more common than mass ejections. Most flares therefore probably originate in a somewhat different environment than the mass ejections, but perhaps by a similar magnetic process. The main differences between the two may just be a matter of physical size, with the compact flares occurring more often than the larger mass ejections.

The mass ejection, with its accompanying erupting prominence, represents a large-scale rearrangement of the Sun's magnetic structure. It blows open the previously closed magnetic structure, like a hot-air balloon that breaks its tether (Fig. 7.17). The magnetic cage is torn asunder, releasing pent-up magnetic energy, perhaps triggered by magnetic reconnection, as solar flares seem to be. Sometimes an associated flare might be formed as the result of the energy released by magnetic coupling of the open field lines as they pinch below the rising prominence. Post-flare loops can be subsequently observed, shining from the newly closed magnetic loops in the flare's thermal afterglow (Fig.7.18).

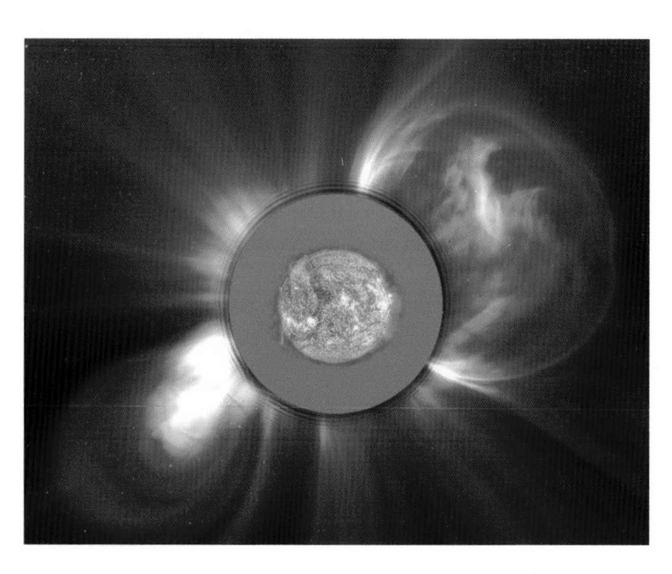

FIG. 7.17 Ejections from the corona A double set of coronal mass ejections seems to be heading in opposite directions from the Sun on 8 November 2000. The dark area corresponds to the coronagraph's occulting disk that blocks the intense sunlight from the photosphere, revealing the surrounding faint corona. An extreme ultraviolet image has been superposed at a location corresponding to the visible solar disk; it was taken on the same day at a wavelength of 30.38 nanometers, emitted by singly ionized helium, denoted He II, at a temperature of about 60,000 kelvin. The coronagraph image was taken with the Large Angle Spectrometric COronagraph, abbreviated LASCO, on the *SOlar and Heliospheric Observatory,* or *SOHO* for short, and the superposed image was taken with the Extreme-ultraviolet Imaging Telescope, abbreviated EIT, on *SOHO.* (Courtesy of the *SOHO* LASCO and EIT consortia and NASA. *SOHO* is a project of international collaboration between ESA and NASA.)

FIG. 7.18 Stitching up the wound An arcade of post-flare loops shines in the extreme ultraviolet radiation of eight and nine time ionized iron, Fe IX and Fe X, at a temperature of about 1.0 million kelvin. This image was taken on 8 November 2000, just after a solar flare occurred in the same active region; at least one coronal mass ejection also accompanied the event (see Fig. 7.17). High-energy particles from the flare entered the Earth's radiation belts, and an associated coronal mass ejection produced a strong geomagnetic storm. This image was taken from the *Transition Region And Coronal Explorer,* abbreviated *TRACE.* (Courtesy of the *TRACE* consortium and NASA; *TRACE* is a mission of the Stanford-Lockheed Institute for Space Research, a joint program of the Lockheed-Martin Solar and Astrophysics Laboratory, or LMSAL for short, and Stanford's Solar Observatories Group.)

7.7 PREDICTING EXPLOSIONS ON THE SUN

Solar flares and coronal mass ejections emit energetic particles, intense radiation, powerful magnetic fields and strong shocks that can have enormous practical implications when directed toward the Earth. They can disrupt radio navigation and communication systems, pose significant hazards to astronauts, satellites and space stations near Earth, and interfere with power transmission lines on the ground. National space environment centers and defense agencies therefore continuously monitor the Sun from ground and space to forecast explosions on the Sun.

The ultimate goal is to learn enough about solar activity to predict when the Sun is about to unleash its pent-up energy. Such space-weather forecasts will probably involve magnetic changes that precede solar flares and coronal mass ejections, especially in the low corona where the energy is released or at lower levels in the photosphere where we can watch internal motions pulling and twisting the magnetism around.

It has been supposed that the energy required for eruptions is stored in stressed magnetic structures, and that the magnetic fields rearrange themselves into a simpler configuration after the event. Solar flares do, in fact, occur in regions of strong magnetic shear in the photosphere, and the low solar corona is in a constant state of agitation and metamorphosis. Coronal loops are magnetically reconfigured as they twist and writhe in response to internal differential rotation and convection motions, and it is likely that coronal loops join and reconnect to trigger solar flares.

Scientists may have discovered one method for predicting the sudden and unexpected outbursts. When the bright, X-ray emitting coronal loops are distorted into a large, twisted sigmoid (S or inverted S) configuration, a coronal mass ejection from that region becomes more likely (Fig. 7.19). In some instances, a coronal mass ejection occurs just a few hours after the magnetic fields have snaked into a sinuous S-shaped feature; the mass ejection arrives at the Earth three or four days later. In the meantime, just after the mass had been expelled from the Sun, the X-ray emitting region dramatically changes shape, exhibiting the telltale, cusp-like signature of magnetic reconnection and an X-ray fading or dimming due to the mass removal. In other words, the magnetism gets stirred up into a complex, stressed and twisted situation before it explodes, and then relaxes to a simpler, less-stressful situation.

However, some regions that exhibit magnetic shear and twist never erupt, so contorted magnetism may be a necessary but not sufficient condition for solar flares or coronal mass ejections. And the Sun's sudden and unexpected outbursts often remain as unpredictable as most human passions. They just keep on happening, and even seem to be necessary to purge the Sun of pent-up frustration and to relieve it of twisted, contorted magnetism.

Since this erratic, unpredictable, impulsive behavior of the Sun is of enormous practical interest to us humans on Earth, it is also critically important to know if the material sent out from the solar outbursts is headed toward our planet. Coronal mass ejections that are expelled from near the visible edge of the Sun will not impact Earth, but threaten other parts of space. Mass ejections are most likely to hit us if they originate near the center of the solar disk, as viewed from the Earth, and are sent directly toward the planet. The outward rush of such a mass ejection appears in coronagraph images as a ring or halo around the occulting disk, but the coronagraph data are unable to determine if the halo-like ejection is traveling toward or away from the observer.

The Earth-directed coronal mass ejections may nevertheless be preceded by coronal activity at extreme ultraviolet and X-ray wavelengths near the center of the solar disk as viewed from the Earth. The mass ejection is itself sometimes associated with

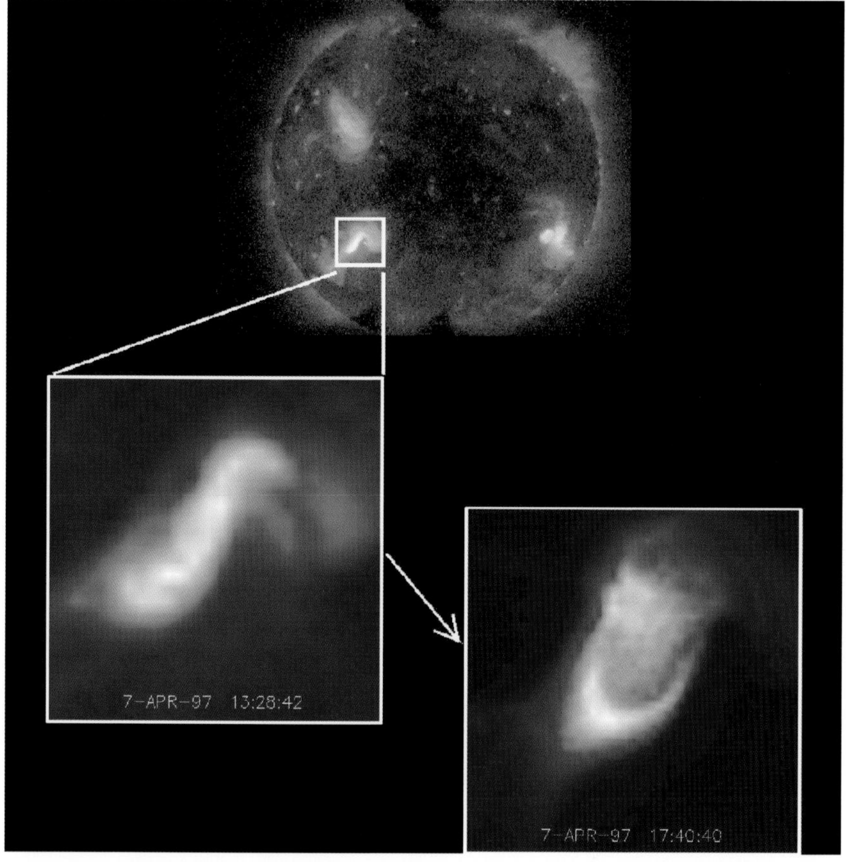

FIG. 7.19 Sigmoid and Cusp The full-disk X-ray image (*top*) shows the Sun with a twisted, sigmoid present on 7 April 1997. It produced a halo Coronal Mass Ejection, abbreviated CME, on the following day. The inset (*bottom left*) shows the soft X-ray sigmoid before eruption of the CME. The other inset (*bottom right*) shows the soft X-ray cusp and arcade formed just after the CME took place. These images were taken with the Soft X-ray Telescope (SXT) on *Yohkoh*. (Courtesy of Richard C. Canfield, Alphonse C. Sterling, NASA, ISAS, LMSAL, the National Astronomical Observatory of Japan and the University of Tokyo.)

dimming X-rays or extreme ultraviolet radiation with associated waves running across the visible disk, like tidal waves or tsunami going across the ocean. Moreover, NASA's *Solar-TErrestrial RElations Observatory*, abbreviated *STEREO* (Fig. 7.20), uses two spacecraft to track coronal mass ejections from the Sun to Earth (Focus 7.2).

The high-energy electrons that accompany solar flares follow the spiral pattern of the interplanetary magnetic field, so they must be emitted from active regions near the west limb and the solar equator to be magnetically connected with the Earth. Solar flares emitted from other places on the Sun are not likely to hit Earth, but they could be headed toward interplanetary spacecraft, the Moon, Mars or other planets.

FIG. 7.20 ***STEREO*** An artist's conception of the *Solar-TErrestrial RElations Observatory,* abbreviated *STEREO,* in which two spacecraft will provide the images for a stereo reconstruction of Coronal Mass Ejections, or CMEs for short. One spacecraft will lead Earth in its orbit and one will be lagging. When simultaneous telescopic images from the two spacecraft are combined with data taken from observations on the ground or in low Earth orbit, the buildup of magnetic energy, the lift off, and the trajectory of CMEs can be traced in three dimensions. (Courtesy of Johns Hopkins, Applied Physics Laboratory.)

FOCUS 7.2
STEREO

NASA's *Solar-TErrestrial RElations Observatory,* abbreviated *STEREO,* mission, scheduled for launch in 2006, will send two identical satellites into solar orbit, one ahead of the Earth in its orbit and one behind. Their coronagraphs and other instruments will give astronomers the first three-dimensional view of coronal mass ejections. Previous observations from a coronagraph aboard a single spacecraft provided an edge on, two-dimensional view.

The instruments aboard the *STEREO* mission will advance our understanding of the origin of coronal mass ejections, their evolution in the interplanetary medium, and their coupling with the Earth's magnetosphere. The mission will also investigate how coronal mass ejections produce energetic particles in space and magnetic storms on Earth.

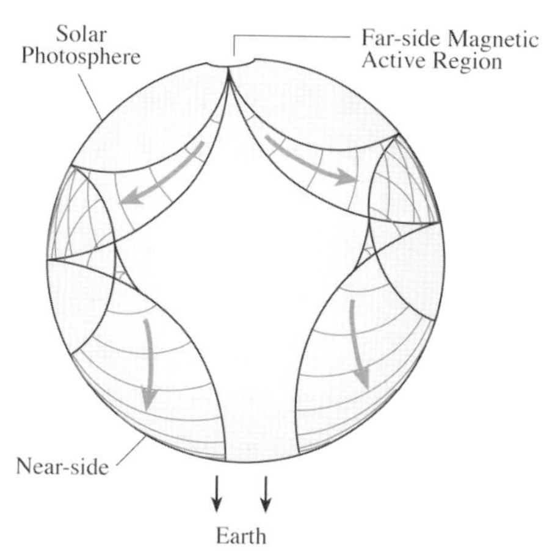

FIG. 7.21 Looking through the Sun The arcing trajectories of sound waves from the far side of the Sun are reflected internally once before reaching the front side, where they are observed with the Michelson Doppler Imager, abbreviated MDI, aboard the *SOlar and Heliospheric Observatory,* or *SOHO* for short. Sound waves returning from a solar active region on the hidden back side of the Sun have a round-trip travel time about twelve seconds shorter than the average of six hours. This timing difference permits scientists to detect potentially threatening active regions on the far side of the Sun before the Sun's rotation brings them around to the front side that faces the Earth. (Courtesy of the *SOHO* MDI/SOI consortium. *SOHO* is a project of international cooperation between ESA and NASA.)

Scientists are also now using sound waves to see right through the Sun to its hidden, normally invisible, backside, describing active regions on the far side of the Sun many days before they rotate onto the side facing Earth. The new technique uses observations of the Sun's oscillations to create a sort of mathematical lens that focuses to different depths, and it is called helioseismic holography because of its similarity to laser holography. A wide ring of sound waves is examined, which emanates from a region on the side of the Sun facing away from the Earth, the far side, and reaches the near side that faces the Earth (Fig. 7.21).

When a large solar active region is present on the backside of the Sun, its intense magnetic fields compress the gases there, making them slightly lower and denser than the surrounding material. So sound waves emerging from an active region on the far side suffer a phase shift, expressed as a change in travel time. A sound wave that would ordinarily take about 6 hours to travel from the near side to the far side of the Sun and back again takes approximately 12 seconds less when it bounces off the compressed active region on the far side. When near-side photosphere oscillations are examined; they can detect the quick return of these sound waves.

This remarkable result, first announced by American solar astronomers Douglas Braun (1961–) and Charles Lindsey (1947–) in 2000, became routine within a few years, with daily far side images obtained from instruments on *SOHO* and shown on the web. Solar astronomers are using this technique to monitor the structure and evolution of large regions of magnetic activity as they cross the far side of the Sun, thereby revealing the regions that are growing in magnetic complexity or strength and seem primed for explosion. Since the solar equator rotates with a period of 27 days, when viewed from the Earth, this can give at least seven day's extra warning of possible bad weather in space before the active region swings into view and pummels the Earth.

Chapter Eight

Energizing Space

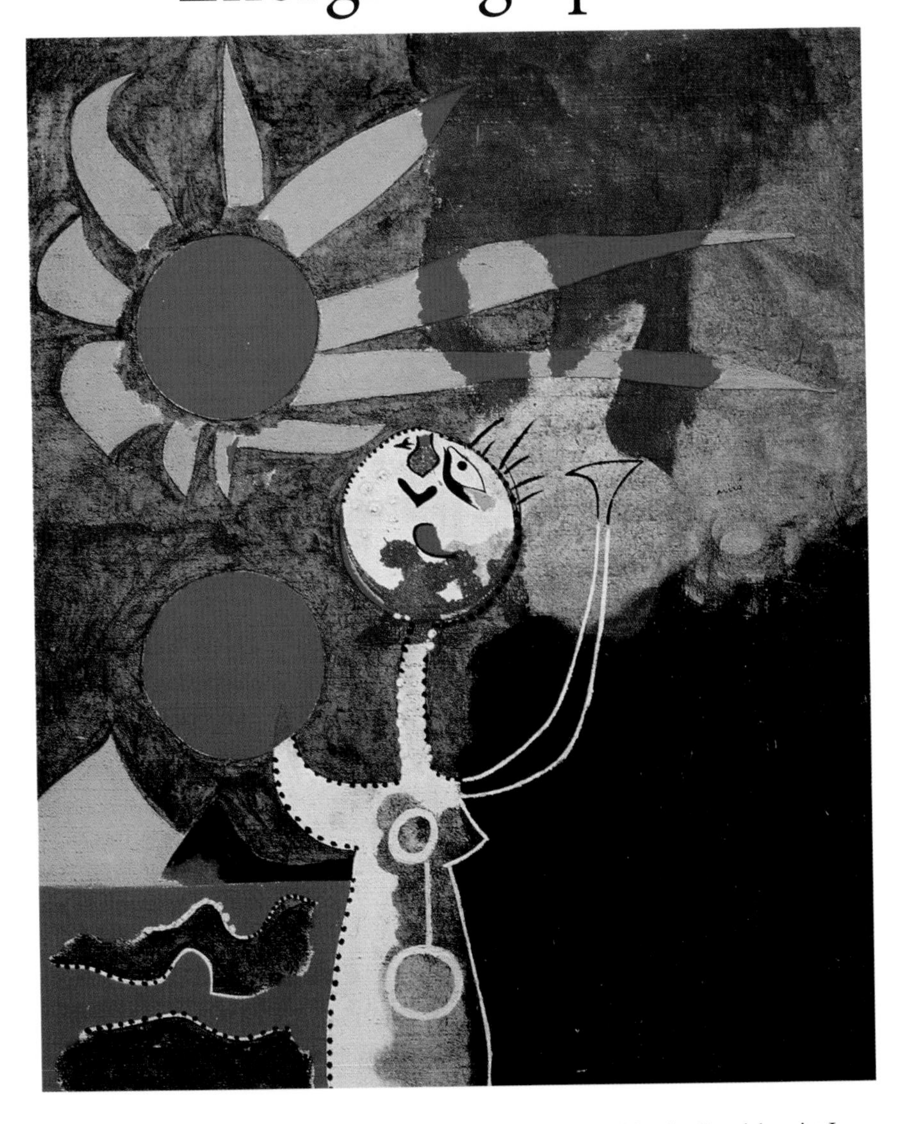

WOMAN IN FRONT OF THE SUN In this painting, executed by the Spanish artist Joan Miró (1893–1983), a tall, erect woman is trying to touch the bright red Sun, while holding a red ball or star in another hand. The luminous Sun sends out rays of hope in the surrounding darkness, linking our life-sustaining star to the Earth. Many of Miró's paintings represent terrestrial forms set against the powerful red disk of the Sun, as well as against the stars of the night sky. (Private collection.)

8.1 THE INGREDIENTS OF SPACE

Several decades ago, before the Space Age, we thought that an imaginary boundary separates the Earth's thin, life-sustaining atmosphere from an empty wasteland called outer space. Back then we visualized our planet traveling in a cold, dark vacuum around the Sun, warmed to life by its radiation. But we now know that the space outside the Earth's atmosphere is not empty! The Earth is immersed in a vast and shifting web of subatomic particles and magnetic fields that come from the Sun.

So, light and heat are not the Sun's only contribution to our environment. The space between the planets, once thought to be an utter void, contains an electrified

FOCUS 8.1
Cosmic Rays

Cosmic rays are extraordinarily energetic elementary particles and atomic nuclei that rain down on the Earth from all directions, traveling at nearly the speed of light. They form a third ingredient of space within our Solar System in addition to the electrified plasma and magnetic fields that emanate from the Sun. Cosmic "rays" were discovered in 1912 when the Austrian physicist Victor Hess (1883–1964) took a Geiger counter, used to study energetic particles emitted by radioactive atoms like uranium, on balloon flights. The measured radioactivity at first decreased with altitude, as would be expected from atmospheric absorption of particles emitted by radioactive rocks, but the Geiger counter's signal unexpectedly increased at even higher distances above the Earth's surface. This meant that the signals were of extraterrestrial origin. By flying his balloon at night and during a solar eclipse, when the high-altitude signals persisted, Hess showed that they could not come from the Sun, but from some other unknown cosmic source.

The new extraterrestrial signals that Hess discovered are now called cosmic rays, distinguishing them from the alpha and beta rays emitted during radioactive decay on the Earth; the alpha and beta rays are now respectively known to be helium nuclei and electrons. The cosmic rays consist mainly of the nuclei of hydrogen atoms, or protons, as well as electrons and the nuclei of heavier atoms. They provide rare samples of matter from outside the Solar System and, although few in number, carry with them the story of the more energetic processes in our Galaxy.

Unlike electromagnetic radiation, the charged cosmic-ray particles are deflected and change direction during numerous encounters with the interstellar magnetic field that wends its way between the stars. By the time they reach the Earth, cosmic rays have therefore lost their orientation. So, we cannot look back along their incoming path and tell where cosmic rays originated; the direction of arrival just shows where they last changed course. We can only speculate that cosmic rays were accelerated to their tremendous energies by shocks associated with the explosions of distant dying stars, called supernovae.

Cosmic rays are also deflected by the Sun's magnetism that is carried into interplanetary space by the solar wind. As first shown by the American physicist Scott Forbush (1904–1984) in 1954, the number of cosmic rays arriving at Earth therefore varies with the 11-year cycle of solar activity, but in the opposite direction to that expected if cosmic rays came from the Sun. At times of maximum solar activity, more charged cosmic-ray particles are deflected away from the Earth by the Sun's magnetic field that stretches out into interplanetary space. Less extensive interplanetary magnetism during a minimum in solar activity lowers the barrier to cosmic particles inbound from the depths of space, and allows more cosmic rays to arrive at the Earth.

gas or plasma made up of more or less equal numbers of positively charged hydrogen nuclei, or protons, and negatively charged free electrons. They are always flowing past the Earth and other planets in a rarefied solar wind, which we cannot see, blowing the Sun away and carrying its magnetic field with them. Thus, one ingredient of space is plasma, and magnetic fields are another.

Interplanetary space also contains a small number of very energetic charged particles of both cosmic and solar origin. Energized particles called cosmic rays travel between the stars, bringing traces of explosive stellar death (Focus 8.1). Powerful solar eruptions also hurl very energetic charged particles into space. We on the surface of the Earth are protected and shielded from these invisible, subatomic particles by our atmosphere and the Earth's own magnetic field, and are therefore normally unaware of them.

8.2 EARTH'S MAGNETIC COCOON

William Gilbert (1544–1603), physician to Queen Elizabeth I (1533–1603) of England, used a spherical magnet, called a *terrella,* or "little Earth", to show why compass needles point north and south. Whenever a compass was placed on the sphere, the needle pointed to its north or south magnetic poles. As suggested by Gilbert in 1600, in his treatise, *De magnete: magnus magnes ipse est globus terrestris,* translated "About magnetism: the terrestrial globe is itself a great magnet", the Earth has a huge dipolar magnetic field with two poles, north and south.

It's as if there were a bar magnet at the center of the planet Earth, with lines of magnetic force emerging from the south pole, looping out through the space near Earth, and re-entering at the north pole (Fig. 8.1). This dipolar magnetic field is generated as in a dynamo by electric currents deep within the Earth's core.

Since the geographic poles are located near the magnetic ones, a compass needle at the equator points north and south, within about 12 degrees. At each magnetic pole, the compass needle would stand upright, pointing into or out of the ground. And in between, at intermediate latitudes, it would point north and south with a downward dip at one end, but not vertically as at a pole.

Invisible powers collide, sometimes violently, in the vast reaches of space between the Earth and the Sun. There the hot, high-speed, magnetized solar wind, which continuously blows from the Sun and rushes toward the Earth, meets the Earth's magnetic field. This turbulent encounter occurs fairly close to home, usually at a distance of about ten times the Earth's radius.

Spacecraft have shown that the terrestrial magnetic field deflects the solar wind away from the Earth and hollows out a cavity in it, forming a protective "cocoon" around the planet. This magnetic cocoon, called the magnetosphere, shields the Earth from the full force of the solar wind, and protects our planet from possibly lethal energetic solar particles. Even though the Earth's magnetism is several times weaker than a toy magnet, it is still strong enough to block and divert the flow of the tenuous solar wind, like a rock in a fast flowing stream of water or a building in a strong wind on Earth.

The magnetosphere of the Earth, or of any other planet, is that region surrounding the planet in which its magnetic field has a controlling influence on, or dominates,

FIG. 8.1 Earth's magnetic dipole The Earth's dipolar (two poles) magnetic field looks like that which would be produced by a bar magnet at the center of the Earth, with the North Magnetic Pole corresponding to the South Geographic Pole and *vice versa.* It originates in swirling currents of molten iron deep in the Earth's core, and extends more than 20 Earth radii, or 126,000 kilometers out into space. Magnetic field lines loop out of the South Geographic Pole and into the North Geographic Pole. A compass needle will always point along a field line. The lines are close together near the magnetic poles where the magnetic force is strong, and spread out above the equator where they are weaker. The magnetic axis is tilted at an angle of 11.7 degrees with respect to the Earth's rotational axis. Notice that the poles of the magnet are inverted with respect to the geographic poles, following the custom of defining positive, north magnetic polarity as the one in which magnetic fields point out, and negative, south magnetic polarity as the place where magnetic fields point in. This dipolar configuration applies near the surface of the Earth, but further out the magnetic field is distorted by the solar wind (see Fig. 8.2).

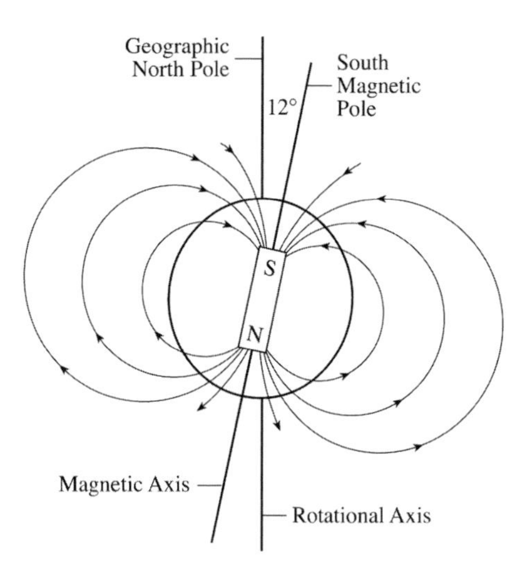

the motions of energetic charged particles such as electrons, protons, or other ions. The basic idea was put forth about a century ago by the Norwegian scientist Kristian Birkeland (1867–1917), who wrote in 1896 that the

> Earth's magnetism will cause there to be a cavity around the Earth in which the [solar wind] corpuscles are, so to speak, swept away.[31]

Birkeland proposed that electrons from the Sun are directed by the Earth's magnetism to the polar regions where they produce auroras, demonstrating his idea in experiments with electrons and magnetic spheres. The Austrian-born, American astrophysicist Thomas Gold (1920–2004) coined the term magnetosphere in 1959, more than half a century after Birkeland's prophetic remarks.

Space probes have now encountered every major planet, showing that six of them are wrapped in their own magnetic sheathing. Mercury, Earth, Jupiter, Saturn, Uranus, and Neptune have dipolar magnetic fields of sufficient strength to deflect the solar wind and form a tear-shaped cavity called a magnetosphere, though the magnetic axes on Uranus and Neptune are tilted by large amounts from their rotation axes. There is no detectable magnetism on Venus. The magnetic fields on Mars are now detected only in surface stripes of alternate magnetic polarity; they are most likely the leftover fossil remnants of a former global dipolar field.

Near the Earth, the magnetic field retains its dipolar configuration, bulging outward into space near the equator and converging inward at its two poles, like the pattern of iron filings scattered near an ordinary bar magnet. But terrestrial magnetism weakens as it extends to greater distances, eventually becoming distorted by the Sun's wind. Far from the Earth, the term sphere in magnetosphere loses its strict geometrical meaning, and instead implies a more general sphere of influence. Out there, the solar wind takes over and continuously molds the Earth's magnetosphere into a changing, asymmetric shape (Fig. 8.2).

Rarefied as it is, the solar wind still possesses the power to bend and move things in its path. It compresses and flattens the outer boundary of the magnetosphere into a blunt-nosed shape on the dayside facing the Sun; on the night side of the Earth away from the Sun, the terrestrial magnetic field is stretched out and turned inward upon itself by the relentless solar wind, forming an invisible magnetotail that always points downwind like a weather vane. The magnetotail contains oppositely directed magnetic field lines derived from the two hemispheres of the terrestrial magnetic field. The magnetic field lines emerging from the southern hemisphere are directed away from the Sun; those originating in the northern hemisphere are directed sunward.

A shock wave forms in front of the Earth's magnetosphere on the dayside facing the Sun; it is called a bow shock because it is shaped like the waves that pile up ahead of

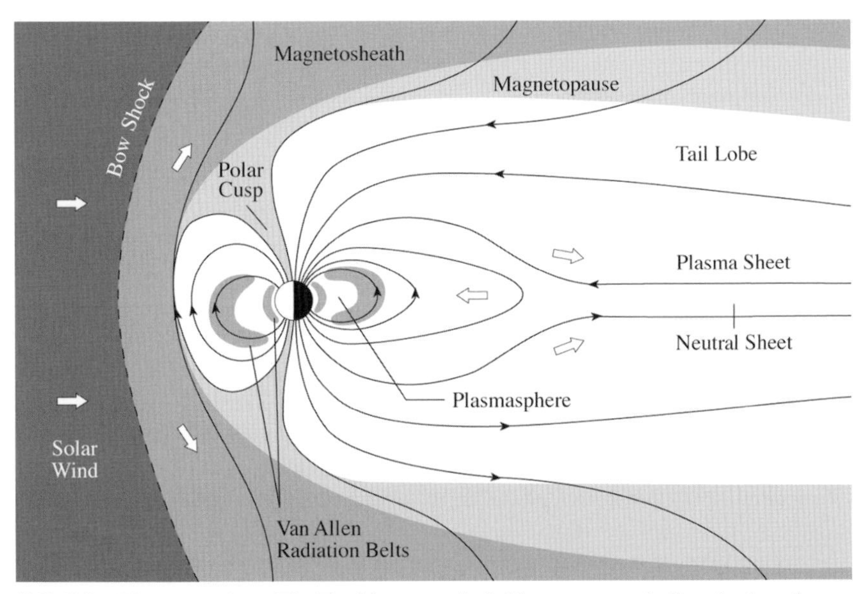

FIG. 8.2 Magnetosphere The Earth's magnetic field carves out a hollow in the solar wind, creating a protective cavity, called the magnetosphere. A bow shock forms at about ten Earth radii on the sunlit side of our planet. The location of the bow shock is highly variable since it is pushed in and out by the gusty solar wind. The magnetopause marks the outer boundary of the magnetosphere, at the place where the solar wind takes control of the motions of charged particles. The solar wind is deflected around the Earth, pulling the terrestrial magnetic field into a long magnetotail on the night side. Plasma in the solar wind is deflected at the bow shock (*left*), flows along the magnetopause into the magnetic tail (*right*), and is then injected back toward the Earth within the plasma sheet (*center*). The Earth, its auroras, atmosphere and ionosphere, and the two Van Allen radiation belts all lie within this magnetic cocoon.

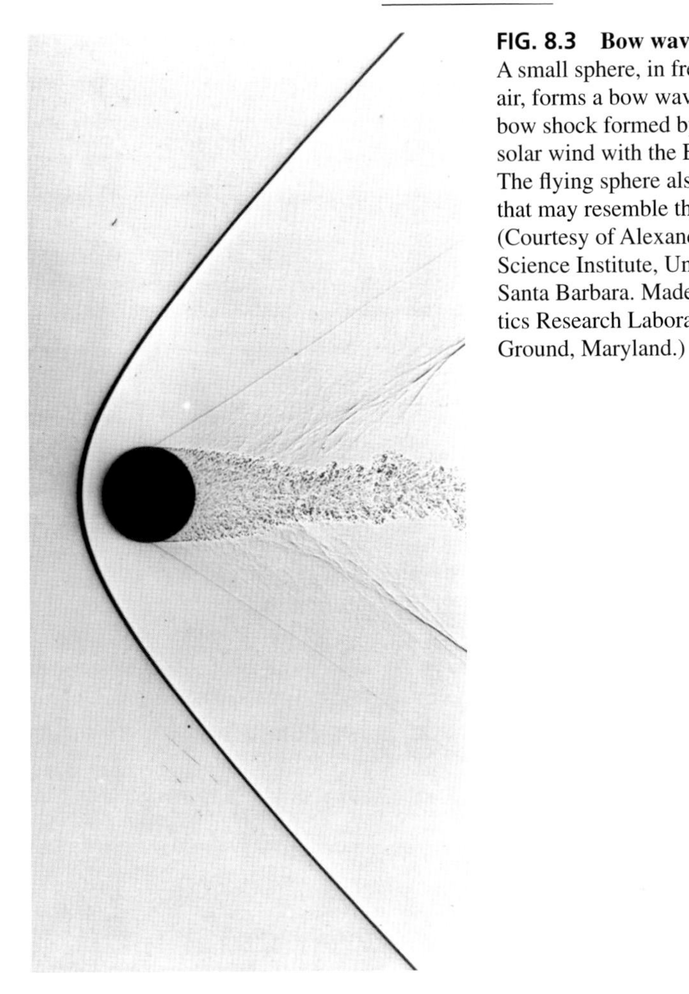

FIG. 8.3 Bow wave of a flying sphere A small sphere, in free flight through the air, forms a bow wave that is similar to the bow shock formed by the interaction of the solar wind with the Earth's magnetic field. The flying sphere also has a turbulent wake that may resemble the Earth's magnetotail. (Courtesy of Alexander C. Charters, Marine Science Institute, University of California, Santa Barbara. Made at U.S. Army Ballistics Research Laboratory, Aberdeen Proving Ground, Maryland.)

the bow of a moving ship (Fig. 8.3). At the bow shock, the solar wind abruptly decelerates to subsonic speed and heats up, like the wheels of a car slamming on its brakes or an ocean wave crashing into foam at the shore. Most of the solar wind is then deflected around the Earth, but some of its particles are reflected back from the bow shock into the onrushing solar wind, like the eddies around a rock in a river.

Like winds on Earth, the solar wind is punctuated by gusts and has its own tempestuous weather, which buffets the magnetosphere. When the solar wind pressure is high, the bow shock moves inward, and when the pressure drops, the Earth's magnetic domain expands; it's like squeezing a rubber ball and letting it go. The entire magnetosphere compresses and expands, changing size constantly as the solar wind varies in density and speed. These variations are frequently caused by violent eruptions on the Sun, which occur more often during the maximum in the 11-year cycle of solar activity.

8.3 PENETRATING EARTH'S MAGNETIC DEFENSE

Particles in the solar wind transport only one ten-billionth the energy of that carried by sunlight, and Earth is protected from the full blast of the dilute, varying solar wind by the terrestrial magnetosphere. The Earth's magnetic shield is so perfect that only 0.1 percent of the mass of the solar wind that hits it manages to penetrate inside. Yet, even that small fraction of the wind particles has a profound influence on the Earth's environment in space. They create an invisible world of energetic particles and electric currents that flow, swirl and encircle the Earth, suspended high above its atmosphere.

As the solar wind brushes past the Earth, the wind carries some of the Sun's magnetic field with it. And since magnetic fields have a direction, the solar wind's magnetism can point toward or away from the direction of the Earth's magnetic field. If the solar field points opposite to that of the Earth when they meet, the two can join each other, just as the opposite poles of two toy magnets stick together. The magnetic field in the solar wind is then broken, and reconnects with the terrestrial magnetic field, plugging the solar wind into the Earth's electrical socket and wiring its magnetosphere to the Sun. This magnetic coupling occurs on the dayside facing the Sun and is dragged downstream all along the length of the Earth's magnetotail. And since the immense magnetic tail forms the bulk of the magnetosphere, it provides the main location for breaching the Earth's magnetic defense.

The passing solar wind is slowed down by the connected fields and decelerates in the vicinity of the tail. Energy is thereby extracted from the nearby solar wind and drives a large-scale circulation, or convection, of charged particles within the magnetosphere (Fig. 8.4). While creating and sustaining the magnetotail, the solar wind brings the oppositely directed tail lobes into close contact, where they can merge together. The magnetotail then snaps like a rubber band that has been stretched too far. The

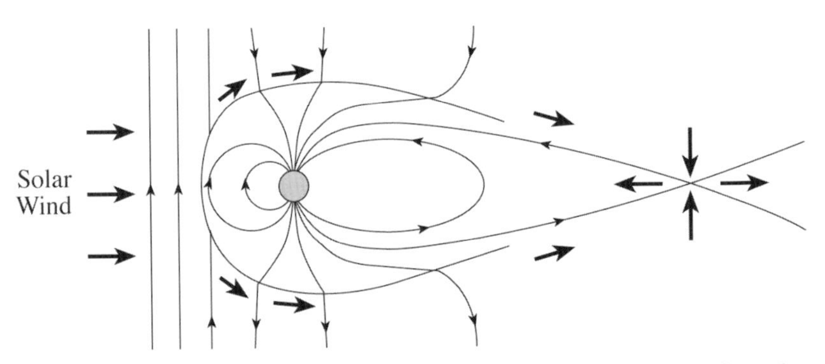

FIG. 8.4 Magnetic connection on the back side The Sun's wind brings solar and terrestrial magnetic fields together on the night side of Earth's magnetosphere, in its magnetotail. Magnetic fields that point in opposite directions (*thin arrows*), or roughly toward and away from the Earth, are brought together and merge, reconnecting and pinching off the magnetotail close to Earth. Material in the plasma sheet is accelerated away from this disturbance (*thick arrows*). Some of the plasma is ejected down the magnetotail and away from the Earth, while other charged particles follow magnetic field lines back toward Earth.

snap catapults part of the tail downstream into space, creating a gust-like eddy in the solar wind.

The other part of the tail, propelled by energy released in the magnetic merging, rebounds back toward our planet. Electrons and ions hurtle along magnetic conduits that are connected to the Earth, linking the solar wind to both equatorial storage regions and down into the polar caps. The electrons that are guided into the poles augment and intensify the northern and southern lights, or auroras. And the tail continues to reform into another, potentially explosive, unstable configuration to await the next magnetic connection.

8.4 STORING INVISIBLE PARTICLES WITHIN EARTH'S MAGNETOSPHERE

Charged particles flowing from the Sun can enter the Earth's magnetic domain and become trapped within it. They can be stored along the stretched dipolar field lines, earthward of the tail magnetic connection site, in a region called the plasma sheet. It acts as a holding tank of electrons and ions, suddenly releasing them when stimulated by the ever-changing Sun. Nearer the Earth particles are stored in the radiation belts, regions of an unexpectedly high flux of high-energy electrons and protons, which girdle the Earth far above the atmosphere in the equatorial regions.

Inspired by Birkeland's experiments with electrons and a dipolar magnetic sphere, Carl Størmer (1874–1957), a young theoretical physicist in Oslo, Norway, studied mathematically the motion of charged particles in the magnetic field of a dipole. Using tedious numerical calculations before the computer age, he showed in 1907 that electrically charged particles could be confined and suspended in space by the Earth's dipolar magnetism, spiraling around the magnetic field lines and bouncing back and forth between the Earth's magnetic poles for long periods if time (Fig. 8.5).

The magnetic field exerts a force on the charged particles that guides them in a helical path; they can move freely along the magnetic field line, but are constrained to a circular motion around it. The spiraling trajectory becomes more tightly coiled in the stronger magnetic fields close to a magnetic pole, until the charged particles are deflected at a mirror point. The stronger magnetism pulls a particle's motion into smaller and smaller circles, slowing the downward motion, stopping it, and pushing the particle back, like throwing a ball against a spring. Most scientists were nevertheless completely surprised by the discovery of donut-shaped belts of electrons and protons trapped within the terrestrial magnetosphere.

The first American satellite *Explorer 1,* launched on 31 January 1958 in response to the Soviet Union's successful *Sputnik* satellites in October and November 1957, had a Geiger counter aboard to measure the intensity of cosmic rays. The instrument recorded the expected cosmic rays near the Earth, but counts rose more than expected at higher altitudes and then disappeared altogether. *Explorer 3* confirmed the effect two months later. (*Explorer 2* went into the ocean.)

It turned out that the satellites had entered a dense region of energetic particles that saturated the Geiger tube, causing the counter to read zero. There were so many high-energy particles coming into the detector that it shut down, as if space itself was radioactive. The instrument showed that space is filled with "radiation"

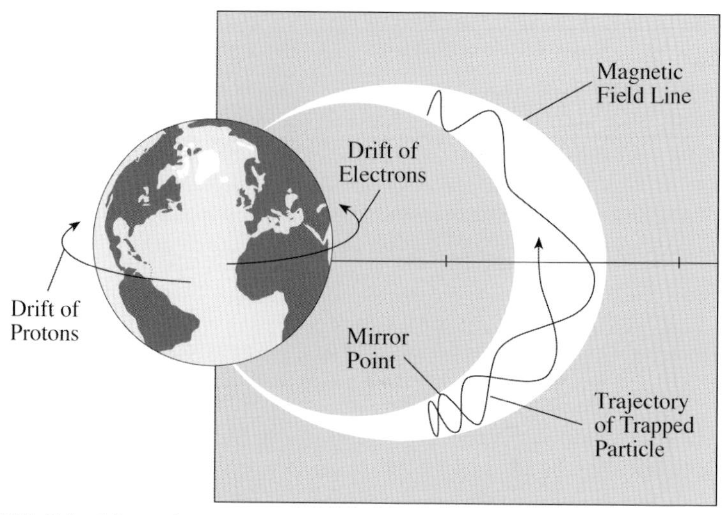

FIG. 8.5 **Magnetic trap** Charged particles can be trapped by Earth's magnetic field. They bounce back and forth between polar mirror points in either hemisphere at intervals of seconds to minutes, and they also drift around the planet on time-scales of hours. As shown by the Norwegian scientist Carl Størmer (1874–1957) in 1907, with the trajectories shown here, the motion is turned around by the stronger magnetic fields near the Earth's magnetic poles. Because of their positive and negative charge, the protons and electrons drift in opposite directions.

of intensities a thousand times greater than expected, suggesting that energetic, charged particles encircle the Earth within donut-shaped regions near the magnetic equator (Fig. 8.6).

These regions are sometimes called the inner and outer Van Allen radiation belts, named after James A. Van Allen (1914–), whose instruments first observed them; they have been dubbed radiation belts since the charged particles that they contain were known as corpuscular radiation at the time of their discovery. The nomenclature is still used today, but it does not imply either electromagnetic radiation or radioactivity.

Van Allen and his colleagues at the University of Iowa were fully aware of Størmer's prior work, as well as related studies by the American physicist Sam Bard Trieman (1925–1999) of cosmic rays interacting with the Earth's atmosphere and magnetic fields, reporting in 1959 that:

> The existence of a high intensity of corpuscular radiation in the vicinity of the Earth was discovered by apparatus carried by satellite *1958 alpha* [*Explorer 1*]. . . . It was proposed in our May 1, 1958 report that the radiation was corpuscular in nature, was presumably trapped in Størmer-Treiman lunes about the Earth, and was likely intimately related to that responsible for aurorae. On the basis of these tentative beliefs it was thought likely that the observed trapped radiation had originally come from the Sun in the form of ionized gas.[32]

FIG. 8.6 Radiation belts Electrons and protons encircle the Earth within two donut-shaped, or torus-shaped, regions near the equator, trapped by the terrestrial magnetic field. These regions are now called the inner and outer Van Allen radiation belts, named after the American scientist James A. Van Allen (1914–) who first observed them with the *Explorer 1* and *3* satellites in 1958. The inner belt's charged particles tend to have higher energies than those in the outer belt. The trapped particles can damage the microcircuits, solar arrays and other materials of spacecraft that pass through the Van Allen radiation belts. A torus of high-energy particles creates a third radiation belt located just inside of the inner Van Allen belt, but not shown here; it contains heavy nuclear ions that originated outside our Solar System and once drifted between the stars.

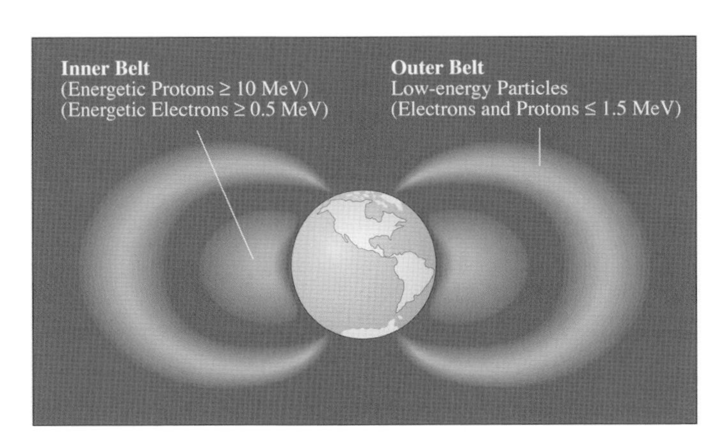

The radiation belts mainly consist of high-energy electrons and protons, with lesser amounts of heavier ions. And since these charged particles are trapped in the Earth's magnetic cage, as Størmer showed, one wonders how they got into the radiation belts in the first place. As we have already shown, some of them can enter through the back door, by magnetic reconnection of the solar and terrestrial magnetic fields in the magnetotail, and once in they can be additionally accelerated to high energies. This coupling can become especially effective when a coronal mass ejection encounters the Earth with the right orientation, supplying more electrons and protons to the magnetosphere and pumping those that are already there up to higher energies.

Particles within the inner magnetosphere close to the Earth can come from the upper terrestrial atmosphere below the radiation belts. Solar extreme ultraviolet and X-ray radiation create the ionosphere in the upper atmosphere, which can vary dramatically with solar activity. Although most of the ionosphere is gravitationally bound to the Earth, some of its particles have sufficient energy to escape and become trapped by the magnetic fields higher up.

In addition, very energetic cosmic rays, entering the atmosphere from outer space, can collide with atoms in our air and eject neutrons from the atomic nuclei. These neutrons travel in all directions, unimpeded by magnetic fields since they have no electrical charge. But once it is librated from an atomic nucleus, a neutron cannot stand being left alone. A free neutron lasts only 10.25 minutes on average before it decays into an electron and proton. Some of the neutrons produced by cosmic rays in our atmosphere move out into the inner radiation belt before they disintegrate, producing electrons and protons that are immediately snared by the magnetic fields and remain stored within them.

8.5 NORTHERN AND SOUTHERN LIGHTS

Dance of the Aurora

The northern and southern lights are one of the most magnificent and earliest-known manifestations of the myriad links between the Sun and the Earth. They illuminate the Arctic and Antarctic skies, where curtains of multi-colored light dance and shimmer across the night sky far above the highest clouds, like giant dragons or snakes (Figs. 8.7, 8.8).

Perhaps because they can be colored red like the rising morning Sun, they are often called the aurora after the Roman goddess of the rosy-fingered dawn; this designation has been traced back to the time of the Italian scientist Galileo Galilei (1564–1642). The auroras seen near the north and south poles have been given the Latin names *aurora borealis,* for "northern lights", and *aurora australis,* for "southern lights."

Aurora activity is not rare; it is almost always present! Residents in far northern locations regularly see the aurora borealis, every clear and dark winter night. Even today, the winter aurora brings solace to circumpolar inhabitants, reminding them of the eventual return of life-sustaining sunlight. But most people never see the awesome lights, for auroras are normally confined to high latitudes in the north or south polar regions with relatively few inhabitants. In addition, only an exceptionally intense aurora can be noticed against bright city lights.

Some of the earliest written accounts of the northern lights are found in Mediterranean countries where schools and libraries flourished long ago. Greek records of the aurora date back to at least Aristotle (384–322 BC), who described an aurora, probably the one occurring in 349 BC, as being blood red in color and with chasm-like or trench-like shapes. However, spectacular auroras rarely extend as far south as Greece, perhaps every 50 to 100 years.

Since civilization began, the multi-colored lights have always been most frequently observed at far northern latitudes, so some of the most vivid and comprehensive accounts come from the Scandinavian countries. The oldest written records by Norwegians concerning the northern lights go back to the Viking period (500 to 1300). In one Norse chronicle, called *Kongespeilet* or *King's Mirror* and written about 1250, it is described by:

> These northern lights have this peculiar nature, that the darker the night is, the brighter they seem, and they always appear at night but never by day, most frequently in the densest darkness and rarely by moonlight. In appearance they resemble a vast flame of fire viewed from a great distance. It also looks as if sharp points were shot from this flame up into the sky, they are of uneven height and in constant motion, now one, now another darting highest; and the light appears to blaze like a living flame . . . It seems to me not unlikely that the frost and the glaciers have become so powerful there that they are able to radiate forth these flames.[33]

FIG. 8.7 Aurora borealis Swirling walls and rays of shimmering green and red light are found in this portrayal of the fluorescent Northern Lights, or Aurora Borealis, painted in 1865 by the American artist Frederic Church (1826–1900). (Courtesy of the National Museum of American Art, Smithsonian Institution, gift of Eleanor Blodgett.)

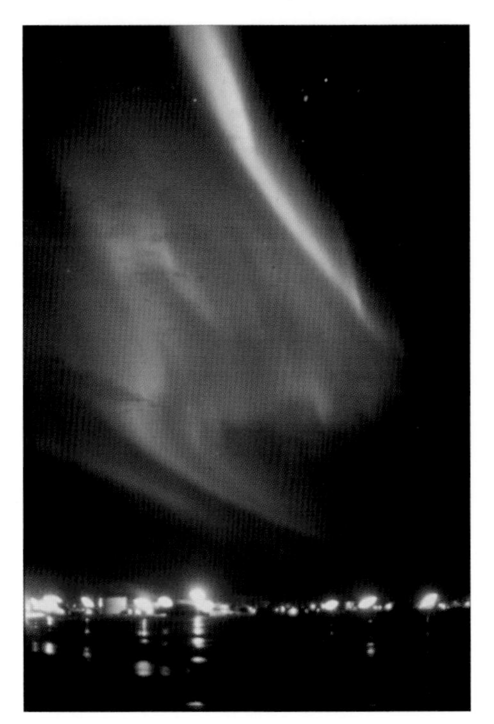

FIG. 8.8 Northern lights Spectacular green curtains of light illuminate the northern sky, like a cosmic neon sign. Forest Baldwin took this photograph of the fluorescent Northern Lights, or Aurora Borealis, in Alaska. (Courtesy of Kathi and Forest Baldwin, Palmer, Alaska.)

In his book *The Fram Expedition*, published in 1897, the Norwegian explorer Fridtjof Nansen (1861–1930) provided this account, written while he was trapped through the long Arctic winter in the frozen pack ice:

> The glowing fire-masses had divided into glistening, many colored bands, which were writhing and twisting across the sky both in the south and north. The rays sparkled with the purest, most crystalline rainbow colors, chiefly violet-red or carmine and the clearest green. Most frequently the rays of the arch were red at the ends, and changed higher up into sparkling green. . . . It was an endless phantasmagoria of sparkling color, surpassing anything that one can dream. Sometimes the spectacle reached such a climax that one's breath was taken away; one felt that now something extraordinary must happen – at the very least the sky must fall.[34]

Nansen, who won the Nobel Peace Prize in 1922, wrote a few more sentences and could not continue; being thinly dressed and without gloves, he had no feeling left in body or limbs.

Auroras occur with the same frequency and simultaneously in both the southern and northern polar regions of the Earth. Indeed, the two auroras are almost mirror images of each other. But the aurora australis have never achieved a renown comparable to the northern lights, probably because the southern ones are not usually located over inhabited land and are instead seen from oceans that are infrequently traveled. In fact, the first recorded sighting of the southern lights did not occur until 1770, by the British Captain James Cook (1728–1779) during the voyage of the *HMS Endeavor.*

Nowadays we can use spacecraft to view both the northern and southern lights from above (Fig. 8.9). The *Space Shuttle* has even flown right through the northern lights, which extend a few hundred kilometers above the ground. While inside the display, astronauts could close their eyes and see flashes of light caused by the charged aurora particles, which ripped through the satellite walls and passed through their eyeballs, making them glow inside.

The view from space is just as magnificent as that from the ground. As the Czechoslovakian astronaut, Vladimir Remek (1948–), expressed it:

> Suddenly, before my eyes, something magical occurred. A greenish radiance poured from the Earth directly up to the [Space] Station, a radiance resembling gigantic phosphorescent organ pipes, whose ends were glowing crimson, and overlapped by waves of swirling green mist.[35]

The play of northern lights in the sky has also been described in the folk lore of Arctic cultures, where they are often interpreted as the spirits of the dead either fighting or playing in the air. The Vikings thought the auroras represented an eternal battle between the spirits of fallen warriors. In Norse mythology, the aurora is a bridge of fire that permits Thor, the God of War, to travel between Heaven and Earth. Eskimos have described the flickering lights as a dance of the dead, amusing themselves in the absence of light from the Sun, or as signals from the deceased trying to contact their living relatives. An Eskimo word for aurora, *aksarnirq,* translates into "ball player." For

FIG. 8.9 Aurora from the shuttle The eerie, beautiful glow of auroras can be detected from space, as shown in this image of the Aurora Australis or Southern Lights taken from the *Space Shuttle Discovery.* The green emission of atomic oxygen extends upward from 90 to 150 kilometers above the Earth's surface, where it is created by beams of high-speed electrons moving down into the atmosphere, exciting the oxygen atoms and making them fluoresce. (Courtesy of NASA.)

Alaskan Eskimos the spirits are playing ball with the heads of children who dared venture outside during the northern lights.

The flickering, colored lights have inspired many poets, such as the English poet Robert Browning (1812–1889), whose vision of Judgment Day was inspired by an aurora, "the final belch of fire like blood, over broke all heaven in one flood", as from a dragon's nostril, and the American poet Wallace Stevens (1879–1955) who wrote of "its polar green, the color of ice and fire and solitude."[36]

In Norway or Sweden it was a common belief that the northern lights were reflections from silvery shoals of herring that flashed light against the clouds when swimming close to the water's surface. According to *The King's Mirror,* the Arctic snow and ice absorbed large amounts of light from the long summer midnight Sun and re-radiated it as the northern lights in wintertime. Almost 400 years later, the French philosopher Rene Descartes (1596–1650) attributed the aurora to sunlight scattered from ice particles found high in the atmosphere at cold northern locations. But all these ideas were eventually shown to be wrong.

Since auroras become more frequent as one travels north from tropical latitudes, it was thought that the northern lights would occur most often at the highest polar latitudes. Early Arctic explorers were therefore surprised to find that their frequency of occurrence did not increase all the way to the poles. In 1860 Elias Loomis (1811–1889),

Professor of Natural Philosophy at Yale University, mapped out their geographic distribution, showing that the northern lights form a luminous ring encircling the North Pole. The Swiss engineer and physicist Herman Fritz (1830–1883) extended Loomis' work, publishing a similar conclusion in 1881 in his then-well-known book *Das Polarlicht.* Loomis and Fritz showed that the intensity and frequency of auroras were greatest in an oval-shaped aurora zone centered on the North Pole with a width of about 500 kilometers and a radius of about 2,000 kilometers. Auroras can occur every night of the year within this zone.

About a century later, in 1957–58, the aurora distribution was mapped out in greater detail using all-sky cameras during the International Geophysical Year. An analysis of hundreds of thousands of photographs, each portraying the sky from horizon to horizon, confirmed that the aurora zone is an oval-shaped band that is centered on the Earth's magnetic pole.

Today spacecraft look down on the aurora oval from high above the north polar region, showing the northern lights in their entirety (Fig. 8.10). They indicate that the luminous aurora oval is constantly in motion, expanding toward the equator or contracting toward the pole, and always changing in brightness. Such ever-changing aurora ovals are created simultaneously in both hemispheres and can be viewed at the same time from the Moon.

Visual auroras normally occur at 100 to 250 kilometers above the ground. This height is much smaller than either the average radius of the aurora oval, at 2,250 kilometers, or the radius of the Earth, about 6,380 kilometers. An observer on the ground therefore sees only a small, changing piece of the aurora oval, which can resemble a bright, thin, windblown curtain hanging vertically down from the Arctic sky.

The Sun's Brightening and Dimming Switch

Loomis and Fritz established a general correlation between the occurrence of sunspots and northern lights. When the number of sunspots is large, bright auroras occur more frequently, and when there are few spots on the Sun the intense auroras are seen less often. So, the frequency of occurrence of the bright auroras tends to follow the 11-year sunspot cycle of solar activity, suggesting that the Sun somehow controls the brilliance of the northern lights. Loomis even suggested a physical connection, noting that an exceptionally intense aurora occurred on 2 September 1859; the day after the first solar flare was discovered.

The Sun's influence was more fully explained in 1896 when the Norwegian physicist Kristian Birkeland (1867–1917) showed that electrons from the Sun might be directed and guided along the Earth's magnetic field lines to the polar regions. Birkeland demonstrated his theory by sending electrons toward his own magnetized sphere, or *terrella,* using phosphorescent paint to show where electrons struck it. An electromagnet was placed in the sphere, creating a dipolar magnetic field, and the entire apparatus was placed in a low-density vacuum that represented outer space. The resulting light indicated that the electrons are curved down toward and around the magnetic poles, and the glowing shapes reproduced many of the observed features of the auroras.

Particle detectors onboard rockets launched into auroras in the early 1960s showed that Birkeland was right, at least in part! The aurora is electrified, principally excited

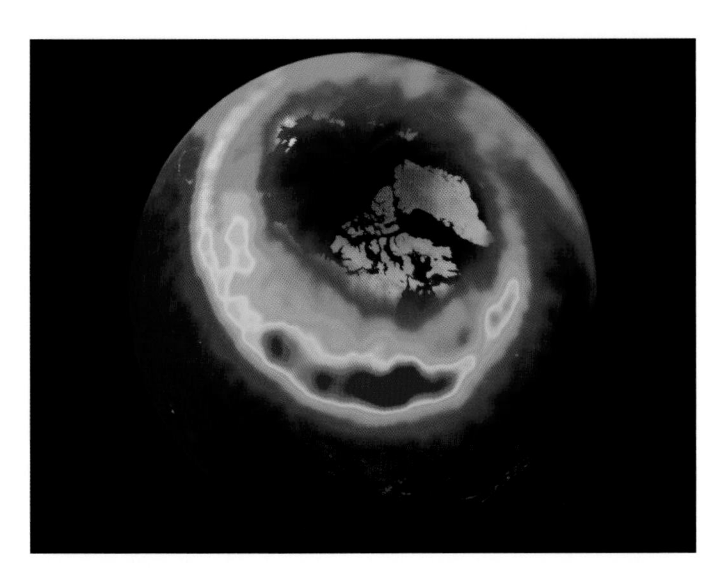

FIG. 8.10 The aurora oval Instruments aboard the *POLAR* spacecraft look down on the aurora from high above the Earth's north polar region on 22 October 1999, showing the northern lights in their entirety. The glowing oval, imaged in ultraviolet light, is 4,500 kilometers across. The most intense aurora activity appears in bright red or yellow, toward the night side of the Earth; it is typically produced by magnetic reconnection events in the Earth's magnetotail. The luminous aurora oval is constantly in motion, expanding toward the equator or contracting toward the pole, and always changing in brightness. Such ever-changing aurora ovals are created simultaneously in both hemispheres. (Courtesy of the Visible Imaging System, University of Iowa and NASA.)

by energetic electrons bombarding the upper atmosphere with energies of about 6 keV, or 6,000 electron volts, and speeds of about 50 kilometers per second. As the electrons cascade down the polar magnetic field lines into the atmosphere, they are slowed down by collisions with the increasingly dense air, exciting the gaseous atoms and causing them to glow like a cosmic neon sign. The luminous aurora shimmers as electrons are injected down from different locations, exciting atoms that shine in long, thin vertical sheets like the folds of a curtain.

Currents can be produced along the aurora ovals that are as strong as a million amperes. These currents flow down from the magnetosphere, through the ionosphere in the upper atmosphere, around the aurora oval, and back out and up to the magnetosphere.

When the electrons slam into the upper atmosphere, they collide with the oxygen and nitrogen atoms there and excite them to energy states unattainable in the denser air below. The most abundant constituents of our atmosphere are oxygen and nitrogen, respectively comprising 21 and 78 percent of the air, and their energized aurora transitions are "forbidden" at high densities. The pumped-up atoms quickly give up the energy they acquired from the electrons, emitting a burst of color in a process called fluorescence. It is similar to electricity making the gas in a neon light shine or a fluorescent lamp glow. The process also resembles the beam of electrons that strikes the screen of your color television set, making it glow in different colors depending on the type of chemicals, or phosphors, that coat the screen.

The colors of the aurora depend on which atoms or molecules are struck by the precipitating electrons, and the atmospheric height at which they are struck. Excited oxygen atoms radiate both green (557.7 nanometers) and red (630.0 and 636.4 nanometers) light; the green oxygen emission appears at an altitude of about 100 kilometers and the red oxygen light at 200 to 400 kilometers. The bottom edge of the most brilliant green curtains are sometimes fringed with the pink glow of neutral, or un-ionized, molecular nitrogen, and rare blue or violet colors are emitted by ionized nitrogen molecules.

So, energetic electrons colliding with oxygen and nitrogen atoms in the air cause the multi-colored aurora light show, but where do the electrons come from and how are they energized? Since the most intense auroras occur at times of maximum solar activity, it was once thought that the aurora electrons came from explosions on the Sun, hurled directly down into the upper atmosphere through the Earth's narrow, funnel-shaped polar magnetic fields. After all, the frequency of the solar explosions also peak at the maximum of the solar magnetic activity cycle.

In another popular theory, solar wind electrons were supposed to be held within the Van Allen radiation belts before being squirted out into the aurora zones, like a squeezed tube of toothpaste, as the result of excessive solar activity. However, solar particles coming in from the polar route apparently do not have enough energy to make all of the auroras, and there are not enough particles in the Van Allen belts.

Even though changing conditions on the Sun may trigger the intense northern and southern lights, we now know that the electrons that cause the auroras arrive indirectly at the polar regions, from the Earth's magnetic tail, and that these electrons can be energized and accelerated locally within the magnetosphere. Changing solar-wind conditions can temporarily pinch off the Earth's magnetotail, opening a valve that lets the solar-wind energy cross into the magnetosphere and additionally shooting electrons stored in the magnetic tail back toward the aurora zones near the poles.

During this magnetic reconnection process, the magnetic fields heading in opposite direction – having opposite north and south polarities – break and reconnect downwind of Earth on its night side. Electrons are pushed up and down the tail, and can be accelerated within the magnetosphere as they travel back toward the Earth and into its polar regions. The electrons that are thrown Earthward follow the path of magnetic field lines, which link the magnetotail to the polar regions and map into the aurora oval.

The entry from the magnetotail also explains why the aurora oval is not completely illuminated when seen from space. It shines mostly on the night side of the Earth, which is magnetically connected to the reconnection site in the magnetotail. Magnetic fields on the sunlit dayside are directly connected to the poles rather than to the magnetotail.

The rare, bright, auroras seen at low latitudes in more clement climates become visible during very intense geomagnetic storms that enlarge the magnetotail. The aurora ovals then intensify and spread down as far as the tropics in both hemispheres. Since these great magnetic storms are produced by solar explosions, known as coronal mass ejections, it is really the Sun that controls the intensity of the brightest, most extensive auroras, like the brightening and dimming switch of a cosmic light, even though auroras are most often fueled by particles accelerated in the magnetosphere.

8.6 GEOMAGNETIC STORMS

Although the gusty solar wind never reaches the Earth's surface, it can cause dramatic changes in the Earth's magnetic field. George Graham (1674–1751), a London watchmaker, first noted the magnetic variation in 1722 as a pronounced swing and rapid fluctuation in the direction that compass needles point. Graham and the Swedish astronomer and physicist Anders Celsius (1701–1744) independently noticed in 1741 that irregular deflections of the compass needle occur when intense auroras are present. Then in the mid-nineteenth century the German explorer and naturalist Baron Alexander von Humboldt (1769–1859) measured variations in the magnetic field during his global trips, giving them the name magnetic storms.

Nowadays we often attach the appellation "geo" to the name, using the term geomagnetic storms to stress their Earthly nature. Unlike localized stormy weather on land or sea, a geomagnetic storm is invisible and silent, undetectable by the human eye or ear, but like our daily weather, geomagnetic storms can sometimes have devastating effects.

The intense geomagnetic storms vary in tandem with the 11-year sunspot cycle. Already in 1852, Colonel Edward Sabine (1788–1883), superintendent of four of the magnetic observatories in the British colonies, was able to show that global magnetic fluctuations are synchronized with this cycle. In response to a letter from the astronomer John Herschel (1792–1871), that called attention to Samuel Heinrich Schwabe's (1789–1875) discovery of the sunspot cycle, Sabine wrote:

> With reference to Schwabe's period of 10 years having a minimum in 1843 and a maximum in 1848, it happens that by a most curious coincidence (if it be nothing more than a coincidence) that in a paper now waiting to be read at the Royal Society, I trace the very same years as those minimum and maximum of an apparent periodical inequality which took place in the frequency and magnitude of the [terrestrial] magnetic disturbances and in the magnitude of the mean monthly range of each of the 3 magnetic elements shown concurrently in the two hemispheres.[37]

Early in the 20th century, the Norwegian physicist Kristian Birkeland (1867–1917) argued that beams of electrons are sent from the Sun to the Earth, causing both auroras and geomagnetic storms. But it was soon realized that the electrical repulsion between the electrons would cause such a solar electron beam to disperse into space before reaching Earth. So in the 1930s, the English geophysicist Sydney Chapman (1889–1970) and his young colleague Vincent Ferraro (1907–1974) reasoned that geomagnetic storms are caused when an electrically neutral plasma cloud is ejected from the Sun and envelops the Earth, generating currents in the magnetosphere and distorting its magnetic field.

We now know that Chapman and Ferraro were close to solving the mystery of geomagnetic storms. The most intense ones are associated with coronal mass ejections, great magnetized bubbles of plasma hurled from the Sun, which can generate currents in the Earth's magnetosphere and energize charged particles in it (Fig. 8.11).

If averaged over both a yearly and global scale, intense geomagnetic storms do vary in step with the sunspot cycle. When the Sun shows more spots, the terrestrial magnetic field is more frequently disturbed by violent storms. But it is not the sunspots themselves that bring about the changes on the Earth. Great magnetic storms are

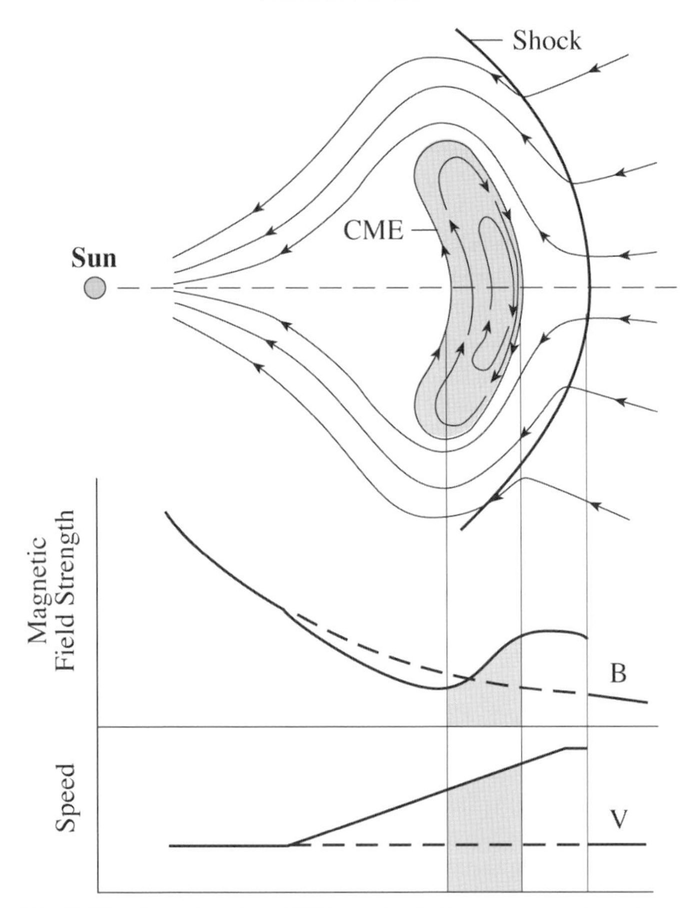

FIG. 8.11 Interplanetary CME shocks As it moves away from the Sun (*top left*) a fast coronal mass ejection (CME, *top right*) pushes an interplanetary shock wave before it, amplifying the solar wind speed, V, and magnetic field strength, B (*bottom*). The CME produces a speed increase all the way to the shock front, where the wind's motion then slows down precipitously to its steady, unperturbed speed. Compression, resulting from the relative motion between the fast CME and its surroundings, produces strong magnetic fields in a broad region extending sunward from the shock. The strong magnetic fields and high flow speeds commonly associated with interplanetary disturbances driven by fast CMEs are what make such events effective in stimulating geomagnetic activity.

caused by coronal mass ejections that occur more often when the Sun is more spotted and active.

The consequences of an exceptionally intense magnetic storm are truly awesome. Several times every solar cycle, a solar eruption of extraordinary energy creates a brief, violent gust in the solar wind that sets the entire magnetosphere reverberating with catastrophic impact. When the high-velocity shocks arrive at the Earth, followed by the magnetic fields, they can compress the dayside magnetosphere down to half its normal size confusing geomagnetic navigational and detection sensors in satellites and disorienting homing pigeons and other migratory animals that depend on the Earth's magnetic field for guidance. High rates of connection between the solar-wind magnetic field and the terrestrial one increase the size of the magnetotail that connects to the poles, and as a result the aurora oval intensifies and spreads eerily beautiful auroras across the sky to tropical latitudes in both hemispheres.

Shortly after the turn of the twentieth century, E. Walter Maunder (1851–1928), a sunspot expert from the Royal Observatory at Greenwich, England, found that many moderate geomagnetic storms occur at intervals of 27 days, corresponding to the apparent rotation period of the Sun as viewed from the Earth. He concluded that this recurrence could be explained:

> by supposing that the Earth has encountered, time after time, a definite stream, a stream which, continually supplied from one and the same area of the Sun's surface, appears to us, at our distance, to be rotating with the same speed as the area from which it rises.[38]

As emphasized by the German scientist Julius Bartels (1899–1964), these recurrent geomagnetic storms sometimes occur when there are no visible sunspots. In 1932 Bartels referred to the sources of geomagnetic storms as M regions and noted that while M regions were sometimes associated with sunspots, in many other cases they seemed to have no visible counterpart. The M probably stood for magnetic, but it might have denoted mysterious; for several decades the elusive source of the moderate, recurrent geomagnetic storms remained one of the great-unsolved mysteries in solar physics.

The recurrent activity is linked to long-lived, high-speed streams in the solar wind that emanate from coronal holes and periodically sweep past the Earth. When the Sun is near a lull in its 11-year activity cycle, the fast wind streams rushing out of coronal holes can extend to the plane of the solar equator. When this fast wind overtakes the slow-speed, equatorial one, the two wind components interact, like two rivers merging to form a larger one. This produces shock waves and intense magnetic fields that rotate with the Sun, producing moderate geomagnetic activity every 27 days. Near solar maximum, at the peak of the 11-year activity cycle, coronal mass ejections dominate the interplanetary medium, producing the most intense geomagnetic storms, and the low-level activity is less noticeable.

Thus, geomagnetic activity shows two components, both connected with the Sun. One, associated with intense geomagnetic storms and coronal mass ejections, varies in tandem with the sunspot cycle of solar magnetic activity. The other, related to moderate geomagnetic storms and recurrent high-speed solar wind streams, is out of phase with the solar cycle, dominating geomagnetic storms when the solar activity cycle is at a minimum.

8.7 SPACE WEATHER

There is danger blowing in the Sun's wind, which contains powerful gusts and squalls. They are the cosmic equivalent of terrestrial blizzards or hurricanes. Down here on the ground, we are shielded from this space weather by the Earth's atmosphere and magnetic fields, keeping us from bodily harm. But out in deep space there is no place to hide, and both humans and satellites are vulnerable.

The storms in space can even kill an unprotected astronaut repairing a space station or walking on the Moon or Mars. The Sun storms can disrupt global radio signals, and disable satellites used for cell phones or pagers, navigation, military reconnaissance and surveillance. Or they might overload electrical grids on the ground, causing massive power blackouts.

Violent explosions on the Sun, known as solar flares and coronal mass ejections, produce space weather, and the most intense storms, producing the greatest damage at Earth, are legendary. There were the record-breaking Halloween storms of October and November 2003, disrupting radio communications across the sunlit side of the Earth; several satellites were also damaged and a power blackout hit Sweden. A couple of years earlier, in March and April 2001, the space weather spawned by solar flares and coronal mass ejections also cut off radio communications, and disrupted or damaged several military and commercial satellites. And damage from the Bastille Day flare (Fig. 8.12), on 14 July 2000, was mitigated by alerts and warnings to industry, the military, and space agencies.

High-Flying Humans At Risk

Although most airplanes fly very low in the atmosphere, well below any satellite, transcontinental flights taking the polar route pass through regions of Earth's magnetic field

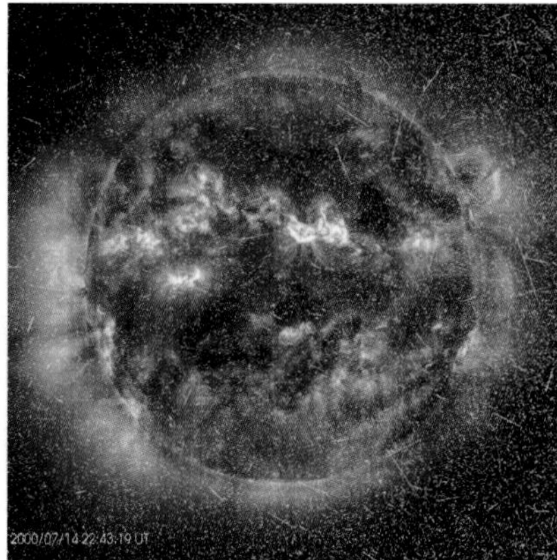

FIG. 8.12 Solar flare produces energetic particle storm A powerful solar flare (*left*), occurring at 10 hours 24 minutes Universal Time on Bastille day 14 July 2000, unleashed high-energy protons that began striking the *SOHO* spacecraft near Earth about 8 minutes later, continuing for many hours, as shown in the image taken on 22 hours 43 minutes Universal Time on the same day (*right*). Both images were taken at a wavelength of 19.5 nanometers, emitted at the Sun by eleven times ionized iron, denoted Fe XII, at a temperature of about 1.5 million kelvin, using the Extreme Ultraviolet Imaging Telescope, abbreviated EIT, on the *SOlar and Heliospheric Observatory,* or *SOHO* for short. (Courtesy of the *SOHO* EIT consortium. *SOHO* is a project of international cooperation between ESA and NASA.)

where solar particles become concentrated. The high-energy particles created by solar explosions can be channeled along the magnetic field and penetrate to low altitudes in the polar regions, exposing airline crews and passengers to elevated levels of particles and radiation from space. The higher the plane is flying and the closer to the poles, the higher the radiation dose. The health risk is small, but highest for frequent fliers, pilots and flight attendants who travel polar routes often. Pregnant women are advised to not take an airplane flying a polar route during a storm on the Sun, to avoid risk of birth defects, but the problem is some women at risk might not even know they are pregnant.

There are even greater hazards aboard spacecraft at higher altitudes. According to an expert at the United States Defense Nuclear Agency, military pilots can be provided with drugs that will make them temporarily survive a lethal dose of solar particles or radiation. It's a matter of patriotism and cost. They are going to die anyway, so why not take the injection, save the spacecraft, and come on home to die. There is no sense in being the first corpse in space.

Go far enough into space and the chemical bonds in your molecules will be broken apart by storms from the Sun, increasing the risk of cancer and errors in genetic information. Space agencies therefore set limits to the exposure to solar energetic particles and radiation an astronaut can have while traveling or working unprotected in space. Because of the potential genetic damage, it is said that astronauts are supposed to have had all their children before flying in outer space; otherwise they might have some very weird offspring. Probably because of hormones, men are more radiation-resistant than women, and the resistance peaks between the ages of forty-five and fifty. So, if genetic harm and other health risks are the dominant factor, most astronauts will be middle-aged men, and they usually are.

Solar energetic particle events can endanger the health and even the lives of astronauts when they are in outer space, unprotected by the Earth's magnetic field. The shielding of a typical spacecraft is not then enough to protect a human from cataracts or skin cancer during a major solar energetic particle event, and high-energy protons from a solar flare or coronal mass ejection can easily pierce a space suit, causing damage to human cells and tissues. They can even kill unprotected astronauts that venture into space (Figs. 8.13, 8.14) to unload spacecraft cargo, construct a space station or walk on the Moon or Mars.

Solar astronomers therefore keep careful watch over the Sun during space missions, to warn of possible activity occurring at just the wrong place or time. Flight controllers can then postpone space walks during solar storms, keeping astronauts within the heavily shielded recesses of a satellite or space station. They would also be told to curtail any strolls on the Moon or Mars, instead moving inside underground storm shelters.

A disaster has so far been avoided because previous stays on the Moon were of short duration (a few days) when no major solar eruption occurred. The manned *Apollo 16* and *17* missions to the Moon in April and December 1972, respectively occurred before and after a large solar flare, on 7 August 1972, which would have caused disabling radiation sickness and possible death to astronauts walking on the Moon.

A longer, future trip to Mars will involve considerable risks. Astronauts would spend six months or more in transit each way, and stay on the Martian surface for as long as a year-and-a-half, until the red planet again moved closest to the Earth. Some

FIG. 8.13 Man in space Astronaut Donald H. Peterson (1933–), on a 50-foot (15-meter) tether line during his 4-hour, 3-orbit space walk, moving toward the tail of the *Space Shuttle Challenger* as it glides around the Earth. Hundreds of kilometers above the Earth, there is no air and astronauts must wear a spacesuit. It supplies the oxygen they need and insulates their body from extreme heat or cold. However, a spacesuit cannot protect an astronaut from energetic particles hurled out from explosions on the Sun. He or she must then be within the protective shielding of a spacecraft or other shelter to avoid the danger. (Courtesy of NASA.)

estimate that every third human cell would be damaged by solar energetic particles during the flight, and others worry about how to keep the astronauts from being irradiated to death. Long exposures to cosmic rays in space also increase the risk of getting cancer, apparently to a forty-percent lifetime chance after a voyage to Mars and far above acceptable thresholds of government agencies. A future return trip to the Moon, with an extended stay, or to explore Mars, must include methods of protection of the crew from the harmful effects of Sun-driven space weather and cosmic rays.

FIG. 8.14 Unprotected from space weather The first untethered walk in space, on 7 February 1984, where there is no place to hide from inclement Sun-driven storms. Astronaut Bruce McCandless II (1937–), a mission specialist, wears a 300-pound (136-kilogram) Manned Maneuvering Unit (MMU) with 24 nitrogen gas thrusters and a 35 mm camera. The MMU permits motion in space where the sensation of gravity has vanished, but it does not protect the astronaut from solar flares or coronal mass ejections. High-energy particles resulting from these explosions on the Sun could kill the unprotected astronaut. (Courtesy of NASA.)

Failing to Communicate

Eight minutes after an energetic solar flare, a strong blast of X-rays and extreme ultraviolet radiation reaches the Earth and radically alters the structure of the planet's upper atmosphere, known as the ionosphere, by producing an increase in the amount of free electrons that are no longer attached to atoms. This can cause the ionosphere to absorb the radio signals it usually reflects, resulting in faded signals and sometimes radio blackouts. During moderately intense flares, long-distance radio communications can be temporarily silenced over the Earth's entire sunlit hemisphere.

The radio blackouts are particularly troublesome for the commercial airline industry, which uses radio transmissions for weather, air traffic and location information, and the United States Air Force and Navy are also concerned about this solar threat to radio communications. A solar flare once blacked out contact with a jet carrying President Ronald Reagan to China; for several hours the country's military Commander in Chief was unable to send or receive messages. The Air Force operates a global system of ground-based radio and optical telescopes and taps into the output of national, space-borne X-ray telescopes and particle detectors in order to continuously monitor the Sun for intense flares that might severely disrupt military communications and satellite surveillance.

Space weather interference with radio communication can be avoided by using short-wavelength signals that pass right through the ionosphere, relaying the transmitted signals by satellite. The communication-satellite industry is nevertheless also threatened by the tempestuous Sun. Solar eruptions apparently incapacitated Pan Am Sat's *Galaxy IV* satellite in May 1998, halting pager service to 45 million customers in North America, including doctors, nurses and irate business people. In addition, radio and television stations, including national public radio, could not distribute their programs. Several new Motorola *Iridium* satellites were disabled at nearly the same time, even before they were put into operation.

And communication satellites are not the only ones whose failure can disrupt our lives. We are children of the Space Age, increasingly dependent on many different kinds of Earth-orbiting satellites.

Satellites in Danger

About 1,000 commercial, military and scientific satellites are now in operation, affecting the lives of millions of people, and the performances and lifetimes of all these satellites are affected by Sun-driven space weather. Geosynchronous satellites, which orbit the Earth at the same rate that the planet spins, stay above the same place on Earth to relay and beam down signals used for cellular phones, global positioning systems and internet commerce and data transmission. They can guide missiles or automobiles to their destinations, enable aviation and marine navigation, aid in search and rescue missions, and permit nearly instantaneous money exchange or investment choices. Other satellites revolve around our planet in closer, low-Earth orbits, scanning air, land and sea for environmental change, weather forecasting and military reconnaissance.

Space weather can noticeably increase the atmospheric friction exerted on satellites in low Earth orbit, at altitudes of 300 to 500 kilometers, causing the satellite orbits to decay more quickly than expected. The enhanced extreme ultraviolet and X-ray radiation from solar flares heats the atmosphere and causes it to expand, and similar or greater effects are caused by coronal mass ejections. The expansion of the terrestrial atmosphere brings higher gas densities to a given altitude, increasing the friction and drag exerted on a satellite, pulling it to a lower altitude, and sometimes causing ground controllers to lose contact with them.

Increased atmospheric friction caused by rising solar activity has sent several satellites to a premature, uncontrollable and fatal spiral toward the Earth, including *Skylab* and the *Solar Maximum Mission*. Both spacecraft were ungratefully destroyed by the very phenomenon they were designed to study – solar flares and coronal mass ejections. Space stations have to be periodically boosted in altitude to higher orbit to avoid a similar fate.

Precise monitoring of all orbiting objects depends on accurate knowledge of atmospheric change caused by storms from the Sun. The U.S. Space Command, for example, often has to recompute the orbits of many hundreds of low-Earth-orbit objects affected by the increase in atmospheric friction.

At higher altitudes, above low-Earth orbit, geosynchronous satellites are endangered by the coronal mass ejections that cause intense geomagnetic storms. These satellites orbit our planet at about 6.6 Earth radii, or about 40,000 kilometers, moving around the Earth once every 24 hours. A coronal mass ejection can compress the magnetosphere from its usual location at about 10 Earth radii to below the satellites' synchronous orbits, exposing them to the full brunt of the gusty solar wind and its charged, energized ingredients.

The Van Allen radiation belts provide a persistent, ever-present threat to high-flying satellites. The energetic electrons trapped in the radiation belts can move right through the thin metallic skin of a spacecraft, damaging the delicate microchip electronics inside. Moreover, when intense solar storms buffet the magnetosphere, they can accelerate the particles trapped in the radiation belts, greatly increasing the amounts of dangerous high-energy electrons. Metal shielding and radiation-hardened computer chips are used to guard against this recurrent hazard, and satellite orbits can be designed to minimize time in the radiation belts, or to avoid them altogether.

Nothing can be done to shield the solar cells used to power nearly all Earth-orbiting satellites; the photovoltaic cells convert sunlight to electricity and therefore have to be exposed to space. The danger was first realized back in 1962, when the United States exploded a 1.4-megaton nuclear bomb, called *Starfish,* about 500 kilometers up in the atmosphere. The explosion increased the energy of the particles in the radiation belts and created new ones. They wiped out the solar arrays of several satellites at the time. As satellites repeatedly pass through the Earth's natural radiation belts, exposure to its energetic particles slowly deteriorates and shortens the useful lives of their solar cells.

The recurrent threat of moving within the radiation belts is particularly acute for satellites in low orbits that pass through the South Atlantic Anomaly, which is caused by a displacement of the Earth's magnetic center by about 500 kilometers from the planet's center. As a result, particles trapped in the inner radiation belt can approach closer to the Earth's surface above the South Atlantic than elsewhere, so this region is anomalous because of its proximity. To avoid malfunctions, the instruments aboard the *Hubble Space Telescope* and some other NASA satellites have had to be shut down each time they repeatedly pass through the South Atlantic Anomaly.

Infrequent, exceptionally large eruptions on the Sun can hurl very energetic protons toward the Earth and elsewhere in space. The solar protons can enter a spacecraft like ghosts, producing erroneous commands and crippling their microelectronics. Such single event upsets have already destroyed at least one weather satellite and disabled several communications satellites. Space weapons can produce a similar effect; so if you didn't know the Sun was at fault, you might think someone was trying to shoot down our satellites.

To put the space-weather threat in perspective, just a few satellites have been lost to storms from the Sun out of thousands deployed. And the U.S. military is more concerned with disruption of radio signals, since they build satellites that can withstand the effects of a nuclear bomb exploded in space. The commercial satellite industry, which constructs satellites that are less expensive and more vulnerable, may not want to recognize the problem, since natural disasters, including those from the Sun, are not covered by insurance policies, but engineering failures are.

A Wired World

The whole Earth has become wired together, first with telegraph wires, then by telephone lines and electrical-power grids. And disabling electrical currents and voltages can be produced in the wires when solar storms produce changes in the Earth's magnetism. This threat is greatest in high-latitude regions where the currents are strongest, such as Canada, the northern United States and Scandinavia.

Even back in the 1840s, when telegraph lines were first deployed, operators noticed extra current whenever overhead auroras signaled the presence of an intense geomagnetic storm. And about a century and a half later, on 13 March 1989, a particularly severe geomagnetic storm, produced by a coronal mass ejection, plunged virtually all of the Canadian province of Quebec into complete darkness without warning and within a few seconds. Six million customers were without electricity for over nine hours, in the middle of a frigid winter night, costing around 500 million dollars counting losses only from unserved demand. The disturbed magnetic fields induced electric currents in the Earth's surface, which in turn created voltage surges on the long-distance power lines, blowing circuit breakers, overheating or melting the windings of transformers, and causing the massive electrical failure.

As demand for electricity increases, utility companies rely more and more on large, interconnected grids of power transmission lines that can span continents, providing rapid response to the diverse energy demands of users scattered throughout the world. In the United States alone, nearly a million kilometers of electrical transmission lines connect more than ten thousand power stations. Such power distribution systems are becoming increasingly vulnerable to severe geomagnetic storms, initiated when a coronal mass ejection with the right magnetic orientation plugs into the magnetosphere. They can plunge major urban centers, like New York City or Montreal, into complete darkness, causing social chaos and threatening safety. The threat doesn't occur very often, perhaps once a year, but the potential consequences are serious enough to employ early warning systems.

Here Comes the Sun

Space weather is here to stay, and the dangers blowing in the Sun's wind are not going away. Humans are spending more and more time in space, while those on the ground become increasingly dependent on satellites that whiz over their heads. In tens of minutes, intense explosions hurl out energetic particles that can endanger astronauts, and their survival may depend on how well one can predict space weather. Forceful solar flares or mass ejections can damage or destroy Earth-orbiting satellites, increase cancer risk for people using commercial airlines over polar routes, and create power surges that can blackout entire cities.

Recognizing our vulnerability, astronomers use telescopes on the ground and *in situ* particle detectors or remote-sensing telescopes on satellites to carefully monitor the

Sun, and government agencies post forecasts that warn of threatening solar activity. This enables evasive action that can reduce disruption or damage to communications, defense and weather satellites, as well as electrical power systems on the ground. Once we know a Sun storm is on its way, the launch of manned space flight missions can be postponed, and walks outside spacecraft or on the Moon or Mars might be delayed. Airplane pilots can be warned of potential radio communication failures. Operators can power down sensitive electronics on communication and navigation satellites, putting them to sleep until the danger passes. Utility companies can reduce load in anticipation of trouble on power lines, in that way trading a temporary "brown out" for a potentially disastrous "black out."

What everyone wants to know is how strong the storm is and when it is going to hit us. Like winter storms on Earth, some of the effects can be predicted days in advance. A coronal mass ejection, for example, arrives at the Earth one to four days after leaving the Sun, and solar astronomers can watch it leave the Sun. The *STEREO* mission will additionally track coronal mass ejections headed for Earth.

Solar flares are another matter. A soon as you see a flare on the Sun, its radiation and fastest particle have already reached us, taking just 8 minutes to travel from the Sun to Earth. One promising technique is to watch to see when the solar magnetism has become twisted into a stressed situation, for it may then be about to release a solar flare. Another one is to look through the Sun and see active regions develop before they rotate to face the Earth. Both of these methods of predicting explosions on the Sun were discussed in Section 7.7.

So we now turn to our life-sustaining atmosphere, which provides the oxygen we breathe and the warmth that keeps the oceans from freezing. Our dynamic atmosphere is being transformed by the Sun above and by humans below.

Chapter Nine

Transforming the Earth's Life-Sustaining Atmosphere

COSMIC SYNCHROMY In this painting, the American artist Morgan Russell (1886–1953) develops the theme found in Robert Delaunay's (1885–1941) portrayal of the Sun and Moon (see frontispiece to Chap. 3), extending arcs of pure color into space. They seem to resonate throughout the Universe like myriads of cosmic rainbows or countless Suns. (Courtesy of Muson-Williams-Proctor Arts Institute, Museum of Art, Utica, New York, Oil on canvas, 16.25 × 13.25 inches.)

9.1 FRAGILE PLANET EARTH – THE VIEW FROM SPACE

About four decades ago astronauts bound for the Moon looked back at our planet, seeing it suspended there all alone, swinging through the chill of outer space. For the first time we saw our home as a glistening blue and turquoise ball, light and round and shimmering like a bubble, flecked with delicate white clouds (Fig. 9.1). This perspective created a worldwide awareness of the Earth as a unique and vulnerable place, a tiny, fragile oasis in space.

The astronomer Fred Hoyle (1915–2001) anticipated such an impact, declaring in 1948 that:

> Once a photograph of the Earth, taken from outside, is available, we shall, in an emotional sense, acquire an additional dimension. . . . Once let the sheer isolation of the Earth become plain to every man whatever his nationality or creed, and a new idea as powerful as any in history will be let loose.[39]

To our ancestors only a few centuries ago, the ocean, land and sky seemed vast and almost limitless. But all that has changed. Today, there is no large unexplored territory on Earth, no new frontier except space itself.

And it is only now that we can see it from space that we realize the magnitude of what we are doing to the Earth. Indeed, it is probably no accident that the advent of the Space Age has coincided with a growing concern about the terrestrial environment and the precarious fate of life within it.

Only from space can we see the living planet Earth as a whole, as a single, unified system that encompasses life on land and in the sea (Fig. 9.2). Spacecraft monitor its vital signs, zooming in with telescopic instruments to take a magnified birds-eye view of the Earth's ever-changing surface (Figs. 9.3, 9.4).

Astronauts view the horizon as a curved line, showing a thin membrane of air that ventilates, protects and incubates us (Fig. 9.5). As astronaut James Buchli (1945–) described it:

> Look at how thin the atmosphere is. . . . Everything beyond that thin blue line is the void of space. And everything below it is what it takes to sustain life. And everything that we do to this environment, and our quality of life, is below that little thin blue line. That's the only difference between what we enjoy here on Earth and the really harsh uninhabitable blackness of space. It's not very wide, is it?[40]

The Earth's diameter is one thousand times greater than the width of the atmosphere surrounding it. Indeed, the distance from the ground to the top of the sky is only about 12 kilometers, or no farther than you might run in an hour. And this narrow, changing atmosphere is linked to the Sun by an energy flow that sustains life.

9.2 CLEAR SKIES AND STORMY WEATHER

Look up! The clear air turns into the blue sky! This is because air molecules scatter blue sunlight more strongly than other colors. The night sky is black because there is no sunlight to illuminate the air. And the sky on the Moon is black even when the Sun shines on it, for the Moon has no atmosphere to scatter sunlight.

FIG. 9.1 The water planet Almost three-quarters of the Earth's surface is covered with water, as seen in this view of the North Pacific Ocean. Earth is the only planet in the Solar System where substantial amounts of water exist in all three possible forms – gas (water vapor), liquid and solid (ice). Here white clouds of water ice swirl just below Alaska; the predominantly white ground area, consisting of snow and ice, is the Kamchatka Peninsula of Siberia. Japan appears near the horizon. From this orientation in space, we also see both the day and night sides of our home planet. (Courtesy of NASA.)

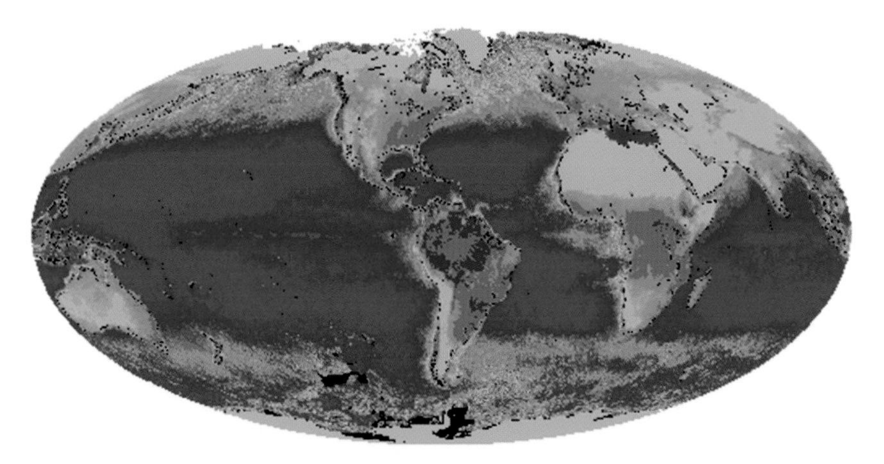

FIG. 9.2 Global biosphere This global view of the Earth's biosphere portrays life on land and sea. The ocean portion shows the chlorophyll concentration arising from phytoplankton, microscopic plants that grow in the upper sunlit portions of the ocean and are the ultimate source of food for most marine life. At sea, the red and orange colors denote the highest concentration of plankton; yellow and green represent areas of moderate plankton concentration, and blue and violet describe the lowest concentrations. The land vegetation image shows rain forests *(dark green areas),* tropical and sub-tropical forests *(light green)* and areas of low vegetation and deserts *(yellow).* This image combines ocean measurements from the Coastal Zone Color Scanner, abbreviated CZCS, aboard the *Nimbus-7* satellite, collected from 1978 to 1986, and three years of land measurements from the Advanced Very High Resolution Radiometer, or AVHRR for short, flown on *NOAA-7*, which was launched on 7 June 1981. (Courtesy of G.C. Feldman and C.J. Tucker, NASA.)

When high in the sky, the Sun is colored yellow. At sunset the Sun's rays pass through a maximum amount of atmosphere; most of the blue light is then scattered out of our viewing direction, and the setting Sun is colored red. Dust in the air also helps redden sunsets.

Over the eons, our planet has been shaped and re-shaped by nature's powerful forces, and "air-conditioned" by several types of cycles. In the oxygen cycle, animals fill their lungs with oxygen, breathing it in from the air, while plants replenish the oxygen in our atmosphere.

In another global cycle, atmospheric carbon dioxide is dissolved in the ocean waters and taken up by plants; animals, volcanoes and burning coal, oil and natural gas return carbon dioxide to the air, completing the cycle. Between 350 million and 65 million years ago, great quantities of carbon were stored deep underground, when plants and animals died and decayed and their remains were compressed into what eventually became coal, oil and natural gas. This carbon was removed from the carbon cycle, until about 200 years ago when humans started using the fossil fuels and releasing carbon dioxide into the atmosphere again.

The most obvious of these grand circulations is the water cycle, which is powered by the Sun's energy and involves interactions between the oceans, atmosphere and

FIG. 9.3 Altamaha river delta, Georgia The history of the Sea Islands in the Altamaha River delta on the coast of Georgia is revealed in this radar image. The outlines of long-lost plantation rice fields, canals, dikes and other inlets are defined. Salt marshes are shown in red, while dense cypress and live oak tree canopies are seen in yellow-greens. Agricultural development of the Altamaha delta began soon after the founding of the Georgia Colony in 1733. The first major crop was indigo, followed by rice and cotton. A major storm in 1824 destroyed much of the town of Darien (*upper right*). The Civil War (1861–1865) ended the plantation system, and many of the island plantations disappeared under heavy brush and new growth pine forests. The Butler Island (*center left*) plantation became a wildlife conservation site growing wild sea rice for migrating ducks and other waterfowl. (AIRborne Synthetic Aperture Radar, abbreviated AIRSAR, image courtesy of NASA, JPL and the University of Edinburgh.)

land. It brings us our daily weather and produces the climatic differences that starkly distinguish one part of the globe from another.

About one-third of the solar energy reaching the Earth's surface as sunlight is expended on the evaporation of seawater. This evaporation releases warm fresh-water moisture into the air and cools the surface of the ocean. The moisture rises high into the cold atmosphere, where clouds are formed. Winds drive the clouds for great distances over sea and land. Rain or snow from the clouds can then fall to land, refreshing streams, lakes and underground reservoirs. The water then runs down to the sea, where the cycle begins once more.

All of the water in the oceans passes through this water cycle once in 2 million years. Yet, the ocean waters are at least 3.5 billion years old, so they have, on average, completed more than a thousand such cycles.

Sunlight illuminates and warms one side of our rotating globe at a time. And because the Earth is spherical, the tropical regions near the equator face the Sun more directly, receiving the greatest amount of heat. At these low latitudes, the Sun's almost vertical rays travel to the ground through the least amount of intervening air. At higher latitudes nearer

FIG. 9.4 New Guinea This Synthetic Imaging Radar, abbreviated SIR, image shows rivers emptying into the sea on the southern coast of the Indonesian half of New Guinea. It was taken from SIR-A aboard the *Space Shuttle Columbia* in 1981, demonstrating the use of imaging radar to acquire and transmit data of different geologic regions. (Courtesy of NASA and JPL.)

the poles, sunlight strikes the ground at a glancing angle and it must also penetrate a greater thickness of absorbing atmosphere, so the ground is heated less than at the equator.

Winds and ocean currents attempt to correct the temperature imbalance, carrying heat from the equator to the poles, or from the warmer to colder places, in both northern and southern directions. Warm air rises at the equator, circulates toward the poles, eventually cools and sinks, and then flows back toward the equator at lower levels near the ground. Two loops of flowing air are formed, one in each hemisphere; and they are named Hadley cells after the English meteorologist George Hadley (1685–1768) who first proposed them in 1735. The two Hadley cells circulate north or south, but they are deflected sideways by the Earth's rotation, forming the low-latitude trade winds that blow westward almost every day of the year.

But climate isn't just determined by how the air circulates. The land and oceans also play a role. Warm water, for example, flows from the tropics to the Arctic and Antarctic, helping move heat around the globe and determining world climate.

And in another of nature's grand transformations, the great landmasses on Earth are continually reorganizing, growing in places and eroding away or breaking apart in others. Over long periods of time, measured in millions of years, entire continents and oceans can be destroyed or created anew, remodeling the entire surface of the Earth.

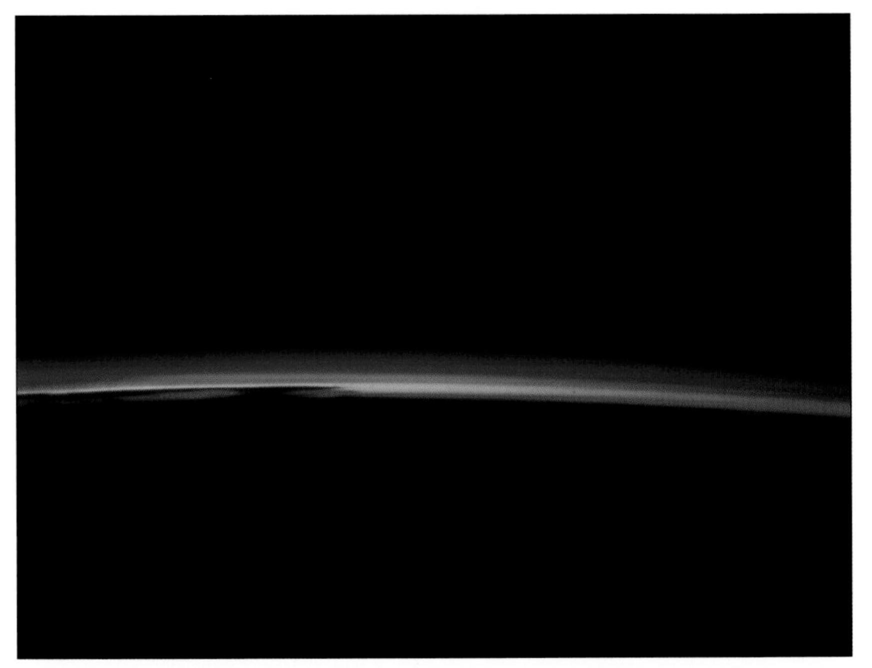

FIG. 9.5 A thin colored line Brilliant red and blue mark the thin atmosphere that warms and protects us, as viewed from space at sunrise over the Pacific Ocean. Without this atmospheric membrane we could not breathe and our lake and ocean waters would freeze. (Courtesy of NASA.)

This results in a changing flow of air and ocean currents that create entirely different global climate patterns.

Nowadays, and in all former times, it is the Sun-driven seasons that dominate our weather. They are created by the tilt of the Earth's rotational axis, together with our planet's annual circuit around the Sun. In each hemisphere, the greatest sunward tilt defines the summer part of its orbit when the Sun is more nearly overhead and its rays strike the surface more directly than in the winter part when that hemisphere is tilted away from the Sun (Fig. 9.6). In fact, the word climate comes from the Greek word *klima,* for "tilt."

This annual change in solar radiation produces large seasonal temperature fluctuations in the Northern Hemisphere where most of the world's land is now found. The difference in surface air temperature, averaged over the northern hemisphere, between winter and summer, is 15 kelvin. Temperatures measured on the more familiar Celsius scale, denoted C, are given by $C = K - 273$, where K denotes the temperature in kelvin units. So a temperature difference of 15 kelvin is also a temperature difference of 15 degrees Centigrade, and it is this difference that brings the climatic seasons, which mark the passage of time.

Without both sunlight and our thin atmosphere, there would be no blue skies or red sunsets, no clouds or rain, and no climates or seasons. And without our atmosphere, the oceans would be frozen solid.

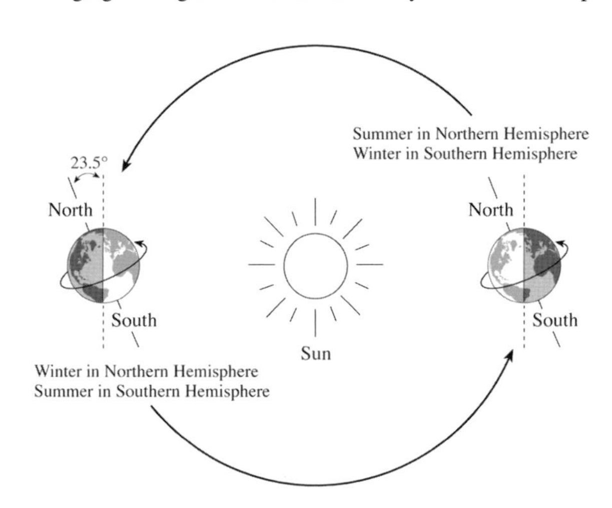

FIG. 9.6 The seasons As the Earth orbits the Sun, the Earth's rotational axis in a given hemisphere is tilted toward or away from the Sun. This variable tilt produces the seasons by changing the angle at which the Sun's rays strike different parts of the Earth's surface. The greatest sunward tilt occurs in the summer when the Sun's rays strike the surface most directly. In the winter, the relevant hemisphere is tilted away from the Sun and the Sun's rays obliquely strike the surface. When it is summer in the northern hemisphere, it is winter in the southern hemisphere and *vice versa.* (Notice that the radius of the Earth and Sun and the Earth's orbit are not drawn to scale.)

9.3 THE LIFE-SAVING GREENHOUSE EFFECT

Our planet's surface is now comfortably warm because the atmosphere traps some of the Sun's heat and keeps it near the surface. The thin blanket of gas acts like a one-way filter, allowing sunlight through to warm the surface, but preventing the escape of some of the heat into surrounding space and raising the temperature of the planet. This warming influence has been dubbed the "natural" greenhouse effect to distinguish it from "enhanced" greenhouse warming caused by "unnatural" human activity.

Right now, the "natural" greenhouse is literally a matter of life and death. Without an atmosphere, the Earth's surface would be warmed by the Sun to just 255 kelvin, which is below the freezing temperature of water at 273 kelvin. But the natural greenhouse effect of the atmosphere keeps the planet 33 kelvin hotter than it would otherwise be, raising the surface temperature to a more hospitable global average of 288 kelvin. Without this greenhouse warming, the oceans would freeze and life as we know it would not exist.

The idea that our atmospheric blanket might warm the Earth was suggested in 1827 by the French mathematician Jean-Baptiste Fourier (1768–1830) and developed by the Irish scientist John Tyndall (1820–1893) in the 1860s. Fourier is now well known for his mathematical investigations of waves and other periodic motion, a branch of mathematics now known as Fourier analysis.

Fourier wondered how the Sun's heat could be retained to keep the Earth hot. And his basic idea, still accepted today, is that the atmosphere lets some of the Sun's radiation in, but it doesn't let all of the radiation back out. Visible sunlight passes through our transparent atmosphere to warm the Earth's land and oceans, and some of this heat is reradiated in infrared form. The longer infrared rays are less energetic than visible ones and do not slice through the atmosphere as easily as visible light.

So our atmosphere absorbs some of the infrared heat radiation from the ground, and some of the trapped heat is reradiated downward to warm the planet's surface and the air immediately above it. Fourier likened the thin atmospheric blanket to a huge glass bell jar, made out of clouds and gases, that holds the Earth's heat close to its surface.

The warming by heat-trapping gases in the air is now known as the greenhouse effect, but this is a misnomer. The air inside a garden greenhouse is heated because it is enclosed, preventing the circulation of air currents that would carry away heat and cool the interior. Nevertheless, the term is now so common that we continue to use it to designate the process by which an atmosphere traps heat near a planet's surface.

Tyndall built an instrument to measure the heat-trapping properties of various gases, examining the transmission of infrared heat radiation through them. He found that the main constituents of our atmosphere – oxygen (21 percent) and nitrogen (78 percent) – were transparent to both visible and infrared radiation. These diatomic, or two-atom, molecules are incapable of absorbing any noticeable amounts of infrared heat radiation.

He also found that water vapor and carbon dioxide, which are minor ingredients of the air, absorb significant heat. As Tyndall realized, these gases are transparent to sunlight, which warms the ground, but partially opaque to the infrared heat rays, which are trapped near the surface and warm our globe. As he wrote in 1861:

> This aqueous vapour, which exercises such a destructive action on the obscure [infrared heat] rays, is comparatively transparent to the rays of light. Hence the differential action, as regards the heat coming from the Sun to the Earth and that radiated from the Earth into space, is vastly augmented by the aqueous vapor of the atmosphere.... Similar remarks would apply to the carbonic acid [carbon dioxide] diffused through the air.[41]

Water vapor and carbon dioxide molecules, which consist of three atoms, are more flexible and free to move in more ways than diatomic molecules, so they absorb the heat radiation.

If the heat-trapping gases were removed from the air for a single night, Tyndall announced, then "the warmth of our fields and gardens would pour itself unrequited into space, and the Sun would rise upon an island held fast in the iron grip of frost."

The Earth might become a frozen globe without the heat-trapping gases, and it could also become noticeably warmer if their amounts increased. The Swedish chemist, Svante Arrhenius (1859–1927) set out to find out what would happen if the amount of carbon dioxide was either cut in half or doubled. After nearly a year of tedious calculations, he concluded that these variations could produce either the ice ages or the warmest intervals between them. In his own words, written in 1896:

> A simple calculation shows that the [ground] temperature in the Arctic regions would rise 8 to 9 degrees Celsius if the carbonic acid [carbon dioxide] increased to 2.5 or 3 times its present value. In order to get the temperature of the ice age, the carbonic acid in the air should sink to 0.65 to 0.55 of its present value, lowering the temperature by 4 to 5 degrees Celsius.[42]

We now believe that changes in the global distribution of sunlight bring on the ice ages. Nevertheless, Arrhenius made a good estimate of the global warming effect of doubling the amount of carbon dioxide in our air, about 5 degrees Celsius or 5 kelvin, without the use of modern calculators or extensive computer models. He also pointed out that industrial activity was then noticeably increasing the amount of atmospheric carbon dioxide, and that humans were therefore altering the temperature of the globe.

Why doesn't the atmosphere just keep heating up until it explodes? Provided that the Sun's radiation output stays constant, and that the composition of the atmosphere does not change, the greenhouse warming rises to a fixed temperature that balances the heat input from sunlight and the heat radiated into space. The level of water in a pond similarly remains much the same even though water is running in one end and out the other.

With greater energy from the Sun, or with more water vapor or carbon dioxide in the air, the balance point will shift to a higher average global temperature. This is why it becomes hotter in the summer when more sunlight directly illuminates the Earth's surface. It's also why cloudy nights tend to be warmer than clear nights; escaping infrared heat radiation is blocked by the water vapor in the clouds, keeping the ground warm at night.

Of course, you can have too much of a good thing, like lying in the Sun all day in the summer or eating too much ice cream too fast. The surface of Venus, under a dense atmosphere of carbon dioxide, has become hot enough to melt lead. The high-temperature world has been boiled dry, like a kettle left on the stove. Nothing could live there. And Mars has only a feeble atmosphere with a meager greenhouse effect. This frozen world is locked in a permanent ice age.

9.4 THE EARTH'S CHANGING ATMOSPHERE AND OCEANS

Nothing in the Cosmos is fixed and unchanging; everything moves and evolves. People live out their changing lives; stars are born and die; the entire Universe is always changing, expanding from its explosive birth long ago. Sometimes the changes are slow and gradual; often they are quick and violent. In fact, the only unchanging thing in the Universe is change itself, and the Earth's atmosphere is no exception.

It is thought that the Earth and other planets formed, together with the Sun, about 4.6 billion years ago. As the body of the newborn Earth cooled on the outside, and warmed up inside, the early atmosphere most likely emerged from ancient volcanoes that spewed lava and released gases that were once trapped in the rock. Cosmic debris, attracted by the gravitational forces of the young Earth, probably slammed into its surface, vaporizing rocks, liberating gas and further enhancing the primitive air.

Because of the energy released by the colliding rocks that merged to form the Earth, it probably began as a molten globe, too hot for any water to accumulate on its surface. Only when the initial bombardment slowed, and the glowing planet cooled, could the oceans form. Their water has remained for billions of years, cycling into clouds and rains and shifting about the globe as continents split open or merged together to open new oceans and close old ones.

Astronomers argue that our oceans were supplied from outside the Earth after it first came together, perhaps by small bodies of rock or ice, similar to today's asteroids and comets, coming in from the cold outer parts of the Solar System. Some geologists reason that the ancient oceans were instead steamed out of the Earth's interior by erupting volcanoes, condensing from water vapor in the primitive atmosphere. No one knows for sure whether external impact or internal volcanism resulted in most of our water. But since the Earth formed by the accumulation of colliding objects, both the atmosphere and the water expelled by the volcanoes had to be originally supplied by cosmic impact.

The Sun began its life shining with only 70 percent of its present luminosity, slowly growing in luminous intensity as it aged. This inexorable increase in the Sun's brightness is a consequence of increasing amounts of helium in the Sun's core; the greater mean density produces higher core temperatures, faster nuclear reactions, and a steady increase in luminosity.

If the Sun were much dimmer eons ago, the Earth's oceans would have been frozen solid even if the present greenhouse effect operated back then. Yet, sedimentary rocks, which must have been deposited in liquid water, date from 3.8 billion years ago. Since fossil evidence suggests that living things have thrived in a warm, liquid environment since that time, the Earth's surface temperature has remained about the same for billions of years, even though the Sun's brightness has increased by some 30 percent since its creation.

The discrepancy between the warm climate record and the predictions of an early cold climate based on the Sun's faint luminosity has come to be known as the faint-young-Sun paradox. It can be resolved if the Earth's early primitive atmosphere contained about a thousand times more carbon dioxide than it does now. The greater heating of the enhanced greenhouse effect could have kept the ancient oceans from freezing. The Earth could subsequently maintain a temperate climate only if its atmosphere, rocks and oceans combined to decrease the amount of carbon dioxide over time, turning down the planet's greenhouse effect as the Sun grew brighter.

Plants and animals also play a role in transforming the atmosphere and determining its composition. Billions of years ago, when the young Earth's volcanic surface was too hot for living things, there must have been very little oxygen in the atmosphere. It was created by the early bloom of life and accumulated slowly. Blue-green algae in the sea began to put oxygen in the air about 2 billion years ago; by about 400 million years ago, they had pumped sufficient oxygen into the atmosphere to sustain animals that breathe oxygen and live on land. Free oxygen is now a major ingredient of the atmosphere, making up about a fifth of our air.

If plants did not continuously replenish the oxygen, living things that breathe it, including humans, would eventually deplete it. Even without animals, the oxygen gas would react chemically with other elements and be locked away in compounds like carbon dioxide and water or within rocks in about 4 million years. However, because ours is a living world, chemically reactive oxygen remains in the air.

Life may even have developed the capability to control its environment, changing the physical world of rocks, oceans and atmosphere to maintain comfortable conditions for living things in spite of threatening changes. This theory, developed by British inventor and scientist James Lovelock (1919–) and the American microbiologist Lynn

Margulis (1938–), has been called the Gaia (pronounced GUY-ah) hypothesis, after the Greek goddess of mother Earth. Lovelock's neighbor, William Golding (1911–1993), author of the book *Lord of the Flies* and recipient of the 1983 Nobel Prize in Literature, suggested the name.

But there is probably no escape in the end. Whatever life on Earth does, its remote future is not secure! As the Sun brightens and turns on the heat, the Earth's oceans will eventually be boiled away, leaving the planet a burned-out cinder, a dead and sterile place billions of years from now.

So, our long-term prospects aren't all that great. Moreover, there are threats on a much shorter time interval that may be caused by human tinkering with the environment. To understand these effects, we must first consider how variable solar radiation transforms our air, heating it up and producing distinct layers within it.

9.5 OUR SUN-LAYERED ATMOSPHERE

The Invisible Atmosphere

On a clear day you can see forever, or at least for a long distance through the air. And on a warm, dry, windless day we are unaware of the touch of the air upon our skin. The drift of a cloud occasionally reminds us of the atmosphere about us, and on cold days we feel the air against our skin. The touch of the wind and the sight of birds and airplanes supported by their motion prove that there is something substantial surrounding us.

Our thin atmosphere is pulled close to the Earth by its gravity, and suspended above the ground by molecular motion. And because air molecules are mainly far apart, our atmosphere is mostly empty space, and it can always be squeezed into a smaller volume. The atmosphere near the ground is compacted to its greatest density and pressure by the weight of the overlying air. Yet, even at the bottom of the atmosphere the density is only about a thousandth of that of liquid water; an entire liter of this air will weigh only one gram.

At greater heights there is less air pushing down from above, so the compression is less and the density of air gradually falls off into the vacuum of space. At a height of 10 kilometers, slightly higher than Mount Everest, the density of air has dropped to 10 percent of its value near the ground. No insects and few birds can fly in such rarefied air. At altitudes above about 50 kilometers, the air is too thin to support a jet airplane.

We can infer the thinning of the air at greater heights by watching a group of hawks circling above a meadow or open field. As air near the warm ground heats, it expands and becomes less dense, rising into the thinner atmosphere above. The rising air carries heat from the ground and distributes it at higher levels, giving free rides to the soaring birds. Hawks riding the currents of heated air sometimes rise so abruptly that it looks as if they were lifted and jerked up by strings.

But why isn't the sky falling, as Chicken Little would have it, if gravity is relentlessly pulling it down? Because the air is warmed by the Sun, its molecules are in continuous motion and collide with each other, producing a pressure that prevents the atmosphere from collapsing to the ground.

The air is thinner at higher altitudes so there are fewer molecular collisions and less air pressure. The decrease in air pressure with height accounts for the rise of balloons. When filled with a light gas, a balloon is buoyed upward by the greater pressure of the air beneath it. If the upward force of buoyancy exactly matches the downward weight of the balloon and its contents, the balloon will remain suspended at the same altitude, moving neither up nor down.

Not only does the atmospheric pressure decrease as we go upward, the temperature of the air also changes, but it is not a simple fall-off. It falls and rises in two full cycles as we move off into space (Fig. 9.7).

The Troposphere and Stratosphere

The temperature decreases steadily with height in the lowest regions of our atmosphere. When warm currents move up from the Earth's surface, the air expands in the lower pressure and becomes cooler. Many of us have experienced the cold temperatures of mountain altitudes. The average air temperature drops below the freezing point of water (0 degrees Celsius or 273 kelvin) at only a kilometer or two above the Earth's surface, and bottoms out at a height of about 12 kilometers above sea level. This is the greatest height achieved by the air currents. All our weather occurs below this altitude, controlled by visible sunlight.

Global atmospheric circulation, driven by differential solar heating of the equatorial and polar surface, create complex, wheeling patterns of weather in this region, leading to the designation troposphere, from the Greek *tropo* for "turning." The average extent of the ground-hugging troposphere varies with latitude, from about 16 kilometers above the warm equator to roughly 8 kilometers over the cold poles.

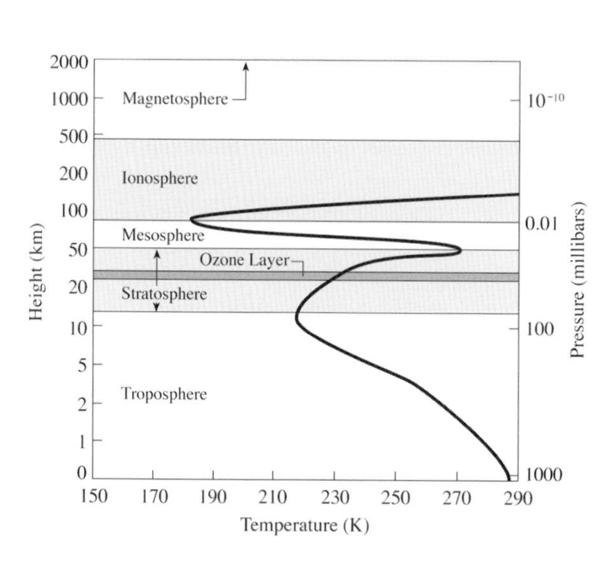

FIG. 9.7 Sun-layered atmosphere The pressure of our atmosphere (*right scale*) decreases with height (*left scale*). This is because fewer particles are able to overcome the Earth's gravitational pull and reach greater heights. The temperature (*bottom scale*) also decreases steadily with height in the ground-hugging troposphere, but the temperature increases in two higher layers that are heated by the Sun. They are the stratosphere, with its critical ozone layer, and the ionosphere. The stratosphere is mainly heated by ultraviolet radiation from the Sun, and the ionosphere is created and modulated by the Sun's X-ray and extreme-ultraviolet radiation. A pressure of 1 millibar is equivalent to 0.001 bar and 100 pascal.

The top of the troposphere was detected near the end of the nineteenth century when Leon Philippe Teisserenc de Bort, (1855–1913), a French meteorologist, launched hundreds of unmanned balloons that carried thermometers and barometers to altitudes as great as 15 kilometers. At this height, the temperature no longer decreased with altitude, and seemed to remain nearly constant. If the temperature was unchanging, the ingredients of the atmosphere above the troposphere might settle down into layers, or strata, depending on their weight, so de Bort named this region the stratosphere.

Contrary to everyone's expectations, the temperature increases at greater heights within the stratosphere, rising to nearly ground-level temperatures at about 50 kilometers above the Earth's surface; but we still use the name stratosphere to designate the layer of the atmosphere that lies immediately above the troposphere. The Sun's invisible ultraviolet radiation is largely absorbed in the stratosphere, where the radiation warms the gas and helps make ozone.

The booming artillery guns of World War I suggested the rise in temperature within the stratosphere. They were heard at unexpectedly large distances, indicating that their sound waves were being bent around by the stratosphere and reflected or mirrored at an angle back down to the ground. Sound travels in much the same way for large distances across the cool surface of a lake in summertime, reflected by the warm air above it.

Above the stratosphere we come to the mesosphere, from the Greek *meso* for "intermediate." The temperature declines rapidly with increasing height in the mesosphere, from temperatures of about 265 kelvin at 50 kilometers altitude to far below freezing at about 85 kilometers, where the temperature reaches the lowest levels in the entire atmosphere.

The mesosphere has been known as the "ignorosphere" because it is too high to be reached by airplanes and too low to be studied by most spacecraft. The air at this height is too thin to support research balloons or aircraft, but thick enough for air drag to cause satellites to decay quickly from orbit. Sounding rockets pass through this region too rapidly to permit detailed study.

The Ionosphere

On 12 December 1901, the Italian electrical engineer Guglielmo Marconi (1874–1937) startled the world by transmitting a radio signal in Morse code across the Atlantic, from England to Newfoundland. Marconi became an international hero, established the American Marconi Company, which later evolved into the Radio Corporation of America, abbreviated RCA, and in 1909 received the Nobel Prize in Physics, jointly with the German physicist Karl Ferdinand Braun (1850–1918) for their development of wireless telegraphy.

Because radio waves travel in straight lines, and cannot pass through the solid Earth, no one expected that Marconi could send a radio signal halfway around the world. Radio waves get around the Earth's curvature by reflection from an electrically charged layer, now called the ionosphere, extending into space from roughly 70 kilometers above the Earth's surface. The atoms in the ionosphere are highly ionized, and many of their electrons have therefore been set free from atomic bonds. These electrons give the ionosphere a high electrical conductivity, which enables them to turn the radio signals back toward the ground, reflecting them like a metal mirror and not allowing the radio signals to pass through.

Arthur E. Kennelly (1861–1939), then at the Harvard School of Engineering, and Oliver Heaviside (1850–1925) in England, independently provided this explanation in 1902. The so-called Heaviside layer in the ionosphere achieved renown after 1981 when the musical *Cats* opened, including Andrew Lloyd Webber's (1948–) song *Journey to the Heaviside Layer.* As a child, Lloyd Webber was fascinated by the American poet T. S. Eliot's (1888–1965) poems about the secret lives of cats, collected in 1939 in *Old Possum's Book of Practical Cats,* and an unpublished poem *Grizabella the Glamor Cat,* presented to Lloyd Weber by the poet's widow Valerie Eliot, described the ascent to Cat Heaven at the Heaviside Layer.

The rapid expansion of radio broadcasting in the 1920s, as well as the concurrent development of pulsed radio signals, helped specify the structure and daily variation of the ionosphere. Edward Appleton (1892–1965) and his students measured the height of the reflecting layer by determining the elapsed time between transmitting a radio pulse and receiving its echo; like all electromagnetic radiation, the radio waves travel at the velocity of light. They showed that there are at least three such reflecting layers, now labeled D, E and F, at respective altitudes of 70, 100 and 200 to 300 kilometers. The three layers collectively account for much of what we today call the ionosphere. In 1947 Sir Appleton received the Nobel Prize in Physics for his investigations of the upper atmosphere, especially for the discovery of the so-called Appleton layers.

The ionosphere does not mirror radio waves unless their wavelength is longer than a certain value. This critical wavelength provides a measure of the electron density in the ionosphere; the shorter it is the larger the density. Radio waves with wavelengths that are less than the critical value can pass right through the ionosphere, because, roughly speaking, they are short enough to pass among the electrons. These shorter wavelengths are used in communications with satellites or other spacecraft in outer space beyond the ionosphere.

The mystery of exactly what produces and controls the ionosphere was not solved until after World War II, when captured German V-2 rockets were brought to the United States. These and subsequent rockets, built by American engineers, were used by the Naval Research Laboratory to loft spectrographs above the atmosphere, showing that the Sun emits very energetic radiation at invisible X-ray and extreme ultraviolet wavelengths. When this radiation reaches the upper atmosphere it breaks the nitrogen and oxygen molecules into their constituent atoms and ionizes them, producing free electrons and atomic ions. The ionosphere above your head therefore develops as the Sun rises and decays as the Sun sets; it lingers on during the night but is not energized then.

The process of ionization by the Sun's invisible rays releases heat to warm the ionosphere, so the temperature rises with altitude in it. In the ionosphere, the temperatures skyrocket to higher values than anywhere else in the entire atmosphere. Indeed, some scientists prefer to call this region the thermosphere, or "hot" sphere. The thermosphere overlaps the E and F regions of the ionosphere, beginning at about 90 kilometers and extending upward to about 500 kilometers.

At higher altitudes, the atmosphere thins out into the *exosphere,* or the "outside sphere." In the exosphere the gas density is so low that an atom can completely orbit the

Earth without colliding with another atom. The temperature is so hot out there, and the atoms move so fast, that some atoms can escape the Earth's gravitational pull and travel into outer space. It is therefore the thermosphere at the top of the ionosphere that caps our Sun-layered atmosphere and provides the Earth's threshold into space.

9.6 THE VANISHING OZONE

Solar Ultraviolet Rays Create the Ozone Layer

What is ozone and how is it created high in the air? When the ultraviolet rays from the Sun strike a molecule of ordinary diatomic oxygen that we breathe, they split or dissociate the molecule into its two component oxygen atoms. Some of the freed oxygen atoms bump into, and become attached to, an oxygen molecule, creating ozone molecules that have three oxygen atoms instead of the normal two. The Sun's ultraviolet rays thereby produce a globe-circling ozone layer in the stratosphere at altitudes between 15 and 50 kilometers. It is a molecular veil that shields Earth by absorbing ultraviolet sunlight.

There isn't much ozone high in the air. The maximum concentration of ozone in the stratosphere is about one part per million by volume or about 5 billion billion, denoted 5×10^{18}, molecules of ozone per cubic meter. This is roughly equivalent to a drop of wine diluted in a swimming pool, but it can still absorb almost all of the Sun's deadly ultraviolet rays before they reach the ground. So this is the good kind of ozone, which protects life below it from dangerous solar ultraviolet radiation. Humans produce a bad kind of ozone closer to the ground (Focus 9.1).

The amount of ozone that is now in our atmosphere is enhanced and depleted, and then enhanced again, when the Sun goes from a highly active state to a relatively calm one, and back to greater activity again, during the 11-year cycle of solar activity. When the Sun brightens during active periods, the increased ultraviolet output produces more ozone in the stratosphere, and when our home star is less active and dims, there is less ultraviolet sunlight and smaller amounts of ozone are produced.

But the varying Sun is not the only cause of the Earth's changing ozone layer, for we have been destroying it.

The Ozone Hole

The amount of ozone in the stratosphere resembles the level of water in a leaky bucket. When water is poured into the bucket, it rises until the amount of water poured in each minute equals the amount leaking out. A steady state has then been reached, and the amount of water in the bucket stops rising and will stay the same as long as you keep pouring water in at the same rate. However, if you pour the water in at a different rate, or punch a few more holes in the bucket, the steady-state level of water in the bucket changes. The amount of ozone in the stratosphere changes in a similar way if there is a variation in the amount of solar ultraviolet streaming into it, or if we punch holes in the ozone layer, as we have been doing with chemicals used in our everyday lives.

A springtime ozone hole forms over Antarctica at the South Pole (Fig. 9.8). It forms within a vast polar vortex that resembles a huge dust devil or the eye of a hur-

FOCUS 9.1

Bad, Ground-Level Ozone, A Harmful Pollutant

Near the Earth's surface, in the air that we breathe, ozone is not so beneficial to life, and it is a main ingredient of eye-burning smog that can damage lung tissue and plants. During smog alerts in large cities, such as Los Angeles and Mexico City, children, elderly people, and jogging enthusiasts are advised to stay indoors because of potential ozone damage to their lungs. Children are especially at risk, because, for their body size, they inhale several times more air than adults, and also since they tend to spend more time outdoors.

In contrast to the "good" naturally occurring ozone in the stratosphere, which is created by incoming solar radiation, the "bad", ground level ozone is produced by humans, primarily from compounds emitted in the exhaust of cars and trucks. Fires have also caused ground-level ozone and smog across fields and grasslands of Brazil and the savannas of southern Africa, with an intensity comparable to the pollution over industrialized regions in Europe, Asia, and the United States. The "bad" ground-level ozone can also be carried hundreds of kilometers by the wind, so the potential damage is not limited to densely populated and heavily industrialized cities.

ricane. Each year the gaping hole opens up during Antarctic spring when the sunlight triggers ozone-destroying chemical reactions. The hole then begins to close up in the early polar fall when the long sunless winter begins. Ozone-depleted air is dispersed globally, and the ozone is slowly restored to fill in the hole until the cycle repeats in the following year.

The ozone layer is itself invisible, so you can't see the ozone hole. Instruments measure the amount of ultraviolet radiation getting through the atmosphere at specific wavelengths or spectral lines. The absorption lines are strengthened when there is more ozone, and weakened if there are lesser amounts.

British scientists first detected a springtime decrease of ozone above the South Pole in 1957–58. They had previously found that the amount of ozone varies annually throughout Europe. When this annual variation was compared to that measured at the Halley Bay station in Antarctica, assuming a six-month difference in the seasons, the ozone values for polar spring were much lower than was expected. It was initially thought that some large mistake had been made, or that the instrument had developed some major fault, but in the polar fall the ozone values jumped to those expected from extrapolation of the European results. According to Gordon Miller Bourne "G.M.B." Dobson (1889–1976):

> It was not until a year later [in 1958–59], when the same type of annual variation was repeated, that we realized that the early results were indeed correct and that Halley Bay showed a most interesting difference from other parts of the world. It was clear that the winter vortex over the South Pole was maintained late into the spring and that this kept the ozone values low. When it suddenly broke up in November both the ozone values and the stratosphere temperatures suddenly rose.[43]

The total ozone concentration was measured above Antarctica for several decades, always showing an anomalous springtime loss that became steadily larger as the years

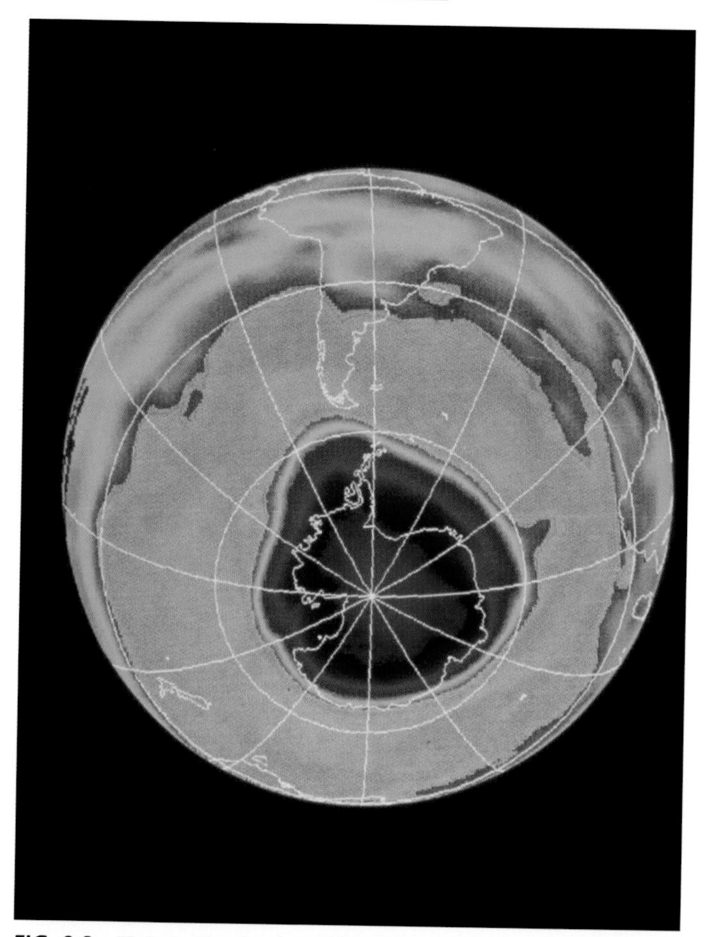

FIG. 9.8 Hole in the sky A satellite map showing an exceptionally low concentration of ozone, called the ozone hole, that forms above the South Pole in the local spring. In October 1990 it had an area larger than the Antarctic continent, shown in outline below the hole. Eventually spring warming breaks up the polar vortex and disperses the ozone-deficient air over the rest of the planet. (Courtesy of NASA.)

went on. By 1985 the thinning of the ozone layer above Antarctica had become so large and dominant in the measurements that it could not be ignored, and the British Antarctic Survey announced their discovery of the so-called ozone hole that is largest and deepest in September and early October each year.

Astounded scientists, who had been monitoring the ozone layer from a satellite, rechecked their data, finding that they had unwittingly recorded the missing ozone for several years. The embarrassed experts had programmed their computers to reject very low values of ozone as bad data, so the now-famous ozone hole had been discarded as an anomaly. Later analysis of the raw satellite data confirmed the discovery of the ozone hole, and demonstrated that it covered most of the Antarctica continent. Up to 70 percent of the ozone normally found over Antarctica was destroyed, resulting in a significant thinning of the ozone layer but not a completely open and empty hole.

Destroying Ozone

What's consuming the ozone layer that protects life on Earth from excessive ultraviolet radiation? In 1974, Mario Molina (1943–) and F. Sherwood Rowland (1927–), two chemists at the University of California at Irvine, showed that man-made chemicals are destroying the protective ozone layer, making it thinner and wasting it away. As they expressed it:

> Chlorofluoromethanes are being added to the environment in steadily increasing amounts. These compounds are chemically inert and may remain in the atmosphere for 40 to 150 years, and concentrations can be expected to reach 10 to 30 times present levels. Photodissociation of the chlorofluoromethanes in the stratosphere produces significant amounts of chlorine atoms, and leads to the destruction of atmospheric ozone.[44]

The ozone-destroyers are man-made gases, invented about a half-century ago and now given the euphonious name "chlorofluorocarbons". This name is a giveaway to their composition, chlorine, fluorine, and carbon, and is often abbreviated as CFC. A number sometimes follows the shorthand CFC notation, providing a complex description of the number of atoms in each molecule, the most widely used being CFC-11 and CFC-12.

The CFCs are synthetic chemicals entirely of human industrial origin with no counterparts in nature, and for several decades they were hailed as wonder chemicals. Their molecules are assembled using the strongest chemical bonds permitted by nature, making them almost impervious to breakage. They are very stable, nontoxic, noncorrosive, nonflammable, and relatively easy to manufacture. The CFCs were once widely used as coolants in refrigerators, freezers and air conditioners, as foaming agents for insulation found in coffee cups and packaging, as propellants in hair spray and deodorants, and as cleaning solvents in the process of manufacturing computer chips.

When initially released in the lower atmosphere, the hardy chemicals don't interact chemically to form other substances that would get removed from the atmosphere naturally. And they're not soluble in water, so they don't get rained out. They are so inert and stable that once released into the atmosphere the CFC molecules can survive for more than a century.

In the lower atmosphere, CFCs are protected from the Sun's intense ultraviolet radiation by the ozone layer, enabling them to migrate intact into the stratosphere. Once there, however, the energetic solar ultraviolet tears the sturdy CFC molecules apart, producing free chlorine atoms, which in turn destroy the ozone. As Molina and Rowland theorized more than three decades ago, the chlorine is the real agent of ozone destruction, being able to destroy thousands of times its own weight in ozone.

Once freed in the stratosphere, a chlorine atom can react with an ozone molecule, taking one oxygen atom to form chlorine monoxide; the ozone is thereby returned to a normal oxygen molecule and its ultraviolet absorbing capability is largely removed. If this were all that happened, there wouldn't be much concern. However, when the chlorine monoxide encounters a free oxygen atom, the chlorine is set free to strike again and again. Because the cycle repeats over and over, a little chlorine goes a long way. In technical terms, the chlorine acts as a catalyst, initiating a series of reactions that destroy ozone, while surviving the process.

A single chlorine atom can disrupt as many as 100,000 ozone molecules before it is captured and locked away in some other molecule. A small amount of chlorine can therefore produce significant changes in the ozone layer, which is itself very rarefied. This discovery that the chlorine in man-made chemicals will destroy enormous amounts of ozone in the atmosphere was so important that Molina and Rowland were awarded the 1995 Nobel Prize in Chemistry, together with the German chemist, Paul J. Crutzen (1933–), for their work in atmospheric chemistry, particularly concerning the formation and decomposition of ozone.

In places, chlorine can destroy ozone at a far faster rate than the gas is replenished naturally. The ozone loss then exceeds the ozone creation, as if water were being poured into a bucket with its bottom cut out. The Sun's powerful ultraviolet rays can then penetrate to the ground, producing widespread damage to living things.

Why doesn't chlorine keep on consuming ozone until there is none left in the stratosphere? The cycle of destruction is eventually broken when a chlorine atom, instead of interacting with ozone, interacts with methane in the atmosphere to form hydrochloric acid. The acid diffuses downward from the stratosphere to the troposphere, and because it's soluble in rain the chlorine is finally washed out, closing the cycle.

The simple scenario of ozone-destroying chlorine, released from CFCs by ultraviolet sunlight, does not explain why severe ozone depletion is limited to the Antarctic region or why it is observed only in the spring. Something must be focusing and intensifying chlorine's destructive power both in space and time.

Although the CFCs were mainly released by human activities in the Northern Hemisphere, where the world's industries are concentrated, the chemicals were redistributed by global winds. Each sunless winter, steady winds blow in a circular pattern over the ocean that surrounds Antarctica, trapping a huge air mass inside. The whirling winds concentrate the ozone-destroying chemicals within a vast, towering polar vortex, confining the most serious ozone depletion to regions above the South Pole and isolating it from the rest of the world for months at a time.

The ozone hole also requires the presence of polar stratospheric clouds of ice crystals. Whipping around the pole, the high-speed vortex winds shut out air from the warmer equatorial regions and keep the temperatures very cold inside. During the long, dark winter, the stratospheric air over Antarctica is therefore colder than elsewhere, allowing ice crystals to form. The crystals provide surfaces or platforms on which chlorine and ozone can alight and interact more readily than if they were free.

Who cares if chemicals are punching a few holes in the sky and letting a little more sunlight reach the ground? You care, and I care, for the solar ultraviolet rays, which are no longer absorbed in the ozone layer, can produce skin cancer and cloud up your eyes with cataracts. It also increases the vulnerability of animals to infections, and can damage or wipe out one-celled plants, called phytoplankton, which live near the surface of the sea and are the base of the ocean food chain. When the

FOCUS 9.2

The Montreal Protocol

Faced with the evidence of vanishing ozone in the Antarctica ozone hole, international diplomats agreed to a treaty, known as *The Montreal Protocol on Substances That Deplete The Ozone Layer,* adopted on 16 September 1987. The international treaty was strengthened by four amendments in 1990, 1992, 1997 and 1999, respectively named the London, Copenhagen, Montreal and Beijing Amendments after the city where they were devised.

The diplomats agreed to a phase out of the production and use of ozone-destroying compounds, such as the CFCs and Halons, in the industrialized countries by the year 1995 and in developing countries by 2010.

The ready acceptance of the *Montreal Protocol* was undoubtedly eased by the development of substitutes for CFCs in its industrial applications, including refrigerators and air conditioners. In fact, the biggest producer, Du Pont, unilaterally stopped making the chemicals even before the *Protocol* required it.

The compounds that consume ozone are now forbidden in the industrialized part of the world. And with the help of the United Nations and the World Bank, use of the dangerous chemicals has decreased considerably throughout the world, and their concentrations in the atmosphere have now leveled off or begun to decline.

plankton population declines, so will the populations of fish that eat plankton, and the mammals that eat the fish – seals, dolphins, whales, and humans.

These concerns, together with the now-famous Antarctic ozone hole, sparked public awareness of the fragile ozone layer and served as a powerful political stimulus to limit and eventually ban the production of CFCs (Focus 9.2).

Owing to the long lifetime of these substances, the healing of the continent-sized ozone hole did not begin until 2003, and complete recovery of the Antarctica ozone layer is not expected until the year 2050 or later. So concerned scientists continue to monitor the ozone using the Total Ozone Mapping Spectrometer, abbreviated TOMS, instrument aboard NASA satellites. By September 2004 the maximum hole size had shrunk to an area of 20 million square kilometers, down nearly 7 million square kilometers from its record size in 2000, but there is still plenty of room for improvement. Measurements of the varying solar ultraviolet output are additionally required to determine whether, and to what extent the Sun is changing the amount of ozone.

9.7 THE INCONSTANT SUN

Solar Radiation Varies Over the 11-Year Activity Cycle

Day after day the Sun rises and sets in an endless cycle, an apparently unchanging ball of heat and light, whose radiation makes life on Earth possible. The total amount of this life-sustaining energy has been called the “solar constant,” because no variations could be reliably detected from anywhere beneath the Earth’s changing atmosphere. Indeed, only a few decades ago, most astronomers and climatologists insisted that the Sun shines steadily.

Yet, reliable as the Sun appears, it is an inconstant companion. The Sun actually fades and brightens in step with the changing level of magnetic activity, and no portion of the Sun's radiation output is invariant. Indeed, the Sun's radiation is variable on all measurable time scales, and this inconstant behavior can be traced to the pervasive role played by magnetic fields in the solar atmosphere.

The discovery of variations in the total solar radiation output, or solar constant, was somewhat unexpected. For instance, a group led by Charles Greeley Abbot (1872–1973), secretary of the Smithsonian Institution from 1919 to 1944, carried out daily measurements of the solar constant for decades, including expeditions to four continents from sea level to mountaintops. And despite Abbot's claim that the Sun is a variable star, his colleagues concluded that any fluctuations noticed by Abbot were due to varying atmospheric absorption of sunlight or to improperly calibrated instruments.

Definite evidence of the changing solar constant was not obtained until the early 1980s, with an exquisitely sensitive detector, created by Richard C. Willson (1937–) of the Jet Propulsion Laboratory in Pasadena, California. He measured the solar constant with his Active Cavity Radiometer Irradiance Monitor, or ACRIM for short, which was placed aboard NASA's *Solar Maximum Mission,* abbreviated *SMM,* satellite, launched in February 1980. The solar constant, defined as the average amount of radiant solar energy per unit time per unit area reaching the top of our atmosphere at the Earth's mean distance from the Sun, was found to be remarkably constant, rarely changing by more than 0.2 percent, or one part in 500, with an average minimum value of about 1365.56 ± 0.01 watts of power per square meter.

But ACRIM determined the solar constant with an incredible precision of 0.01 percent or one part in 10,000, when averaged over one day, and at this level of accuracy, the Sun's total radiation output is almost always changing, at amounts of up to a few tenths of a percent and on all time scales from 1 second to 10 years. Similar variations were also detected by the Earth Radiation Budget (ERB) radiometer aboard the *Nimbus 7* satellite, launched in October 1978, with an even longer record, but with less precision.

Hugh S. Hudson (1939–), a solar physicist at the University of California at San Diego, teamed up with Willson to help him identify the irradiance variations with well-known features on the Sun. The largest, relatively brief, downward excursions correspond to, and are explained by, the rotation of a large group of sunspots across the face of the Sun. The concentrated magnetism in sunspots acts as a valve, blocking the energy outflow, which is why sunspots are dark and cool, producing reductions in the solar constant of up to 0.3 percent that last a few days.

Bright patches on the visible solar surface cause the increased luminosity. They are called *faculae,* from the Latin for "little torches." On the short-term scale of minutes to days, sunspot blocking dominates facular brightening; on the long-term scale of months and years, faculae are dominant because they have significantly longer lifetimes than sunspots and cover a larger fraction of the solar disk.

The decade-long variations in the Sun's total radiation output are the result of a competition between dark sunspots, in which radiation is depleted, and bright faculae, which are sources of enhanced radiation, with the faculae winning out. As magnetic activity increases, they both appear more often, but the excess radiance from bright faculae is greater than the loss from sunspots. This was demonstrated by Peter V. Foukal (1945–), of Cambridge Research and Instrumentation, Inc., in Massachusetts, and Judith Lean (1953–) of the Naval Research Laboratory in Washington, DC. They showed that there is a good correlation between the long term, solar constant variations and facular emission from the entire solar disk, once the effects of sunspot dimming have been removed.

More than five instruments on different spacecraft have made precise measurements of the solar constant for more than three decades, including the second Active Cavity Radiometer Irradiance Monitor, or ACRIM II, on the *Upper Atmosphere Research Satellite,* abbreviated *UARS* and launched in 1991, and the Variability of IRradiance and Gravity Oscillations, or VIRGO, instrument on the *SOlar and Heliospheric Observatory,* or *SOHO* for short and launched in 1995. They indicate that during recent epochs of high solar activity, near maxima in the 11-year solar cycle, the mean levels of the solar constant increase by about 0.7 percent, relative to solar cycle minimum levels (Fig. 9.9).

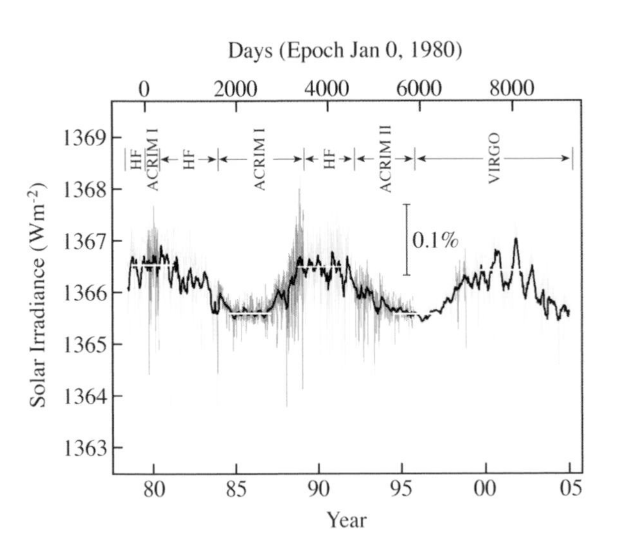

FIG. 9.9 Variations in the solar constant Observations with very stable and precise detectors on several Earth-orbiting satellites from 1978 to 2005 show that the Sun's total radiation input to the Earth, termed the solar irradiance, is not a constant, but instead varies over time scales of days and years. Here the total irradiance just outside our atmosphere, also called the solar constant, is given in units of watts per square meter, abbreviated W m^{-2}, where one watt is equivalent to one joule per second. The observations show that the Sun's output fluctuates during each 11-year sunspot cycle, changing by about 0.1 percent between maximums (1979, 1990 and 2001) and minimums (1987 and 1997) in magnetic activity. The average minimum value is 1365.560 ± 0.009 W m^{-2}, and the cycle amplitudes are 0.934 ± 0.019, 0.897 ± 0.020 and 0.829 ± 0.017 W m^{-2} above the average minimum value. Temporary dips of up to 0.3 percent and a few days' duration are due to the presence of large sunspots on the visible hemisphere. The larger number of sunspots near the peak in the 11-year cycle is accompanied by a rise in magnetic activity that creates an increase in luminous output that exceeds the cooling effects of sunspots. The capital letters given at the top are acronyms for the different radiometers. (Courtesy of Claus Fröhlich.)

The entire spectrum of the Sun's radiation is modulated by solar activity. Overwhelmingly the greatest part of solar radiation is emitted in the visible part of the spectrum where the 11-year variations are a relatively modest 0.1 percent. In contrast, radiation at the short-wavelength, invisible parts of the solar spectrum change significantly during the solar cycle, though contributing only a tiny fraction of the Sun's total radiation. The ultraviolet radiation is at least ten times more variable than visible radiation, and the Sun's X-ray emission changes even more, by at least a factor of one hundred.

The Earth's upper atmosphere acts as a sponge, soaking up the unseen ultraviolet and X-ray radiation. At times of high solar activity, the Sun pumps out much more of the invisible rays, the air absorbs more of them, and our upper atmosphere heats up; when solar activity diminishes the high-altitude air absorbs less and cools down.

The X-rays are absorbed at high altitudes in the terrestrial atmosphere, where the global mean temperature can double between the minimum and maximum of solar activity (Fig. 9.10). The Sun's variable ultraviolet radiation modulates the vital ozone

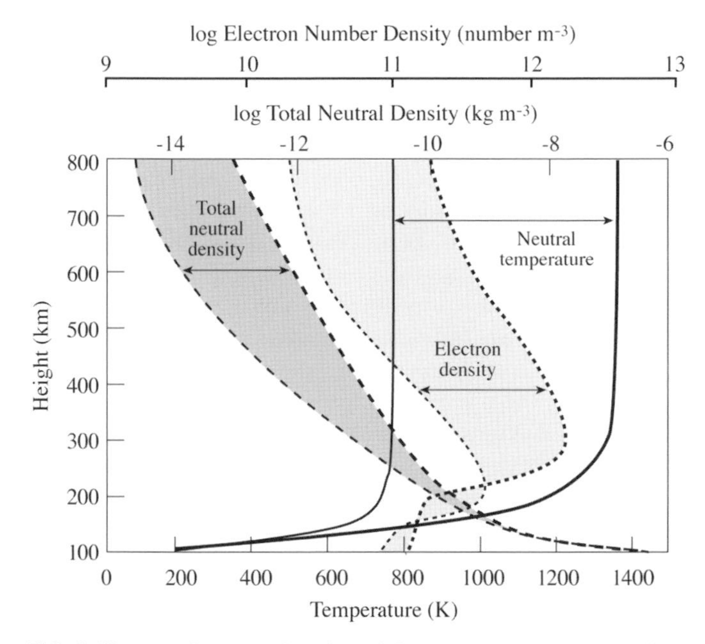

FIG. 9.10 Varying solar heating of the upper atmosphere During the Sun's 11-year activity cycle, the upper-atmosphere temperatures fluctuate by factors of two, and neutral (un-ionized atom) and electron densities by factors of ten. The bold lines (*right side*) register maximum values and the less bold (*left side*) the minimum values. Enhanced magnetic activity on the Sun produces increased ultraviolet and X-ray radiation that heats the Earth's upper atmosphere and causes it to expand, resulting in higher temperatures and greater densities at a given altitude in our atmosphere. (Courtesy of Judith Lean.)

layer lower in the atmosphere. Only the least variable, visible portion of the solar spectrum penetrates through to the relatively placid lower atmosphere, but this sunlight might still vary enough to warm and cool the air.

Moreover, the 11-year cycle that clocks the rise and fall in magnetic activity on the Sun does not repeat with the same strength at maximum. It instead seems to be modulated over 100-year time intervals.

Solar Variability Over the Centuries

Despite diligent observations by European astronomers, very few dark spots were found on the Sun between 1645 and 1715, a 70-year period that included the reign of France's "Sun King" Louis XIV (1638–1715). The German astronomer Gustav Spörer (1822–1895) called attention to the 70-year absence in 1887. An indirect consequence of the missing sunspots was reported much earlier, in 1733, by the French author Jean Jacques D'Ortous de Mairan (1678–1771), as a decrease in the number of auroras seen on Earth, but he was ridiculed for thinking that the northern lights could be related to increases in the number of sunspots.

E. Walter Maunder (1851–1928), at the Royal Greenwich Observatory in England, fully documented the dearth of sunspots using extensive historical records covering hundreds of years. His accounts, entitled *A Prolonged Sunspot Minimum,* were presented to the Royal Astronomical Society in 1890 and 1894, but they remained largely ignored until the 1970s when the American solar astronomer John A. "Jack" Eddy (1931–) provided further evidence for the "Maunder Minimum", as he called it, using the growth rings of trees.

As the trees lay down their rings each year, they record the amount of atmospheric carbon dioxide captured in the process of photosynthesis. The carbon intake comes in two varieties, or isotopes, stable carbon 12 and radioactive carbon 14, and the radioactive type tells us how active the Sun was at the time. Cosmic rays from outer space produce radioactive carbon 14 when they strike atoms in the air. Because the cosmic rays are deflected away from the Earth by the Sun's magnetic fields during high solar-activity levels, there is less radioactive carbon in the air when the Sun is very active, and more of it at times of low activity on the Sun.

An analysis of the world's longest-lived trees, the bristle cone pines, suggests that the Sun's output has been turned low for several extended periods in the past millennia. Eddy used the technique to read the history of solar activity all the way back to the Bronze Age, and showed that the tree-ring data are supported by other evidence such as the ancient sightings of terrestrial auroras (Fig. 9.11). In an important paper entitled "The Maunder Minimum", published in *Science* on 18 June 1976, Eddy concluded that the Sun has spent nearly a third of the past two thousand years in a relatively inactive state. He pinpointed several periods of low activity with significantly more radioactive carbon 14, each about a century long, naming them the Maunder, Spörer and Wolf minima.

Two periods of missing sunspots, the Spörer minimum from 1460 to 1550 and the Maunder Minimum from 1645 to 1715, occurred during the Little Ice Age (1400–1850) when Europe and North America endured a period of extremely harsh winters and colder than normal summers. During that time, alpine glaciers expanded, and the river Thames in London and the canals of Venice regularly froze over. Owing to the apparent

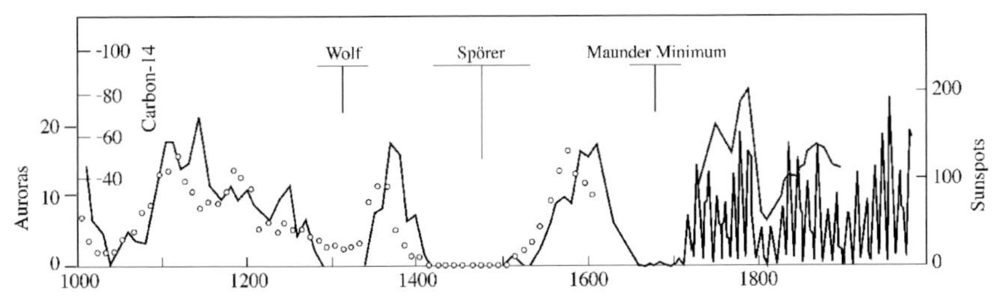

FIG. 9.11 Long periods of solar inactivity Three independent indices demonstrate the existence of prolonged decreases in the level of solar activity. The observed annual mean sunspot numbers (*scale at right*) also follows the 11-year solar activity cycle after 1700. The curve extending from A.D. 1000 to 1900 is a proxy sunspot number index derived from measurements of carbon-14 in tree rings. Increased carbon-14 is plotted downward (*scale at left-inside*), so increased solar activity and larger proxy sunspot numbers correspond to reduced amounts of radiocarbon in the Earth's atmosphere. Open circles are an index of the occurrence of auroras in the Northern Hemisphere (*scale at left-outside*). The pronounced absence of sunspots from 1645 and 1715 is named for the English astronomer E. Walter Maunder (1851–1928), who fully documented it, and another noticeable lack of solar activity is named for the German astronomer Gustav Spörer (1822–1895) who previously called attention to the prolonged absence between 1645 and 1715. The third prolonged absence of sunspots is named for the Swiss astronomer Johann Rudolf Wolf (1816–1893) who investigated the connection of the 11-year sunspot cycle with geomagnetic activity and devised what is now known as the Wolf sunspot number. (Courtesy of John A. "Jack" Eddy.)

lack of solar activity, the Sun was probably noticeably dimmer, suggesting that it might be related to the colder temperatures during the Little Ice Age. But many uncertainties preclude a definitive understanding of the Sun's role in the Little Ice Age, and no one is certain what brought it about some 500 years ago.

9.8 THE HEAT IS ON

Changing Temperatures

Over the past century, the planet as a whole has warmed up in fits and starts by almost half a degree Celsius or 0.5 kelvin. The world heated up by about that amount between 1920 and 1940, for example, but it entered a cooling phase after that. From 1940 to 1970 temperatures dropped so much that some experts predicted a coming ice age. In subsequent decades, temperatures began to rise again, and the world became unusually hot on average. So there is no smooth increase or decrease in the global temperature record.

Detailed examinations indicate that the Earth's climate could have been controlled by the Sun for much of the past 400 years. Historical records of the varying solar irradiance of Earth show that the Sun's changing brightness dominated our climate for two centuries, from 1600 to 1800. In the 19th century, major volcanoes, such as Mount Tambora in 1815 and Mount Krakatoa in 1885, cooled the Earth by throwing hazy dust and other particles high above the ground, forming an invisible, umbrella-like shield that blocks some of the incoming solar radiation and causes temporary global cooling. But the Sun noticeably warmed the world for another century, from 1870 to 1970.

In 1991 the Danish scientists Eigil Friis-Christensen (1944–) and Knud Lassen (1921–) found an intriguing link between solar activity and temperatures here on Earth. The global variations in the air temperature over land fluctuate in synchronism with the length of the solar cycle for 130 years, beginning about 1860. Cycles a few months shorter than 11 years in duration are characteristic of greater solar activity that apparently warms our planet, while cycles a few months longer than 11 years signify decreased activity on the Sun and cooler times at the Earth's surface.

These temperature variations may be caused by the Sun's 11-year modulation of the amount of cosmic rays, which would otherwise reach the Earth and help produce clouds. Interplanetary magnetic fields associated with greater solar activity cut off the Earth-directed cosmic rays, resulting in fewer clouds that would have reflected sunlight, and more sunlight therefore reaches the ground to warm it.

Surface temperatures, cloud cover, drought, rainfall, tropical monsoons, and forest fires all show a definite correlation with solar activity, but the climate experts usually cannot explain why. They just do not know how or why the climate responds to the solar variability. And most recently it looks more and more like humans are helping to turn up the global thermostat.

Global mean temperatures have been inferred for the past 1,000 years using the widths of tree rings, the isotopic composition of layers found in ice cores, and the chemical composition of corals. When these results are linked to much more precise thermometer readings from the past 100 years, the data indicate that the Earth became hotter in the late 20th and early 21st centuries than at any time during the previous millennium. The dramatic rise in temperature coincides with the unprecedented release of carbon dioxide and other heat-trapping gases into the Earth's atmosphere, suggesting that human activity is taking control of the world's temperature, boosting it to unheard of levels.

Some scientists have called attention to the large uncertainties in determining temperatures over the past 1,000 years, and to the possibility of large climate swings during this time. Even skeptical climate scientists nevertheless agree that recent decades are still the warmest of the past millennium, and that the temperature rise is probably explained by human activity together with natural influences.

Carbon Dioxide and Other Heat-Trapping Gases

Because carbon dioxide is colorless, odorless and disperses immediately in the atmosphere, few realized how much of the potentially dangerous, heat-trapping gas enters into the air. Half a century ago, no one even knew if any of the carbon dioxide stays in the atmosphere or if it was all being absorbed in the forests and oceans. Then in 1958, Charles Keeling (1928–2005) began measurements of its abundance in the clean air at the Mauna Loa Observatory in Hawaii. It is located at a remote, high-altitude site on top of a volcano in the midst of a barren lava field, far from cars and people that produce carbon dioxide and from nearby plants that might absorb it.

The most striking aspect of the carbon dioxide curve (Fig. 9.12) is its smooth, systematic increase over the past half century, when the atmospheric concentrations of carbon dioxide have increased by almost 20 percent. Both the increase and its accelera-

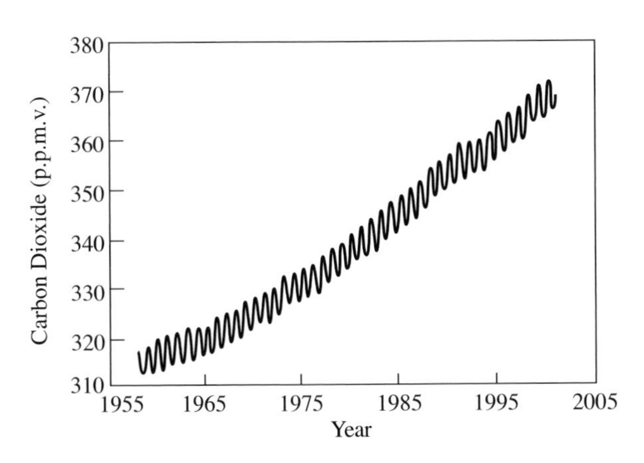

FIG. 9.12 Rise in atmospheric carbon dioxide The average monthly concentration of atmospheric carbon dioxide, denoted CO_2, in parts per million by volume, abbreviated ppmv, of dry air plotted against time in years observed continuously since 1958 at the Mauna Loa Observatory, Hawaii. It shows that the atmospheric amounts of the principal waste gas of industrial societies, carbon dioxide, have risen steadily for nearly half a century. The up and down fluctuations, that are superimposed on the systematic increase, reflect a local seasonal rise and fall in the absorption of carbon dioxide by trees and other vegetation. Summertime lows are caused by the uptake of carbon dioxide by plants, and the winter highs occur when the plants' leaves fall and some of the gas is returned to the air. Carbon dioxide ice-core data indicate that the exponential increase in the amount of carbon dioxide has been continuing for the past two and a half centuries. (Courtesy of Dave Keeling and Tim Whorf, Scripps Institution of Oceanography.)

tion have been rock steady – as inexorable as the expansion of the world's population, human industry and pollution.

Ice core studies indicate that the buildup of carbon dioxide has been going on since the beginning of the industrial revolution in the mid-18th century. Air is sealed off in bubbles when the ice is laid down, and can be extracted from cores drilled deep within layered glaciers.

The ice cores show that the carbon-dioxide levels stood at about 280 parts per million per volume, abbreviated ppmv, in 1860, and that they were at this level thousands of years earlier, at the times of Julius Cesar and well before that. Then about 200 years ago, industrial combustion of fossil fuels, such a coal, oil and natural gas, and accelerating deforestation resulted in an unprecedented release of carbon dioxide into the atmosphere at an ever-increasing rate, and no end to the buildup is in site.

The amounts of carbon dioxide in our air rose gradually at first, and then much more rapidly, rising from 280 to 315 ppmv in a century, from 1860 to 1960, and jumping to 360 just 30 years later. If current trends continue, atmospheric carbon dioxide will reach 500 ppmv – nearly double pre-industrial levels – around the middle of the 21st century.

There is more than 750 billion tons of carbon in the atmosphere in the form of carbon dioxide. And each person now on Earth is, on average, dumping about a ton of carbon dioxide into the air every year, releasing an annual total of about 7 billion tons of carbon dioxide. Each American produces about 4.5 tons every year, on average, more than a citizen of any other country. And a single automobile generates between

50 and 100 tons of carbon dioxide in its lifetime. About half of all this gas remains in the atmosphere; the rest is absorbed in the sea and consumed by forest trees.

During the last few decades, other heat-trapping gases, as well as carbon dioxide, have been accumulating noticeably in the atmosphere (Focus 9.3). They now contribute about as much to global warming as carbon dioxide. However, industrial and volcanic emissions of sulfur in the form of small particles, called sulfate aerosols, may be reflecting solar radiation back into space, thereby masking the greenhouse effect of these gases over some parts of the Earth.

Climatologists are, to put it mildly, concerned about the possibility of global warming by the greenhouse effect if this irreversible buildup of carbon dioxide and other heat-trapping gases continues unabated. Two American chemists, Roger Revelle (1909–1991) and Hans Suess (1909–1993), noticed the danger nearly half a century ago, stating in 1957 that:

> During the next few decades the rate of combustion of fossil fuels will continue to increase, if the fuel and power requirements of our worldwide industrial civilization continue to rise exponentially. . . . Thus human beings are now carrying out a large-scale geophysical experiment of a kind that could not have happened in the past nor be reproduced in the future. Within a few centuries we are returning to the atmosphere and oceans the concentrated organic carbon stored in sedimentary rocks over hundreds of millions of years.[45]

FOCUS 9.3

New Heat-Trapping Gases in Our Atmosphere

Increased carbon dioxide may be responsible for only about half the "unnatural" greenhouse effect. The other half of the warming may be due to methane, nitrous oxide, and chlorofluorocarbons, abbreviated CFCs.

Like carbon dioxide, the atmospheric concentration of methane, also known as natural gas, has more than doubled since the start of the industrial revolution. Although atmospheric methane remains about 100 times less abundant than carbon dioxide, each incremental molecule of methane is about 20 times more effective in trapping heat than each additional molecule of carbon dioxide. Methane is produced by some of the bacteria that thrive in oxygen-free places like swamps, rice paddies, landfills, garbage dumps, termite colonies, and in the stomachs of cows. Some methane also escapes from coal mines, natural gas wells and leaky pipelines. Methane in swamps, known as marsh gas, sometimes ignites spontaneously, producing flickering blue flames called will-o'-the-wisps. It is hard to believe, but the United States Environmental Protection Agency once spent half a million dollars to find out if cattle belch enough methane to contribute to global warming.

Nitrous oxide, or laughing gas, is also building up in the air, although not as rapidly as methane. The current rate of increase is about 0.2 percent a year, primarily as the result of nitrogen-based fertilizers but also from burning of fossil fuels in cars and power plants.

The addition of one CFC molecule can have the same greenhouse effect as the addition of 10,000 molecules of carbon dioxide to the present atmosphere, but fortunately the warming effect of these industrial chemicals may soon be leveling off since they have been banned on the basis of their ozone-destroying capability. Contrary to popular misconceptions, the thinning of the ozone layer does not by itself make the Earth's surface hotter.

Without remedial action, the levels of atmospheric carbon dioxide in this century will become twice those of the previous century, and this will eventually raise the temperature of the Earth's surface.

Signs of Global Warming

The heat is on, and the temperatures are rising. The evidence comes from direct measurements of rising surface air temperatures and subsurface ocean temperatures and from phenomena such as retreating glaciers, increases in average global sea levels, and changes to many physical and biological systems.

Nearly every major glacier in the world is shrinking at an accelerating pace, from Montana to Alaska, from the Bolivian Andes to Patagonia, Chile, and from Tibet to Greenland. Some glaciers are retreating so quickly that they will shrivel up and vanish in a few decades.

Huge glaciers in Antarctica are thinning, and ice shelves the size of American states are either disintegrating or retreating, allowing the inland ice to increase the pace of its motion to the sea. And sea ice in the Arctic is also shrinking, by two hundred and fifty million acres since 1979, an area the size of California and Texas combined. The Arctic sea ice could well disappear by the end of the century.

Most of the Arctic and nearly a quarter of the land in the Northern Hemisphere are underlain by frozen ground known as permafrost, which the rising temperatures are also melting, undermining support for roads and buildings. The snow and ice in the frozen ground reflect a lot of sunlight; in technical terms they have a high *albedo,* a Latin word for "whiteness." When the warming climate melts the permafrost, darker, less reflective land or water, of low albedo, is exposed. And because bare ground and seawater absorb sunlight more efficiently than ice does, the warming process tends to be self-reinforcing with a positive feedback, giving rise to further melting and warming at an accelerated pace.

Rising seas provide additional evidence for a hotter world. As the water in the sea gets warmer, it expands as most substances do when heated. The sea then ascends to higher levels, in much the same way that heating the fluid in a thermometer causes the fluid to expand and rise up. This is because warm water or other fluids occupy a greater volume than cold ones.

Rapidly shrinking glaciers are also significant contributors to rising seas, as is the melting of ice covering land in the Arctic and Antarctica. But contrary to popular belief, the melting of floating icebergs will not raise the level of the surrounding sea. When ice cubes in your drink at home melt, they similarly do not cause any change in its level, for the melted ice produces the same volume of water that the ice cube displaced.

Current global warming is also signaled by warmer nights, and a smaller difference between day and night temperatures. Spring is arriving earlier than it used to, by a week or so, autumn lasts a week or two longer, and we now have longer, hotter summers than many of us can remember. And the growing season is longer, benefiting both crops and weeds. Animals are shifting their ranges poleward, birds are nesting earlier, and plants are blooming days, and in some cases weeks, earlier than they used to.

Other warning signs of global warming include record rainfalls and severe snowstorms; extraordinarily destructive hurricanes and tornadoes; heat waves unique in weather annals; widespread droughts and devastating forest fires; and some of the worst floods in recorded weather history.

So why isn't more being done to curtail the ongoing global warming? One reason is that the predictions of future consequences are complicated by uncertain factors.

Uncertain Computations

To assess the effects of global warming, one assumes that the dominant cause of climate change during this century will be the release of heat-trapping greenhouse gases into the atmosphere by humans, and that the increasing rate of atmospheric build up of the waste gases will continue unabated. Supercomputer models then evaluate all the factors that push and pull the climate, setting the global thermostat. But these models of global warming involve large grids that blur distinctions between land and sea, mountains and plains, or clouds and ice, and there are great uncertainties due to the many climatic variables and nonlinear, interactive feedback mechanisms that can either amplify (positive feedback) or reduce (negative feedback) small changes.

An example is the ocean temperature. Water absorbs more carbon dioxide when it is colder, which is why you should always keep a carbonated beverage cold. When the oceans get warmer, their waters will absorb less carbon dioxide, leaving more of it in the air, strengthening the greenhouse effect, and further warming the oceans in a positive feedback. Yet, a precise knowledge of how carbon dioxide is buried deep within the oceans, and how the gas is released from them, is not available.

An incomplete knowledge of how clouds affect the Earth's temperature also causes uncertainty in the warming forecasts. At any given moment, clouds now cover at least half the area of the planet, and increased warming should produce even more clouds. The high-flying clouds can cool the planet by reflecting more incident sunlight back into space. You may have noticed this cooling effect when a large cloud passed overhead. Clouds also produce a warming effect by absorbing some of the infrared heat radiation emitted by the ground. This heat is reradiated downward, keeping the planet warm. That accounts for warmer nights on a cloudy day. The net temperature effect of clouds depends on which effect dominates and how strong it is, but that is the difficulty, for no one seems to know for sure.

Because of the various uncertainties, scientists have to generate a wide range of possible futures, some very threatening and others less so. Not all of these outcomes are likely to be true, and none is definitive, but people tend to latch onto those that fit their preconceptions. Especially the extremists, who choose alarming doom and gloom or cheerfully favor good times for all, at least for now.

But these uncertain details should not distract from the overall, global problem. If current emissions of carbon dioxide and other greenhouse gases go unchecked, the increased heat and violent weather will drastically change the climate we are used to. And once in the atmosphere, carbon dioxide stays there for centuries, so future generations will have to contend with the consequences of our present actions. The invisible waste gases that we have already dumped in the air will slowly change the climate of the

Earth regardless of future actions, and sometime in the future a lot of people might be feeling like the world is melting down in a pool of sweat.

Likely Future Consequences of Overheating the Earth

When the Earth reaches the inevitable high-temperature stage, our life will certainly be changed, but the predicted consequences depend on how quickly it occurs and whether it happens at the bottom or top of the predicted range, of 1.5 to 4.5 kelvin when the greenhouse gases double.

If the warming is in the lower range of the prediction, it could bring benefits. Increased rainfall would make some areas of the world more fertile. Crops and plants would be enriched and invigorated by rising concentrations of atmospheric carbon dioxide, but some studies show that carbon dioxide promotes the growth of invasive weeds far more than it stimulates crops. The growing season would lengthen in the north, with longer summers and shorter winters, and residents of cities like Boston would have to shovel less snow and perhaps go skiing less often. The capacity of humans to adapt is extraordinary, and most of us would probably welcome a little more heat or a move to a warmer place.

Moreover, people already experience extremes of temperature greater than that of any predicted global warming. Seasonal temperature changes averaged over the Northern Hemisphere between winter and summer are 15 kelvin. The average temperature differential between New York City and Atlanta, Georgia, or between Paris and Naples, is as large as the most extreme predictions of global warming, and there is little evidence of greater risk to people who live in the warmer southern climates.

There is nevertheless a long list of possible catastrophes that might be caused by significant global warming. The associated rise in sea level will flood coastal cities; hurricanes, which draw their energy from warm waters of the ocean, will become stronger and wetter; water supplies will be reduced and forest fires will become more common; more species will become extinct; drought will be intensified within the interiors of many continents; the American Midwest might become a colossal dust bowl; power companies will be unable to air condition our sweltering cities; and extreme heat waves will cause great human stress and more deaths, particularly among the poor, elderly, and weak or those with cardiovascular and respiratory disease.

More than a hundred million people worldwide live within a meter of mean sea level. And when all the world's glaciers experience complete meltdown, in about 100 years or less at the present rate, all of these people will have their homes destroyed. Some of the poorest regions on Earth will be hit hardest, such as densely populated Bangladesh. The resultant flooding will disrupt major cities like Bangkok, Boston, New York City, Shanghai and Tokyo. Venice and Alexandria will be inundated. And residents of South Florida will not have to worry about the sweltering heat; their homes will be flooded with seawater.

The boundary between salt water and fresh water at the mouths of rivers will move several kilometers inland, so the Nile, Yangtze, Mekong, and Mississippi deltas are all at risk. Island nations will suffer severe coastal flooding or disappear altogether under the rising waters; they include Cyprus and Malta in the Mediterranean, many of the Caribbean islands and several Archipelagos around the Pacific Ocean.

Even a modest rise in sea level will wipe many of the world's beaches out of existence. Flooding isn't the problem; it's the removal of sand by waves. Just a 0.3-meter rise in sea level will create wave action that erodes away up to 5 meters of horizontal sandy beach shoreline. So, people who live by the seashore on Caribbean islands had better sell their homes before they die; for their grandchildren can forget pleasant winter vacations there.

Although the most severe consequences of global warming are not likely to be noticed by you or your children, we've already initiated changes that will affect future generations. No matter what we do about dumping increasing amounts of heat-trapping gas into the air, global temperatures and the sea level will continue to rise for many decades to come, owing to the carbon dioxide already circulating in the atmosphere and the considerable inertia of the climate system.

Even if worldwide emission of carbon dioxide were capped at today's amounts, its concentrations in the atmosphere would continue to increase. And if we stopped burning fossil fuels altogether, the atmosphere would not immediately recover. The carbon dioxide already released in the air will stay there for at least a century, and there is a built in delay before the full impact of atmospheric changes that have already started. This is because heating the oceans, which requires tremendous amounts of energy, involves a lot of thermal inertia.

Most scientists therefore support prudent steps to curb the continued buildup of heat-trapping gases, even asserting that the evidence warrants a sense of urgency.

Doing Something About Global Warming

The majority of scientists are now in agreement that we ought to do something to curtail human emissions of carbon dioxide and other heat-trapping gases. On 8 June 2005, for example, the world's most influential scientific academies warned world leaders that they can no longer ignore the "clear and increasing" threat posed by global warming, and that "the scientific understanding of climate change is now sufficiently clear to justify nations taking prompt action." The unprecedented joint statement included the heads of the scientific academies of Brazil, Canada, China, France, Germany, India, Italy, Japan, Russia, the United Kingdom, and the United States.

In contrast, the political establishment in the United States has repeatedly described global warming as a matter of ongoing scientific debate, stating that we cannot say with any certainty what constitutes a dangerous level of warming, and therefore what level must be avoided. For many years, President George W. Bush (1946–) emphasized uncertainties about the relationship between rising global temperatures and rising concentrations of heat-trapping gases. A White House official, with former ties to the oil industry, has even edited climate research reports to play down the links between such emissions and global warming, including removal and adjustments of statements already approved by the government scientists and their supervisors.

Senator James M. Inhofe (1934–) of Oklahoma has been even more outspoken in his criticism, calling global warming "the greatest hoax ever perpetuated on the American people." In 2005, at about the same time that scientific leaders were calling for urgent action to avert further global warming, Inhofe was citing new evidence that "makes a mockery" of the notion that human-induced warming is occurring. In support of his claims, the Senator repeatedly cited the fiction writer Michael Crichton (1942–), whose novel *State of Fear* uses charts and references to argue that global warming is a scientific delusion. Inhofe chairs the Senate's Environment and Public Works Committee, which oversees legislation concerning emissions of greenhouse gases, and once the voting public is persuaded that there is no scientific consensus about global warming, mandates for limits may become very difficult to legislate in the United States.

In contrast, many other industrial countries have now endorsed the *Kyoto Protocol*, agreeing to legally binding limits to the emission of heat-trapping gases (Focus 9.4).

FOCUS 9.4

The Kyoto Protocol

In December 1997, at an International Climate Summit in the ancient Japanese capital of Kyoto, more than 100 nations agreed to reduce the emissions of heat-trapping gases that can warm the planet. Known as *The Kyoto Protocol to the United Nations Framework Convention on Climate Change*, or just the *Kyoto Protocol* for short, the agreement calls for mandatory reductions in the emissions of greenhouse gases, such as carbon dioxide and methane, by industrialized nations.

In order for the protocol to take effect, it had to be approved by countries responsible for at least 55 percent of the industrialized world's 1990 emissions linked to global warming. Since the United States withdrew from the negotiations, and by itself accounts for 36 percent of 1990 emissions, it looked for many years like the treaty would never be implemented. But Britain, Canada, France, Germany, Italy and Japan all signed on, and in 2004 the threshold was crossed when Russia, with 17.4 percent of the 1990 emissions, ratified the treaty. With the collapse of Communism, the Russian economy had collapsed and the country's gas emissions fell far below 1990 levels anyway.

The treaty, which began on 16 February 2005 and expires in 2012, commits industrialized countries that signed the agreement to reduce emissions of six greenhouse gases, including carbon dioxide, by 2012 to an average of 5.2 percent below emission levels in 1990. Different nations have slightly different obligations; the reduction of the European Union nations is set at 8 percent, and Japan at 6 percent, below 1990 levels. Industrialized nations can also meet their targets, in part, by buying or selling emission credits with other countries or by investing in "clean development" projects in other nations.

Developing countries are exempt from the agreement. Such a rich-poor, "two-world" difference was employed successfully in the *Montreal Protocol* for curtailing the production of ozone-destroying chemicals. And for the *Kyoto Protocol* it was defended by the fact that the average person in the rich, industrial countries eats more food, consumes more energy and pollutes the air more than the average person in poor, developing countries. And the industrial countries, it is argued, became rich largely by burning coal and oil for about two hundred years, producing most of the heat-trapping carbon dioxide that is now in the atmosphere. So, the poorer nations say, it is only right that the wealthy countries be the ones to cut back on their emissions, while the poorer economies grow, especially since such growth is needed for the very survival of their people.

But the *Kyoto Protocol* will not by itself solve the problem. The United States has not signed it. And just between 1990 and 2002, the carbon-dioxide emissions of the United States increased by more that the combined cut of the emissions that will be achieved when all agreeing countries hit the *Kyoto Protocol* targets. So the agreement can just serve as a hopeful beginning, perhaps a framework, for future international treaties that effectively curtail the emissions of heat-trapping gases.

The United States, which did not sign the *Kyoto Protocol* and is not affected by the treaty, has been cast as the wealthy villain, the most greedy, selfish and irresponsible of all. It has the world's biggest economy and is by far the biggest single producer of heat-trapping gases, both in total output and on a per capita basis, contributing about 18 percent of the total emissions with just 4 percent of the world's population. In a year, the average American produces the same greenhouse gas emissions as two Europeans, or four and a half Mexicans, or ten to fifteen Chinese or Indians, or ninety-nine Bangladeshis. Why, the poor nations argue, should anyone have the right to pollute the world more than anyone else?

But the United States continues to find the *Kyoto Protocol* unacceptable because it unfairly requires only the industrialized countries to cut emissions. They insist that warming is of global concern, since emissions from any country quickly enter the atmosphere and spread across the entire world. All countries, they argue, must share in the solution, particularly since some of the developing countries' emissions are expected to surpass even the unrestrained emissions of the richer nations in a few decades. China, for example, is not bound by the treaty restrictions, but it is rapidly increasing its consumption of coal and oil and is expected to overtake the United States as the world's largest carbon-dioxide emitter around the year 2025. Moreover, the *Kyoto Protocol* will not curtail the total emissions of greenhouse gases, since the United States and China are not participating.

Preventive legislation in the United States might hamper economic growth and interfere with the free market. No one has yet found a cheap substitute for coal and oil, so staggering costs of trillions of dollars are foreseen if alternate energy sources are used to curb the emission of carbon dioxide. The average American is not about to give up his or her comfortable life style for an uncertain future, or to pay the estimated thousands of dollars of annual costs per family to implement the *Kyoto Protocol* in the United States. And the typical Asian is not about to condone restraints that might slow his country's growth just to feed an American's insatiable appetite.

State and national governments can nevertheless blunt future global warming by setting carbon-dioxide emission limits on cars. They can adopt energy policies that shift from coal and oil to gas, and eventually to wind, water, solar or nuclear power. Countries can avoid the clearing of their forests and plant a lot more trees, and farmers can use plants to pull carbon out of the air, while not plowing the soil which releases the carbon back into the air. By protecting existing forests and planting new ones, or by practicing no-till farming, countries might offset up to 20 percent of the expected carbon dioxide build up during this century.

While waiting for the next governmental action, individuals can take steps to help. They can reduce their consumption of coal, oil or natural gas that electrify and heat their homes, offices and schools, power their vehicles, and fuel their factories. Ordinary people can use energy-efficient appliances, reduce their daily electricity use, drive their cars less, and insulate their buildings.

Cars account for about 21 percent of the world's total carbon-dioxide emission. A sports utility vehicle emits twice as much carbon dioxide as most other cars, and so does a pickup truck. So everyone should avoid the gas-guzzling cars, buy fewer cars, reduce the number of cars per family, and increase the number of passengers in every car. People should start using a hybrid car that combines internal combustion with battery power, or better yet ride a bicycle to work.

And if government and individual action doesn't solve the global-warming problem, nature will. In perhaps 100 years or even less, we will completely exhaust oil supplies, and the entire world will run out of gas. Once that happens, the Earth's climate should cool gradually, as the deep ocean waters slowly absorb the carbon dioxide pumped into the air during the current frenetic pulse of activity. Then a long-overdue ice age might be on its way.

9.9 THE ICE IS COMING

During the past two million years, huge ice sheets have advanced across the Northern Hemisphere and retreated again more than twenty times. The great, extended ice sheets last roughly 100,000 years, keeping the climate cold and the sea level low. The warm periods that punctuate the cold spells have been exceptional, each lasting roughly 10,000 years. Their scientific name, the interglacials, reflects their unusual nature.

The most recent advance, called the Wisconsin, started about 120,000 years ago in Canada, Scandinavia, and Siberia. By the time the ice had spread to its maximum southern extent, most of northern Europe, New England and the Midwestern United States were buried under ice a kilometer thick. The sea level had fallen to about 100 meters lower than it is today, and it would have been possible to walk from England to France, from Siberia to Alaska, and from New Guinea to Australia.

The present interglacial, called the Holocene, began about 10,000 years ago, when the world became warmer and wetter, about 5-kelvin warmer on average. The ice sheets melted and shrank back to their present-day configurations, leaving only the glacial ice in Greenland and parts of arctic Canada, as well as the massive ice sheets of Antarctica, and the sea level rose rapidly around the world.

This warm spell enabled civilization to flourish. Agriculture, and technology were developed. And during all that time populations grew throughout the world.

The ponderous ebb and flow of the great continental glaciers are affected by three astronomical rhythms that slowly alter the distances and angles at which sunlight strikes the Earth (Fig. 9.13). They are sometimes called the Milankovitch cycles, after the Serbian engineer Milutin Milankovitch (1879–1959) who described how variations in the planet's orbit, wobble and tilt could influence the pattern of incoming solar radiation at different locations on the globe. Joseph Alphonse Adhémar (1797–1862), a French mathematician, previously suggested in 1842 that the ice ages might be due to variations in the way the Earth moves around the Sun, and James Croll (1821–1890), a self-taught Scotsman, took up the idea in greater detail in 1864, showing how long periodic variations in the Earth's distance from the Sun might change the terrestrial climate. But the theory received its fullest mathematical development from 1920 to 1941 by Milankovitch .

The shortest astronomical rhythm is a periodic wobble, or precession, in the Earth's rotation axis that is repeated in periods of 23,000 years. It determines whether the seasons in a given hemisphere are enhanced or weakened by orbital variations. A longer periodic variation, of the Earth's axial tilt from 21.5 degrees to 24.5 degrees and

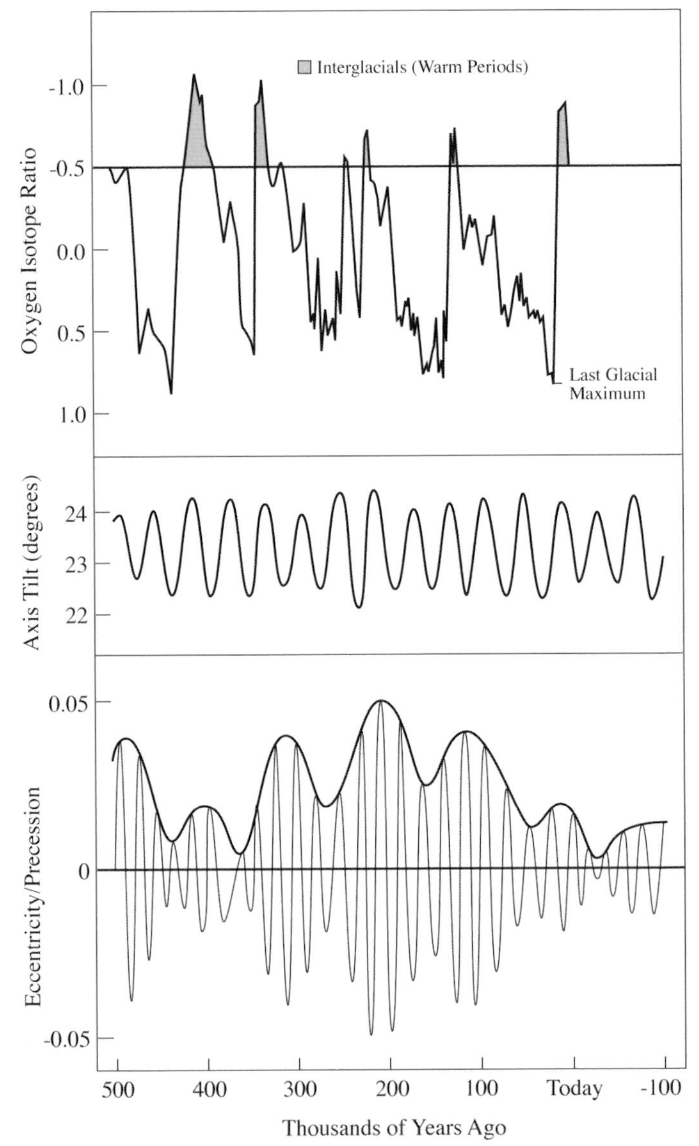

FIG. 9.13 Astronomical cycles cause the ice ages The advance and retreat of glaciers are controlled by changes in the Earth's orbital shape or eccentricity, and variations in its axial tilt and wobble. They alter the angles and distances from which solar radiation reaches Earth, and therefore change the amount and distribution of sunlight on our planet. The global ebb and flow of ice is inferred from the presence of lighter and heavier forms of oxygen, called isotopes, in the fossilized shells of tiny marine animals found in deep-sea sediments. During glaciations, the shells are enriched with oxygen-18 because oxygen-16, a lighter form, is trapped in glacial ice. The relative abundance of oxygen-18 and oxygen-16 (*top*) is compared with periodic 41,000-year variations in the tilt of the Earth's axis (*middle*) and in the shape, or eccentricity – longer 100,000-year variation, and wobble, or precession, of the Earth's rotation axis – shorter 23,000-year variation (*bottom*).

back again, occurs every 41,000 years. It is currently 23.5 degrees, and accounts for our yearly seasons. The greater the tilt is, the more intense the seasons in both hemispheres, with hotter summers and colder winters.

The third and longest cycle is due to a slow periodic change in the shape of the Earth's orbit every 100,000 years. As the orbit becomes more elongated, the Earth's distance from the Sun varies more during the year, intensifying the seasons in one hemisphere and moderating them in the other.

The astronomical theory for the recurring ice ages was not strongly supported until 1976 when American climate scientists James D. Hays (1933–) and John Imbrie (1925–) and English geophysicist Nicholas J. "Nick" Shackleton (1937–) published a key paper on the subject, entitled "Variations in the Earth's Orbit: Pacemaker of the Ice Ages" on 10 December 1976 in *Science.* They used an analysis of different types, or isotopes, of oxygen atoms in deep-sea sediments to infer the proportion of the world's water frozen within the glacial ice sheets at different times, revealing all three astronomical rhythms, with a dominant 100,000-year one. Analysis of ice cores drilled in Antarctica and Greenland between 1985 and 2005 confirmed that the major ice ages are initiated every 100,000 years by orbital-induced changes in the intensity and distribution of sunlight arriving at Earth.

It was somewhat surprising that the glaciers have advanced and retreated in synchronism with this longer rhythmic stretching of the Earth's orbit. The shorter cycles have a greater, direct effect on the seasonal change in sunlight, but apparently produce smaller changes in ice volume than the longer one that has a weaker seasonal effect. By itself the 100,000-year cycle does not appear strong enough to bring about direct alterations of the terrestrial climate, so it must be leveraged by some other factor.

Successive layers of frozen atmosphere have been laid down in Greenland and Antarctica, providing a natural archive of the Earth's climate over the past 420,000 years. Bubbles of air trapped in falling snowflakes and entombed in ice are deposited every year, building up on top of each other like layers of sediment. When extracted in deep ice cores, the air bubbles trapped in the ancient ice indicate that the Antarctic air temperatures and atmospheric heat-trapping gases rose and fell in tandem as the glaciers came and went and came again.

The temperatures go up whenever the levels of carbon dioxide and methane do, and they decrease together as well (Fig. 9.14). Scientists cannot yet agree whether an increase in greenhouse gases preceded or followed the rising temperatures, but the heat-trapping-gas increase does answer the riddle of why the largest climate variations occur every 100,000 years. Although the orbital variation in the intensity of incident solar radiation is far too small to directly create the observed temperature changes, the build up of greenhouse gases apparently amplifies effects triggered and timed by the rhythmic 100,000-year orbital change in the distance between the Earth and the Sun.

Since the present period of interglacial warmth has lasted about as long as any other interglacial, another ice age might soon be on its way. In fact, humans were most likely contributing significant quantities of greenhouse gases for thousands of years before the industrial revolution, offsetting the natural cooling trend of an approaching ice age. Clearing forests to increase tillable land would have produced carbon dioxide, and methane must have been released by irrigating extensive rice fields in Asia. The extra gases probably offset declines associated with variations in the Earth's orbit, and

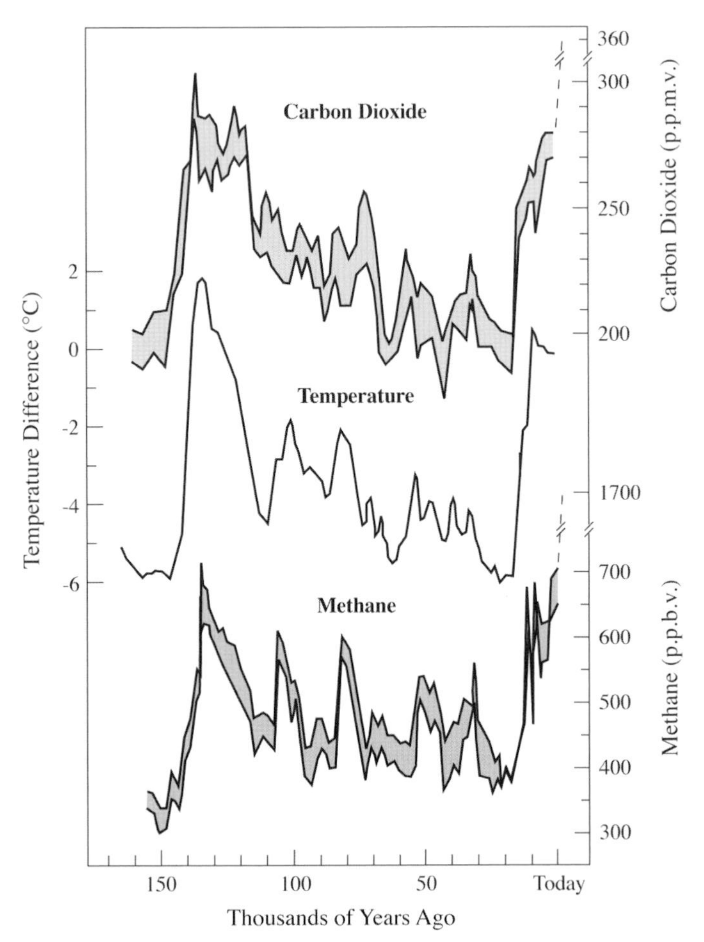

FIG. 9.14 Ice age temperatures and greenhouse gases Ice-core data indicate that changes in the atmospheric temperature over Antarctica closely parallel variations in the atmospheric concentrations of two greenhouse gases, carbon dioxide and methane, for the past 160,000 years. When the temperature rises, so does the amount of these two greenhouse gases, and *vice versa.* The strong correlation has been extended by a deeper Vostok ice core, to 3,623 meters in depth and the past 420,000 years. The carbon dioxide (parts per million per volume) and methane (parts per billion per volume) increases may have contributed to the glacial-interglacial changes by amplifying orbital forcing of climate change. The ice-core data does not include the past 200 years, shown as dashed and broken lines at the right. They indicate that the present-day levels of carbon dioxide and methane are unprecedented during the past four 150,000-year glacial-interglacial cycles. (Adapted from Claude Lorius, *EOS,* Vol. 69, No. 26, 1988.)

the associated distribution of sunlight on the globe, which would have otherwise made the planet colder and perhaps triggered an ice age.

Because the current level of greenhouse gases, recently deposited in our atmosphere by humans, far surpasses any natural fluctuation of these substances recorded during past ice ages (Fig. 9.14), we are not sure if the next ice age will dampen future global warming or whether recent global warming is delaying the coming of the ice. But when we completely exhaust oil supplies, in 100 years or less, and when our sources of coal and natural gas are also depleted, perhaps in a few hundred years, the ice should be on its way.

It is very hard to predict the distant future over much longer intervals of millions of years. We now live at times when polar ice caps exist, and the recurrent ice ages occur. But on a time scale of 100 million years, drifting continents can remove land from the polar regions, and permanent glaciers will not form there. A hundred million years ago, when the dinosaurs roamed the Earth, the climate was some 15-kelvin warmer everywhere than it is today, and there were no polar ice caps. Indeed, over most of its 4.6 billion years the Earth has probably been free of ice, even at its poles.

During the past 200 million years, powerful internal forces have reshaped the Earth, producing steady, irreversible changes in our global climate. Drifting continents collided to create towering mountain ranges, or split open to make way for new oceans, altering the flow of air or sea, which influence the climate, and paving the way for the growth of continental ice sheets. All of these forces continue today.

So, we really don't know if the distant future is one of fire or ice. I would side with the American poet Robert Frost (1874–1963) who wrote:

Some say the world will end in fire. Some say in ice.
From what I've tasted of desire
I hold with those who favor fire.
But if it had to perish twice,
I think I know enough of hate
To say that for destruction ice
Is also great
And would suffice.[46]

Quotation References

1. An incantation from Ptolemaic Egypt (373–30 BC). Quoted by Carl Sagan in *Cosmos.* Random House, New York 1980, p. 217.
2. Friedrich Nietzsche (1844–1900): *Thus Spoke Zarathustra – Of Immaculate Perception.* Translated by R. J. Hollingdale. Penguin Books, Harmondsworth 1961, p. 146.
3. Francis William Bourdillon (1852–1921): Among the Flowers (1878). In: *The Oxford Dictionary of Quotations,* Fourth Edition (Angela Partington, ed.). Oxford University Press, New York 1992, p. 138. Also in John Bartlett's *Familiar Quotations,* Sixteenth Edition (Justin Kaplan, ed.). Little, Brown and Company, Boston 1992, p. 563.
4. Letter written by Robert William Bunsen (1811–1899) to the English chemist Henry Enfield Roscoe (1833–1915) in November 1859. Quoted by Roscoe in: *The Life and Experiences of Sir Henry Enfield Roscoe,* London 1906, p. 71. It is reproduced by A. J. Meadows in his article: The Origins of Astrophysics, *The General History of Astronomy, Vol. 4, Astrophysics and Twentieth-Century Astronomy to 1950, Part A* (Owen Gingerich, ed.), Cambridge University Press, New York 1984, p. 5. Also see Gustav Robert Kirchhoff (1824–1877): On the Chemical Analysis of the Solar Atmosphere, *Philosophical Magazine and Journal of Science* **21,** 185–188 (1861). Reproduced by A. J. Meadows in: *Early Solar Physics,* Pergamon Press, Oxford 1970, pp. 103–106; Gustav Kirchhoff and Robert Bunsen: Chemical Analysis of Spectrum Observations, *Philosophical Magazine and Journal of Science* **20,** 89–109 (1860), **22,** 329–349, 498–510 (1861).
5. William Thomson (1824–1907): On the Age of the Sun's Heat, *Macmillan's Magazine,* March 5, 288–293 (1862). *Popular Lectures* **I,** 349–368. William Thomson is better known today as Lord Kelvin. Also see J. D. Burchfield: *Lord Kelvin and the Age of the Earth.* Science History Publications, New York 1975.
6. Arthur Stanley Eddington (1882–1944): The internal constitution of the stars. *Nature* **106,** 14–20 (1920), *Observatory* **43,** 341–358 (1920). Reproduced by Kenneth R. Lang and Owen Gingerich in: *A Source Book in Astronomy and Astrophysics, 1900–1975.* Harvard University Press, Cambridge, MA 1979, pp. 281–290.
7. Arthur Stanley Eddington (1882–1944): ibid., reference 6.
8. Jean Baptiste Perrin (1870–1942): Atomes et lumiere. *La Revue du Mois* **21,** 113–166 (1920).
9. Arthur Stanley Eddington (1882–1944): *The Internal Constitution of the Stars.* Dover, New York 1959, p. 301 (first edition 1926).
10. Herman Melville (1819–1891): *Moby Dick,* Marshall Cavendish Paperworks, London 1987; a reproduction of the 1922 edition, p. 141. According to Captain Ahab, all visible objects are but pasteboard masks, and man must strike through the mask. He must break through the prison wall. To Captain Ahab, the white whale was that wall. Nowadays, I am reminded of the song by the rock group The Doors entitled: *Break on Through to the Other Side.*
11. Paul Adrien Maurice "P.A.M." Dirac (1902–1984): Quantized Singularities in the Electromagnetic Field. *Proceedings of the Royal Society of London* **A133,** 60–72 (1931). This paper makes the first mention of the anti-electron. Carl David Anderson (1905–1991) discovered it in cosmic-ray cloud chamber tracks in 1932, and in the following year proposed the name positron. Carl David Anderson: The Positive

Electron. *Physical Review* **43,** 491–494 (1933)]. However, Dirac's theory apparently played no part whatsoever in the discovery of the positron. Also see Helge Kragh: *Dirac – A Scientific Biography.* Cambridge University Press, New York 1990.
12. Madonna: *Material Girl* (1985). Written by Peter Brown and Robert Rans, published by Candy Castle Music, BMI.
13. John Updike (1932–): Cosmic Gall. Originally published in: *The New Yorker,* 17 December 1960, p. 36. Reproduced in John Updike (1932–): *Telephone Poles and Other Poems.* Alfred A. Knopf, New York 1979, p. 5, and John Updike (1932–): *Collected Poems 1953-1993.* Alfred A. Knopf, New York 1993, p. 315.
14. Wolfgang Pauli (1900–1958): Remarks at the Seventh Solvay Conference, October 1933. Reproduced in the original French in: *Collected Scientific Papers of Wolfgang Pauli,* Vol. 2 (Ralph Kronig and Victor F. Weisskopf, eds.). Wiley Interscience, New York 1964, p. 1319. Quoted in English by Christine Sutton in: *Spaceship Neutrino.* Cambridge University Press, New York 1992, p. 19.
15. Frederick Reines (1918–1998) and Clyde L. Cowan (1919–1974): Telegram to Pauli dated 14 June 1956. Quoted in: Proceedings of the International Colloquium on the History of Particle Physics. *Journal de Physique* **43,** supplement C8, 237 (1982). Also quoted by Christine Sutton in: *Spaceship Neutrino.* Cambridge University Press, New York 1992, p. 44.
16. Yoji Totsuka (1942–): Recent results on solar neutrinos from Kamiokande. *Nuclear Physics* **B19,** 69–76 (1991).
17. John N. Bahcall (1934–2005) and Hans A. Bethe (1906–2005): A solution to the solar neutrino problem. *Physical Review Letters* **65,** 2233–2235 (1990).
18. Arthur D. "Art" McDonald (1943–): First results from the Sudbury Neutrino Observatory explain the missing solar neutrinos and reveal new neutrino properties. News Release of the Sudbury Neutrino Observatory on 18 June 2001.
19. Robert B. Leighton (1919–1997): Considerations on localized velocity fields in the solar atmosphere. *Aerodynamic phenomena in stellar atmospheres – International Astronomical Union Symposium No. 12. Supplemento del Nuovo Cimento* **22,** 321–325 (1961).
20. Robert B. Leighton (1919–1997), Robert W. Noyes (1934–) and George W. Simon (1934–): Velocity fields in the solar atmosphere. Preliminary report. *Astrophysical Journal* **135,** 497 (1962).
21. Edward N. Frazier (1939–): A spatiotemporal analysis of velocity fields in the solar photosphere. *Zeitschrift fur Astrophysik* **68,** 345 (1968).
22. Samuel Heinrich Schwabe (1789–1875): Sonnen 'Beobachtungen im Jahre 1843. *Astronomische Nachrichten* **20,** No. 495, 233–236 (1844). English translation: Solar Observations During 1843 in A. J. Meadows: *Early Solar Physics.* Pergamon Press, Oxford 1970, pp. 95–97.
23. Manuel John Johnson (1805–1859): Address Delivered by the President, M. J. Johnson, Esq., on presenting the medal of the society to M. Schwabe. *Monthly Notices of the Royal Astronomical Society* **16,** 129 (1857). Reproduced in part by H. H. Turner in: *Astronomical Discovery.* Edward Arnold, London 1904, pp. 156–176.
24. Horace W. Babcock (1912–2003): The topology of the Sun's magnetic field and the 22-year cycle. *Astrophysical Journal* **133,** 572–587 (1961).
25. Francis Bailey (1774–1844): Some remarks on the total eclipse of the Sun, on July 8th 1842. *Monthly Notices of the Royal Astronomical Society* **5,** 208–214 (1842).
26. Ludwig Franz Biermann (1907–1986): Solar corpuscular radiation and the interplanetary gas. *Observatory* *77,* 109–110 (1957). Reproduced by Kenneth R. Lang and Owen Gingerich in: *Source Book in Astronomy and Astrophysics, 1900–1975.* Harvard University Press, Cambridge, Massachusetts 1979, p. 148. Also see Eugene N. Parker (1927–): Dynamics of the interplanetary gas and magnetic fields. *Astrophysical Journal* **128,** 664–676 (1958).
27. Dante Alighieri (1265–1321): *The Comedy of Dante Alighieri, Cantica I Hell (L'Inferno).* Translated by Dorothy L. Sayers in: Basic Books, New York 1948, Canto XXVI, p. 236.
28. Joachim du Bellay (1522–1560): Les Regrets (1559). Quoted in *John Bartlett's Familiar Quotations, Sixteenth Edition* (Justin Kaplan, ed.). Little, Brown and Co., Boston 1992, p. 144.

29. Richard C. Carrington (1826–1875): Description of a singular appearance seen in the Sun on September 1, 1859. *Monthly Notices of the Royal Astronomical Society* **20,** 13–15 (1860). Richard Hodgson's (1804–1872) account "On a curious appearance seen in the Sun" is given on the following pages 15–16. Both articles are reproduced by A. J. Meadows in: *Early Solar Physics.* Pergamon Press, Oxford 1970, pp. 181–185.
30. Richard C. Carrington (1826–1875): ibid., reference 29.
31. Kristian Birkeland (1867–1917): Sur les rayons cathodiques sons l'action de forces magnetiques. *Archives des Sciences Physiques et Naturelles* **1,** 497 (1896). Birkeland's analysis of simultaneous magnetic data at several northern stations led him to the conclusion that large currents flowed along the Earth's magnetic field lines during auroras; see Kristian Birkeland (1867–1917): *The Norwegian Aurora Polaris Expedition 1902–1903. Vol. 1: On the Cause of Magnetic Storms and the Origin of Terrestrial Magnetism.* H. Aschehoug Co., Christiania 1908, 1913.
32. James Alfred Van Allen (1914–), Carl E. McIlwain (1931–), and George H. Ludwig (1927–): Radiation observations with satellite 1958 epsilon. *Journal of Geophysical Research* **64,** 271–280 (1959). Reproduced by Kenneth R. Lang and Owen Gingerich in: *Source Book in Astronomy and Astrophysics, 1900–1975.* Harvard University Press, Cambridge, Massachusetts 1975, pp. 150–151.
33. *King's Mirror* (1250): Quoted by Robert H. Eather in: *Majestic Lights – The Aurora in Science, History, and the Arts.* American Geophysical Union, Washington, D.C. 1980, p. 42. From *Kongespeilet (The King's Mirror)* translated by L. M. Larson, New York, Twayne Publishing 1917.
34. Fridtjof Nansen (1861–1930): Quoted by Robert H. Eather in: *Majestic Lights – The Aurora in Science, History, and the Arts.* American Geophysical Union, Washington, D.C. 1980, p. 205. From Fridtjof Nansen (1861–1930): *The Fram Expedition – Nansen in the Frozen World.* A. G. Holman, Philadelphia, 1897.
35. Vladimir Remek (1948–): Quoted in: *The Home Planet* (Kevin W. Kelly, ed.), with photograph number 31. Addison-Wesley, New York 1988.
36. Robert Browning (1812–1889) *Easter Day,* and Wallace Stevens (1879–1955) *The Auroras of Autumn.*
37. Edward Sabine (1788–1883): Letter to John Herschel 16 March 1852. *Herschel Letters No. 15.235* (Royal Society). Quoted by A. J. Meadows and J. E. Kennedy in: The origin of solar- terrestrial studies. *Vistas in Astronomy* **25,** 419–426 (1982).
38. Edward Walter Maunder (1851–1928): Magnetic disturbances, 1882 to 1903, as recorded at the Royal Observatory, Greenwich, and their association with sunspots. *Monthly Notices of the Royal Astronomical Society* **65,** 2–34 (1905). For the M regions also see Julius Bartels (1899–1964): Terrestrial magnetic activity and its relation to solar phenomena. *Journal of Geophysical Research* **3,** 1–52 (1932). The role of coronal mass ejections in creating intense geomagnetic storms is discussed in John "Jack" T. Gosling (1938–): The solar flare myth. *Journal of Geophysical Research* **98,** 18937–18949 (1993).
39. Fred Hoyle (1915–2001): Lecture in 1948. Quoted by Jon Darius in: *Beyond Vision.* Oxford University Press, New York 1984, p. 142.
40. James Buchli (1945–): Quoted in the narration for: *The Blue Planet* (1990).
41. John Tyndall (1820–1893): On the absorption and radiation of heat by gases and vapours, and on the physical connexion of radiation, absorption, and conduction. *Philosophical Magazine and Journal of Science* **22A,** 276–277 (1861).
42. Svante August Arrhenius (1859–1927): On the influence of carbonic acid in the air upon the temperature of the ground. *Philosophical Magazine and Journal of Science* **41,** 237–268 (1896). Arrhenius' ideas were taken up in greater detail by Thomas Chrowder Chamberlin (1843–1928): An attempt to frame a working hypothesis of the cause of the glacial periods on an atmospheric basis. *The Journal of Geology* 7, 545–584 (1899). Here Chamberlin drew attention to the absorption of atmospheric carbon dioxide by the ocean, the variable depletion of carbon dioxide by weathering on globally-changing land area, and to the amplification of the greenhouse effect by

increased water vapor evaporated from hotter seas.

43. Gordon Miller Bourne "G. M. B." Dobson (1889–1976): Forty years research on atmospheric ozone at Oxford: A history. *Applied Optics* 7, 387–405 (1968).
44. Mario J. Molina (1943–) and F. Sherwood Rowland (1927–): Stratospheric sink for chlorofluoromethanes. Chlorine atomic-catalysed destruction of ozone. *Nature* **249,** 810–812 (1974).
45. Roger Revelle (1909–1991) and Hans E. Suess (1909–1993): Carbon dioxide exchange between atmosphere and ocean and the question of an increase of atmospheric carbon dioxide during the past decades. *Tellus* **9,** 18–27 (1957).
46. Robert Frost (1874–1963): Fire and Ice, in: *The Poetry of Robert Frost* (Edward Connery Lathem, ed.). Holt, Rinehart and Winston, New York 1979, pp. 220–221.

GLOSSARY

A

Absolute luminosity: The total radiant energy of a celestial object emitted per unit time, and a measure of its intrinsic brightness. Denoted by L, the amount of energy radiated per unit time, measured in units of watts. One Joule per second is equal to one watt of power, and to ten million erg per second. The absolute luminosity of the thermal radiation from a blackbody is given by the Stefan-Boltzmann law in which $L = 4\pi\sigma R^2 T_{eff}^4$, where $\pi = 3.14159$, the Stefan-Boltzmann constant, $\sigma = 5.67051 \times 10^{-8}$ Joule $m^{-2}\,°K^{-4}\,s^{-1}$, the radius is R and the effective temperature is T_{eff}.

Absolute temperature: The temperature as measured on a scale whose zero point is absolute zero, the point at which all motion at the molecular level ceases. The unit of absolute temperature is the degree kelvin, denoted by °K or just K and written kelvin without a capital K. The freezing temperature of water is 273 °K and the boiling temperature of water is 373 °K. Absolute zero is 0 °K. The equivalent temperature in degrees Celsius is °C = °K – 273, and the equivalent temperature in degrees Fahrenheit is °F = (9 °K/5) – 459.4. *See* Celsius, Fahrenheit and Kelvin.

Absolute zero: The lowest possible temperature, –273 degrees Celsius and 0 degrees on the Kelvin scale. *See* absolute temperature.

Absorption: The process by which the intensity of radiation decreases as it passes through a material medium. The energy lost by the radiation is transferred to the medium.

Absorption line: A "dark" line of decreased radiation intensity at a particular wavelength of the electromagnetic spectrum, produced when radiation from a distant source passes through a gas cloud closer to the observer. An absorption line can be formed when a cool, tenuous gas, between a hot, radiating source and an observer, absorbs electromagnetic radiation of that wavelength, as in the spectra of stars whose hot internal radiation is absorbed by the cooler, outer stellar layers. This spectral feature looks like a line when the radiation intensity is displayed as a function of wavelength; such a display is called a spectrum. Different atoms, ions and molecules produce characteristic patterns of absorption lines, and observations of these lines enable identification of the chemical ingredients of the gas between the light source and the observer. Absorption lines can also be used to determine the velocity of the source, using the Doppler effect. The absorption lines in the Sun's spectrum are often called Fraunhofer lines. *See* Doppler effect, Fraunhofer lines, spectroheliograph and spectrum.

Acoustic waves: Sound waves that propagate through the interior of the Sun. They are produced when there is a small displacement of the gas and pressure is the dominant restoring force. Convective turbulence in the Sun and some other stars acts as a source of acoustic waves. Sound waves in the Sun produce oscillations of its visible photosphere with periods of about five minutes; observations of these oscillations can be used to infer the interior constitution of the Sun. Acoustic waves also play a role in heating the chromosphere. *See* five minute oscillations and helioseismology.

Active region: A region in the solar atmosphere, from the photosphere to the corona, that develops when strong magnetic fields emerge from inside the Sun. The magnetized realm in, around and above sunspots is called an active region. Radiation from active regions is enhanced, when compared to neighboring areas in the chromosphere and corona, over the whole electromagnetic spectrum, from X-rays to radio waves. Active regions may last from several hours to a few months. They are the sites of intense explosions, called solar flares, which last a few minutes to hours. The number of active regions varies in step with the 11-year sunspot cycle. *See* solar flare, solar activity cycle, and sunspot cycle.

Alfvén velocity: The velocity of propagation of disturbances in a magnetized plasma propagates along the magnetic field lines. The Alfvén velocity is given by $B/(4\pi\rho)^{1/2}$ for a magnetic field strength of B and a gas mass density of ρ.

Alfvén wave: A wave motion occurring in a magnetized plasma in which the magnetic field lines oscillates transversely to the direction of propagation without a change in magnetic field strength. The tension in the field lines acts as a restoring force. These magnetohydrodynamic waves propagate at the Alfvén velocity in the direction of the magnetic field lines.

Alpha particle: The nucleus of a helium atom, consisting of two protons and two neutrons. The helium nucleus has a charge twice that of a proton and a mass just 0.007, or 0.7 percent, less than the mass of four protons. The alpha particle is a helium ion. *See* helium and ion.

Ångström: A unit of wavelength designated by the symbol Å, and equal to 10^{-10} meters, or 0.00000001 of a centimeter and 0.1 nanometers. An Ångström is on the order of the size of an atom. Blue light has a wavelength of about 4400 Å, yellow light 5500 Å, and red light 6500 Å. The unit is named after the Swedish astronomer **Anders Jonas Ångström (1814–1874)** who in 1868 published *Researches of the Solar System,* in which he presented measurements of the wavelengths of more than a hundred dark absorption lines in the Sun's spectrum using this unit of measurement.

Antielectron: The antiparticle of an electron, also called a positron.

Antiparticle: A particle equal in mass and most other properties to an elementary particle, but opposite in charge. The antiparticle of an electron is the positron or "positive electron", and the antiparticle of the neutrino is an antineutrino. When a particle meets with its antiparticle, they annihilate each other, producing energetic radiation called gamma rays.

Arcade: A series of arches formed by magnetic loops that confine hot gas.

Arc second: *See* Second of arc.

Astronomical unit: Abbreviated AU, a unit of astronomical distances especially within the Solar System, equal to the mean distance between the Earth and the Sun, about 150 million kilometers or 92.8 million miles. The astronomical unit has a more exact value of $1.495\,978\,706\,1 \times 10^{11}$ meters, which is equal to 499.012 light-seconds. The astronomical unit is also defined as the distance from the Sun at which a mass-less particle in an unperturbed orbit would have an orbital period of 365.256 898 3 days.

Atmosphere: The gases surrounding the surface of a planet or natural satellite, and the outermost gaseous layers of a star, held near them by their gravity. Since gas has a natural tendency to expand into space, only bodies that have a sufficiently strong gravitational pull can retain atmospheres. The ability of a planet or satellite to retain an atmosphere depends on the gas temperature, determined by its distance from the Sun, and its escape velocity, which increases with the object's mass. Hotter, lighter gas molecules move faster, and are more likely to escape from a planet or satellite; less massive objects such as our Moon are less likely to retain an atmosphere. We use the term atmosphere for the tenuous outer material of the Sun because it is relatively transparent at visible wavelengths. The solar atmosphere includes, from the deepest layers outward, the photosphere, the chromosphere, and the corona. The Earth's atmosphere is transparent and consists mainly of molecular nitrogen (78 percent) and molecular oxygen (21 percent) with trace amounts of carbon dioxide (0.033 percent) and water vapor (variable, a few times 0.0001 percent).

Atmospheric window: A wavelength band in the electromagnetic spectrum for which radiation is able to pass through the Earth's atmosphere with relatively little attenuation by absorption, scattering or reflection. There are two main windows – the optical window of visible light and the radio window.

Atom: The smallest particle of a chemical element that still has the characteristics and properties of that element. An atom is composed of a dense and massive nucleus, containing protons and neutrons, surrounded by electrons. The hydrogen atom has one proton and no neutrons in its nucleus. A neutral atom has as many electrons as there are protons, and so is without net charge. An ionized atom has fewer electrons than protons, and therefore a positive charge. The atomic number is equal to the number of protons in the atom's nucleus, while the atomic mass number is equal to the total number of protons and neutrons in the nucleus.

AU: Abbreviation for astronomical unit. *See* astronomical unit.

Aurora: A display of rapidly varying, colored light usually seen from magnetic polar regions of a planet. The light is given off by collisions between charged particles and the atoms, ions and molecules in a planet's upper atmosphere or ionosphere. Auroras are visible on Earth as the aurora borealis, or northern lights, near the North Pole, and the aurora australis, or southern lights, near the South Pole. They include the green and red emission lines from oxygen atoms and nitrogen molecules that have been excited by high-speed electrons from the Sun. Terrestrial auroras occur at heights of around 100 to 250 thousand kilometers, which is within the Earth's ionosphere.

Auroral oval: One of the two oval-shaped zones around the Earth's geomagnetic poles in which the auroras are observed most often. They are located at latitudes of about 67 degrees north and south, and are about 6 degrees wide. Their position and width both vary with geomagnetic and solar activity. The auroral ovals are detected from Earth-orbiting satellites. Ground-based observers detect only a section of an oval, as shimmering ribbons and curtains of light.

B

Baily's beads: A string of bright lights observed at the extreme edge of the Sun's disk, seen during a total eclipse of the Sun just before or after totality. They are caused by sunlight shining through the valleys on the Moon's edge, or limb, while mountains on the limb block other rays of sunlight. The English astronomer **Francis Baily (1774–1844)** first described them in his account of the total eclipse of the Sun on 15 May 1836, and hence the name Baily's beads.

Balmer series: A mathematical formula devised in 1885 by the Swiss mathematics teacher **Johann Balmer (1825–1898)** to give the wavelengths, or frequencies, of the visible spectral lines of the hydrogen atom seen in the light of the Sun and other stars. The first Balmer line is called the red hydrogen alpha line, designated Hα, at a wavelength of 656.3 nanometers, which is used to detect the chromosphere across the face of the Sun. *See* Bohr atom.

Beta decay: The disintegration of a radioactive nucleus, in which a neutron turns into a proton by ejecting an electron, historically called a beta particle, and an antineutrino. The resulting nucleus has a charge one greater than the initial one.

Beta particle: An electron emitted from an atomic nucleus as a result of a nuclear reaction or in the course of radioactive decay. Such particles were originally called beta rays. *See* neutrino.

Billion: One thousand million, written 1,000,000,000 or 10^9.

Blueshift: The Doppler shift in the wavelength of a spectral line toward a shorter, bluer wavelength, caused when the source of radiation and its observer are moving toward each other. The greater the blueshift is, the faster an object is moving toward us. *See* Doppler shift.

Bohr atom: In 1913 the Danish physicist **Niels Bohr (1885–1962)** proposed a model, now known as the Bohr Atom, in which the lone electron in a hydrogen atom revolves about the atomic nucleus, a proton, is specific orbits with definite, quantized values of energy, and an electron only emits or absorbs radiation when jumping between these allowed orbits. The model explains the wavelengths of the visible radiation emitted by the hydrogen atoms in the Sun, described in the mathematical formula derived in 1885 by the Swiss mathematics teacher **Johann Balmer (1825–1898).** *See* Balmer series.

Bow shock: The boundary on a planet's sunlit side where the solar wind is deflected and there is a sharp decrease in the wind's velocity. The plasma of the solar wind is heated and compressed at the bow shock. The stream of particles in the solar wind is deflected around the planet, like water flowing past the bow of a moving ship.

Bremsstrahlung: Radiation that is emitted when an energetic electron is deflected and accelerated by an ion. It is also called a free-free transition because the electron is free both before and after the encounter, remaining in an unbound hyperbolic orbit without being captured by the ion. *Bremsstrahlung* is German for "braking radiation."

Brightness: The amount of light coming from an object, usually meaning the luminous intensity or amount of energy per second. *See* luminosity.

Burst: A sudden, transient increase in solar radio emission during a solar flare, caused by energetic electrons. They are probably caused by the sudden release of large amounts (up to 10^{25} Joule) of magnetic energy in a relatively small volume in the solar corona.

Butterfly diagram: A plot of the heliographic latitudes of sunspots over the course of the 11-year sunspot cycle. At the start of a cycle there are sunspots at solar latitudes of up to 35 to 45 degrees north and south, but very few near the equator. As the cycle progresses, sunspots occur nearer the equator. This graphical representation of sunspot latitudes as a function of time was first plotted in 1922 by Edward Walter Maunder and is also known as the Maunder diagram. The plot resembles butterfly wings, which gives the diagram its popular name. *See* heliographic latitude and sunspot cycle.

C

Calcium H and K lines: Spectral lines of ionized calcium, denoted by Ca II, in the violet part of the spectrum at 396.8 nanometers (H) and 393.4 nanometers (K). They are conspicuous features in the spectra of many stars, including the Sun. The designations H and K were given by the German physicist **Joseph von Fraunhofer (1787–1826)** in1814, and are still commonly used. The Sun's chromosphere and its magnetic chromospheric network are detected across the face of the Sun using the calcium H and K lines with a spectroheliograph. *See* chromosphere, chromospheric network, Fraunhofer lines, H and K lines, plage and spectroheliograph.

Carbon dioxide: A molecule, symbolized by CO_2, composed of one atom of carbon, denoted by C, and two atoms of oxygen, each abbreviated by O. Carbon dioxide gas is the main constituent of the atmospheres of Venus and Mars, but only a fraction of the Earth's atmosphere. Carbon dioxide is a greenhouse gas that traps solar energy and warms the surfaces of the Earth and Venus. *See* greenhouse effect, and global warming.

Celsius: A unit of temperature denoted by C or by °C for degrees Celsius. Absolute zero on the Celsius temperature scale is at −273 °C, the freezing point of water is at 0 °C, and the boiling point of water is at 100 °C. The equivalent temperature in degrees kelvin is given by °K = °C + 273. The equivalent temperature in degrees Fahrenheit is given by °F = 9 °C/5 + 32. The temperature unit is named for the Swedish astronomer **Anders Celsius (1701–1744),** who proposed the temperature scale in 1742. *See* absolute temperature, Fahrenheit, and Kelvin.

Centigrade: An older name for the Celsius unit of temperature, derived from the fact that there are 100 degrees between the freezing and boiling temperatures of water. *See* Celsius.

Charged particles: Fundamental components of subatomic matter, such as protons, other ions and electrons, which have electrical charge. The charged subatomic particles are surrounded by electrical force fields that attract particles of opposite charge and repel those of like charge. Charged particles are either negative (electron) or positive (proton) and are responsible for all electrical phenomena.

Cherenkov radiation: Light or other forms of electromagnetic radiation emitted when a charged particle has a velocity that exceeds the velocity of light in a medium. A neutrino can be detected by the cone of blue Cherenkov light emitted when a neutrino enters a large underground tank of water and accelerates an electron by collision to a velocity larger than the velocity of light in water. The Cherenkov effect is named for the Russian physicist **Pavel A. Cherenkov (1904–1990),** who first observed it in 1934. *See* Kamiokande, Sudbury Neutrino Observatory, and Super-Kamiokande.

Chlorine experiment: The subterranean neutrino detector located in the Homestake Gold Mine near Lead South Dakota, in which a neutrino strikes a chlorine nucleus in a huge tank of cleaning fluid to produce a nucleus of radioactive argon. The American physicist **Raymond Davis Jr. (1914–)** used the experiment to detect neutrinos from the Sun for more than a quarter of a century, beginning in 1967, always detecting about one-third of the expected amount of solar neutrinos. *See* Homestake, and solar neutrino problem.

Chromosphere: The layer or region of the solar atmosphere lying above the photosphere and beneath the transition region and the corona. The Sun's temperature rises to about 10 000 kelvin in the chromosphere. The name literally means a "sphere of color". The chromosphere is normally invisible because of the glare of the photosphere shining through it, but it is briefly visible near the beginning or end of a total solar eclipse as a spiky red rim around the Moon's disk. At other times, the chromosphere can be studied by spectroscopy, observing it across the solar disk in the red light of the hydrogen-alpha line or the calcium H and K lines. Solar filaments, plage, prominences and spicules are all seen in the chromosphere. A thin transition region separates the chromosphere and the corona. *See* calcium H and K lines, chromospheric network, corona, filament, hydrogen-alpha line, photosphere, plage, prominence, spicule, and transition region.

Chromospheric network: A large-scale cellular pattern visible in spectroheliograms taken in the calcium H and K lines. The network appears at the boundaries of the photosphere supergranulation cells, and contains magnetic fields that have been swept to the edges of the cells by the flow of material in the cells. *See* calcium H and K lines, spectroheliograph, and supergranulation.

Climate: The average weather conditions of a place over a period of years.

CO_2: *See* carbon dioxide.

Continuous spectrum: An unbroken distribution of radiation over a broad range of wavelengths.

Continuum: That part of a spectrum that has neither absorption nor emission lines, but only a smooth wavelength distribution of radiant intensity. *See* continuous spectrum, and spectrum.

Convection: The transport of energy from a lower, hotter region to a higher, cooler region by the physical upwelling of hot matter. A bubble of gas that is hotter than its surroundings expands and rises vertically, resulting in transport and mixing. When it has cooled by passing on its extra heat to its surroundings, the bubble sinks again. Convection can occur when there is a substantial decrease in temperature with height such as the Earth's troposphere, the Sun's convective zone or a boiling pot of water.

Convective zone: A layer in a star in which convection currents, or mass motions, are the main mechanism by which energy is transported outward. In the Sun, a convective zone extends from 0.713 of the solar radius to just below the photosphere. The opacity in the convective zone is so large that energy cannot be transported by radiation.

Core: The central region of a planet, star or other celestial object. In solar and stellar astronomy, the core is the central location where energy is generated by nuclear reactions. *See* fusion, and proton-proton chain.

Corona: A shimmering halo of pearl-white light seen momentarily at the limb, or apparent edge, of the Sun during a total solar eclipse, called the *corona,* from the Latin word for "crown." The outermost, high-temperature region of the solar atmosphere, above the chromosphere and transition region, consisting of almost fully ionized gas contained in closed magnetic loops, called coronal loops, or expanding out along open magnetic field lines to form the solar wind. In 1941 the Swedish astronomer **Bengt Edlén (1906–1993)** and the German astronomer **Walter Grotrian (1890–1954)** showed that emission lines from the solar corona were forbidden transitions of atoms that are highly ionized at an unexpectedly high, million-degree temperature. In 1958 the American physicist **Eugene N. Parker (1927–)** showed that the million-degree electrons and protons in the corona will overcome the Sun's gravity and accelerate to supersonic speeds, naming the resultant radial outflow the solar wind. Even near the Sun, the corona is a highly rarefied, low-density gas, with electron densities of less than 10^{16} electrons per cubic meter, heated to temperatures of millions of degrees. The corona is briefly visible to the unaided eye during a total eclipse of the Sun, for at most 7.5 minutes; at other times it can be observed in visible white light by using a special instrument called a coronagraph. The corona can always be observed across the solar disk at X-ray and radio wavelengths. The shape of the corona is determined by the distribution of solar magnetic fields, which varies with the solar activity cycle. *See* coronagraph, coronal hole, coronal loop, coronal mass ejection, solar activity cycle, and solar wind.

Coronagraph: An instrument, used in conjunction with a telescope, that makes it possible to mask the light from the Sun or other star, in order to observe gas, dust or larger objects very close to the star. A coronagraph is used to observe the faint solar corona in white light, or in all the colors combined, at times other than during a solar eclipse. The bright light of the Sun's photosphere is blocked out by an occulting disk, providing an artificial eclipse, with additional precautions for removing all traces of stray light. The French astronomer **Bernard Lyot (1897–1952)** invented the coronagraph in 1930. Even with a coronagraph located at a high site where the sky is very clear, scattering of light by the Earth's atmosphere is a problem. Coronagraphs therefore work best when placed on satellites above the Earth's atmosphere.

Coronal green line: An emission line of Fe XIV at 530.3 nanometers, the strongest visible line in the solar corona.

Coronal hole: An extended region in the solar corona where the density and temperature are lower than other places in the corona, and the coronal emission lines and the extreme ultraviolet and soft X-ray coronal emission are abnormally faint or absent. The weak, diverging and open magnetic field lines in coronal holes extend radially outward and do not immediately return back to the Sun. The high-speed part of the solar wind streams out from coronal holes. The low density of the gas makes these parts of the corona appear dark in extreme ultraviolet and soft X-ray images of the Sun, as if there were a hole in the corona. Coronal holes are nearly always present near the solar poles, and can also occur at lower solar latitudes.

Coronal loop: A magnetic loop that passes through the corona and joins regions, called footpoints, of opposite magnetic polarity in the underlying photosphere. Coronal loops can have exceptionally strong magnetic fields, and they often contain the dense, million-degree coronal gas that emits intense X-ray radiation. *See* footpoint.

Coronal mass ejection: Abbreviated CME, the transient ejection of plasma and magnetic fields from the Sun's corona into interplanetary space, detected by white-light coronagraphs. A large body of magnetically confined, coronal material being released from the corona. A coronal mass ejection contains 5 billion to 50 billion tons, or 5 million million to 50 million million kilograms, of gas, and can travel through interplanetary space at a high, supersonic speeds up to 1200 kilometers per second. It is often associated with an eruptive prominence, and sometimes with a strong solar flare. Coronal mass ejections produce intense shock waves, accelerate vast quantities of energetic particles, grow larger than the Sun in a few hours, and when directed at Earth can cause intense geomagnetic storms, disrupt communications, damage satellites and produce power surges on electrical transmission lines. *See* geomagnetic storm, and prominence.

Coronal streamer: A magnetically confined, loop-like coronal structure in the low corona straddling a magnetic neutral line on the solar photosphere. These high-density, bright coronal structures have ray-like stalks that extend radially outward to large distances in the outer corona. Near the minimum in the 11-year solar activity cycle, coronal streamers are located mostly near the solar equator and appear to be the source of the slow-speed solar wind. *See* helmet streamer, solar activity cycle and solar wind.

Coronal transient: *See* coronal mass ejection.

Coronium: A supposedly unknown chemical element emitting unidentified emission lines in the spectrum of the solar corona. In 1941, the Swedish astronomer **Bengt Edlén (1906–1993)** and the German astronomer **Walter Grotrian (1890–1954)** showed that the mysterious lines are produced by highly ionized forms of known elements, such as iron, in the unexpectedly high, million-degree temperature of the solar corona.

Corpuscular radiation: Charged particles, mainly protons and electrons, emitted by the Sun. The stream of electrically charged particles was hypothesized in the 1950s and later renamed the solar wind. *See* solar wind.

Cosmic rays: High-energy, charged, subatomic particles that enter the Solar System from interstellar space, moving at speeds approaching the speed of light and attaining energies greater than a million eV, typically a billion, or 10^9, eV and as large as 10^{20} eV. Cosmic rays were discovered in 1912 by the Austrian physicist **Victor F. Hess (1883–1964).** In 1925 the American physicist **Robert Millikan (1868–1953)** conclusively showed that they come from outer space, giving them the name cosmic rays. In 1933 the American physicist **Arthur H. Compton (1892–1962)** demonstrated that the "rays" are charged particles. Cosmic rays can smash into atoms and split them apart, creating lighter elements, such as lithium, beryllium, and boron. Protons are the most abundant kind of cosmic rays, but they include lesser amounts of heavier atomic nuclei, such as alpha particles or helium nuclei, and electrons. Cosmic rays beyond the Earth's atmosphere are known collectively as primary cosmic rays. When they enter the Earth's atmosphere, cosmic rays collide with atmospheric atoms and produce various atomic and subatomic particles, called secondary cosmic rays, which can be detected at ground level. Sometimes the term galactic cosmic ray is used to distinguish those coming from interstellar space from high-energy, charged particles coming from the Sun. The latter have historically been called solar cosmic rays.

Coulomb barrier: The electric field repulsion experienced during the approach of two charged subatomic particles, each of positive charge or both with negative charge. When the Coulomb barrier of two protons is overcome by the tunnel effect, nuclear reactions that produce the Sun's energy and radiation can occur. *See* tunnel effect.

Current sheet: The two-dimensional surface that separates magnetic fields of opposite polarities or directions.

D

Degree: The unit of angular measurement, equal to 1/360 of a full circle. There are 90 degrees from the zenith to the horizon, and the angle subtended by the Sun and the Moon is about one half of a degree. A degree is also a unit of temperature, as in degrees Celsius, Fahrenheit or Kelvin.

Density: An object's mass divided by its volume, also known as the mass density. *See* mass density.

Deuterium: The rare, heavy isotope of hydrogen containing both a proton and a neutron in its nucleus, also known as heavy hydrogen. An atom of light hydrogen, which is much more abundant than deuterium, contains one proton and no neutron in its nucleus. Deuterium was discovered in 1931 by the American chemist **Harold Clayton Urey (1893–1981).**

Deuteron: The nucleus of a deuterioum atom. This nucleus contains one neutron and one proton.

Diameter: The length of an imaginary straight line passing through the center of a spherical object, equal to twice the radius of the sphere.

Diamond-ring effect: An intense point of light seen during a total eclipse of the Sun, just before or after totality, when just the bright central part of the edge of the Sun's disk is visible, giving the appearance of a diamond ring.

Differential rotation: Rotation of the different parts of a non-solid body or collection of bodies at different rates. The outer layers of the giant planets and the Sun exhibit differential rotation, with their middle equatorial regions moving faster than the polar ones. Stars and interstellar gas also rotate differentially, revolving about the galactic center in independent orbits at speeds that can differ. The particles in the rings of Saturn revolve around the planet at speeds that decrease with increasing distance, just as more distant planets move about the massive Sun with slower speeds and longer orbital periods. Stars that are nearer the galactic center also revolve about it at faster speeds than more distant stars.

Diffraction grating: A metal or glass surface on which a large number of equidistant parallel lines have been ruled, typically at 100 to 1000 per millimeter. Light striking a grating is dispersed by diffraction into a high-quality spectrum, which is reflected by a metal grating and transmitted by a glass one.

Dipole: Magnetic fields that include both a north and a south magnetic pole.

Disk: The visible part of the Sun or other cosmic object.

D layer: The lowest part of the Earth's daytime ionosphere at heights between about 50 and 90 kilometers. This is the layer that reflects radio waves at wavelengths longer than about 10 meters, or frequencies less than about 30 MegaHertz.

Doppler effect: The change in the observed wavelength or frequency of sound or electromagnetic radiation due to the relative motion between the observer and the emitter along the observer's line of sight. The change is to longer wavelengths when the source of waves and the observer are moving away from each other, and to shorter wavelengths when they are moving toward each other. The Doppler effect produces the change in pitch of a siren as an ambulance speeds past. In astronomy, the Doppler effect is used to detect and measure relative motion along the line of sight, determining the radial velocity from the Doppler shift in wavelength. The effect is named after the Austrian physicist **Christian Doppler (1803–1853),** who first described it in 1842. *See* blueshift, radial velocity, and redshift.

Doppler shift: A change in the apparent wavelength or frequency of the radiation emitted from a moving source, caused by its relative motion along the line of sight, either toward or away from the observer. An object moving away from the observer will appear to be emitting radiation at a longer wavelength, or lower frequency, than if at rest or non-moving. A Doppler shift in the spectrum of an astronomical object is commonly described as a redshift when it is towards longer wavelengths (object receding) and as a blueshift when it is towards shorter wavelengths (object approaching). A measurement of the Doppler shift makes it possible to determine the radial velocity, or velocity along the line of sight. *See* blueshift, radial velocity, and redshift.

Dynamo: An electric generator that employs a spinning magnetic field to produce electricity. The dynamo mechanism generates magnetic fields through the interaction of the convection of conducting matter with rotation and rotational shear. A dynamo converts the kinetic energy of a moving electrical conductor to the energy of electric currents and a magnetic field. It uses the motion of a conducting, convecting fluid to generate or sustain a magnetic field. The terrestrial magnetic field is supposed to be generated by such a dynamo, located in the Earth's molten core. The Sun's magnetic field is also sustained by an internal dynamo, perhaps located at a region of rotational shear just below the convective zone.

E

Earth: The third planet from the Sun, which we live on. The Earth has a mean radius of 6371 kilometers, a mean mass density of 5515 kilograms per cubic meter, and an age of 4.6 billion years. The mean distance between the Earth and the Sun is one Astronomical Unit, denoted by AU. The Earth's atmosphere is mainly composed of molecular nitrogen, at 77 percent, and molecular oxygen, at 21 percent. The greenhouse effect, caused by relatively small amounts of atmospheric carbon dioxide and water vapor, warms the surface of the Earth to an average temperature of about 288 kelvin (15 degrees Celsius and 59 degrees Fahrenheit), and without this effect the surface temperature would be 255 kelvin, below the freezing point of water at 273 kelvin. Global warming produced by human activity, such as the burning of coal, gas and oil to produce carbon dioxide, will continue to increase the Earth's surface temperature by the greenhouse effect. Liquid oceans cover 71 percent of the Earth's surface, but new oceans form and old ones close as the continents disperse and then reassemble.

Eclipse: The partial or total obscuration of the light from an astronomical object by another such object. In a solar eclipse, the Moon passes between the Sun and the Earth, blocking the Sun's light. In a lunar eclipse, the Earth passes between the Moon and the Sun, blocking the solar light that the Moon reflects. The eclipse of a star by the Moon or by a planet or other body in the Solar System is called an occultation. A total solar eclipse can occur only at the new Moon, and it has a maximum duration of 7.5 minutes. During the brief moments of a total solar eclipse, darkness falls, and the outer parts of the Sun, the chromosphere and the corona, are seen. At any given point on the Earth's surface, a total solar eclipse occurs, on the average, once every 360 years. *See* solar eclipse.

E layer: A daytime layer of the Earth's ionoshere roughly between altitudes of 95 to 140 kilometers.

Electromagnetic radiation: Radiation that carries energy and moves through vacuous space in periodic waves at the speed of light, propagating by the interplay of oscillating electrical and magnetic fields. The velocity of light is usually designated by the letter c and has a value of 299 792.458 kilometers per second. Electromagnetic radiation includes visible light, radio waves, infrared radiation, ultraviolet radiation, X-rays and gamma rays. Electromagnetic radiation, in common with any wave, has a wavelength, denoted by λ, and a frequency, denoted by ν; their product is equal to the velocity of light, or $\lambda\nu = c$, so the wavelength decreases when the frequency increases and *vice versa*. The energy associated with the radiation increases in direct proportion to frequency, and this energy, known as the photon energy and denoted by E, is given by $E = h\nu = hc/\lambda$, where Planck's constant $h = 6.6261 \times 10^{-34}$ Joule second. There is a continuum of electromagnetic radiation – from long-wavelength radio waves of low frequency and low energy, through visible-light waves, to short-wavelength X-rays and gamma rays of high frequency and high energy. *See* electromagnetic spectrum, gamma-ray radiation, radio radiation, ultraviolet radiation, visible radiation, and X-ray radiation.

Electromagnetic spectrum: All types and wavelengths of electromagnetic radiation, from the most energetic and shortest waves, the gamma rays, to the least energetic and longest ones, the radio waves. From short wavelengths to long ones, the electromagnetic spectrum includes gamma rays, X-rays, ultraviolet radiation, visible light, infrared radiation and radio waves; visible light comprises just one small segment of this much broader spectrum. *See* gamma-ray radiation, optical spectrum, radio radiation, ultraviolet radiation, visible radiation, and X-ray radiation.

Electron: An elementary, negatively charged, subatomic particle that surrounds the positively charged nucleus of an atom, but can exist in isolation outside an atom. The English physicist **Joseph John "J. J." Thomson (1856–1940)** discovered the negatively charged "corpuscles" in 1897 while investigating cathode rays. The Dutch physicist **Hendrik Lorentz (1853–1928)** named them "electrons"; a term previously introduced by the Irish physicist **George Johnstone Stoney (1826–1911).** The interactions between electrons of neighboring atoms create the chemical bonds that link atoms together as molecules. A neutral, or uncharged, atom has as many electrons as positively charged protons, which reside in the nucleus. An ionized atom is usually positively charged and has fewer electrons than protons. Free electrons have broken away from their atomic bonds and are not bound to atoms. At the hot temperatures inside the Sun and other stars, and in the solar corona and solar wind, the electrons have been set free of their atomic bonds. The mass of an electron is 9.1094×10^{-31} kilograms, or 1/1836 of the mass of a proton. *See* ion, and proton.

Electron neutrino: The type of neutrino that interacts with the electron. It is the only kind of neutrino produced by the nuclear reactions that make the Sun shine. *See* neutrino, and solar neutrino problem.

Electron volt: A unit of energy, denoted by eV, often used for measuring the energies of particles and electromagnetic radiation. An electron volt is defined as the energy acquired by an electron when it is accelerated through a potential difference of 1 volt in a vacuum. 1 eV $= 1.602177 \times 10^{-19}$ Joule, where one Joule is equal to 10 million erg. Radiation with an energy of one electron volt has a wavelength of 1240 nanometers and a frequency of 2.42×10^{14} Hertz. The energies of X-rays are expressed in thousands of electron volts, abbreviated keV. Millions and billions of electron volts, respectively denoted by MeV and GeV, are used as units for very energetic charged particles, such as cosmic rays. An electron with a kinetic energy of a few MeV is traveling at almost the velocity of light.

Element: A chemically homogeneous substance that cannot be decomposed by chemical means. The atoms of a particular element all have the same number of protons in their nucleus, but the number of neutrons can vary, giving rise to different isotopes of the same element. There are 92 naturally occurring elements, with the number of protons, or atomic number, ranging from 1 for hydrogen to 92 for uranium.

Emission line: A bright spectral feature at a particular wavelength or narrow bands of wavelengths, emitted directly by a hot, luminous gaseous source, such as an emission nebula or planetary nebula, revealing by its wavelength a chemical constituent of its gas.

Emission spectrum: A series of bright emission lines. *See* emission line, and forbidden line.

Equator: An imaginary line around the center of a body where every point on the line is an equal distance from the poles. The equator defines the boundary between the northern and southern hemispheres.

Erg: A unit of energy equal to the work done by a force of 1 dyne acting over a distance of 1 centimeter. An energy of one Joule is equal to 10^7, or 10 million, erg.

EUV: Acronym for Extreme Ultra-Violet, a portion of the electromagnetic spectrum from approximately 10^{-8} to 10^{-7} meters, or 100 to 1,000 Ångströms. *See* extreme ultraviolet radiation.

eV: Abbreviation of electron volt. *See* electron volt.

Evershed effect: The radial flow of gases within the penumbra of a sunspot, discovered in 1908 by the English astronomer **John Evershed (1864–1956)** from the Doppler shift in the spectrum of a sunspot. The flow is outward at the level of the photosphere, but inward and downward at higher levels in the solar atmosphere. *See* photosphere, and sunspot.

Extreme ultraviolet radiation: Abbreviated EUV, a portion of the electromagnetic spectrum from approximately 10^{-8} to 10^{-7} meters, or 10 to 100 nanometers and 100 to 1,000 Ångströms.

F

Faculae: Bright regions of the photosphere, associated with sunspots, seen in white light and visible only near the limb of the Sun. They are brighter than the surrounding medium due to their higher temperatures and greater densities. Faculae appear some hours before the associated sunspots, in the same place, but can remain for months after the sunspots have gone. The word *facula* is Latin for "little torch." A plage appears in the chromosphere just above a facula. *See* plage.

Fahrenheit: A unit of temperature denoted by F or by °F for degrees Fahrenheit. It is named after the Polish-born Dutch physicist **Gabriel Daniel Fahrenheit (1686–1736),** who invented the first accurate thermometer and devised the Fahrenheit scale of temperature. The freezing temperature of water is 32 °F and the boiling temperature of water is 212 °F. The equivalent temperature in degrees kelvin is °K = (5 °F/9) + 255.22, and the equivalent temperature in degrees Celsius is °C = 5(°F −32)/9. *See* Celsius, and Kelvin.

Faint-young-Sun paradox: The discrepancy between the Earth's warm climate record and a dimmer Sun billions of years ago. Geological evidence indicates the presence of liquid water on Earth from 3.8 billion years ago. Yet, the Sun began shining 4.5 billion years ago with about 70 percent of the brightness it has today; assuming an unchanging atmosphere comparable to that of today, the oceans would have been frozen more than 2 billion years ago.

Fibrils: Dark elongated features seen in hydrogen-alpha spectroheliograms of the chromosphere, forming a linear pattern thought to delineate the magnetic field. *See* chromosphere, hydrogen-alpha line, and spectroheliograph.

Filament: A mass of relatively cool and dense material suspended above the photosphere in the low corona by magnetic fields, generally along a magnetic inversion, or neutral, line separating regions of opposite magnetic polarity in the underlying photosphere. Filaments appear as dark, elongated features when observed on the disk in the light of hydrogen alpha or ionized calcium. A dark prominence is seen in projection against the bright solar disk. When detected above the limb of the Sun, a filament is seen in emission against the dark sky, and it is then called a prominence. *See* chromosphere, hydrogen-alpha line, and prominence.

Fission: The splitting, or breaking apart, of a heavy atomic nucleus into two or more lighter atomic nuclei, releasing energy. Fission is spontaneous during the decay of a radioactive element, and it can be induced by particle bombardment. Fission powers atomic bombs, and occurs in man-made nuclear power reactors on Earth.

Five-minute oscillations: Vertical oscillations of the solar photosphere with a period of five minutes, interpreted in terms of trapped sound waves. The five-minute oscillations are detected in the Doppler effect of solar absorption lines formed in the photosphere and in light variations of the photosphere. The American physicist **Robert Leighton (1919–1997)** and his colleagues discovered localized five-minute oscillations while investigating motions of the photosphere. Observations of the oscillations have been used to infer the composition, motions, structure and temperature of the solar interior through the technique of helioseismology. *See* acoustic waves, Doppler effect, Global Oscillation Network Group, helioseismology, photosphere, ***SOlar and Heliospheric Observatory,*** spicules and supergranulation.

Flare: A sudden and violent release of matter and energy from the Sun or another star in the form of electromagnetic radiation, energetic particles, wave motions and shock waves, lasting minutes to hours. The frequency and intensity of flares on the Sun increase near the maximum of the solar activity cycle, and they occur in solar active regions. Many red dwarf stars also emit flares, and they are called flare stars. Solar flares accelerate charged particles into interplanetary space. The impulsive or flash phase of flares usually lasts for a few minutes, during which matter can reach temperatures of hundreds of millions of degrees. The flare subsequently fades during the gradual or decay phase lasting about an hour. Most of the radiation is emitted as X-rays, but flares are also observed at visible hydrogen-alpha wavelengths and radio wavelengths. They are probably caused by the sudden release of large amounts of magnetic energy, up to 10^{25} Joule, in a relatively small volume in the solar corona.

Flash spectrum: The emission line spectrum of the solar chromosphere that is seen for a few seconds just before and after totality during an eclipse of the Sun. *See* chromosphere, emission line, and emission spectrum.

F layer: The upper layer of the ionsphere, approximately 140 to 1,500 kilometers in height. The densiest part of the F layer is at heights between 200 and 600 kilometers.

Footpoint: Intersection of a magnetic loop with the photosphere. The lowest visible portion of the magnetic loop. *See* coronal loop.

Forbidden lines: Emission lines not normally observed under laboratory conditions because they have a low probability of occurrence, often resulting from a transition between a metastable excited state and the ground state. Under typical conditions on Earth, an atom in a metastable state will lose energy through a collision before it is able to decay to the ground state and emit line radiation. Under astrophysical conditions, including the solar corona and highly rarefied nebulae, the metastable state can last long enough for the "forbidden" lines to be emitted. *See* emission line.

Fraunhofer lines: The dark absorption features in the solar spectrum, caused by absorption at specific wavelengths in the cooler layers of the Sun's atmosphere, including the photosphere and chromosphere. They were first observed in 1802 by the English chemist and physicist **William Hyde Wollaston (1766–1828).** They were first carefully studied from 1814 by the German physicist and optician **Joseph von Fraunhofer (1787–1826),** who catalogued the wavelengths of more than 300 of them, assigning Roman letters to the most prominent, such as the H and K lines of ionized calcium, the D line of neutral sodium, the E line of iron, and the C and F lines of hydrogen. *See* absorption line, H and K lines, hydrogen-alpha line, and spectrum.

Free electron: An electron that has broken free of its atomic bond and is therefore not bound to an atom.

Frequency: The number of crests of a wave passing a fixed point each second usually measured in units of Hertz, where one Hertz is equal to one oscillation per second. *See* Hertz, Hz and MHz.

Fusion: The joining together of two or more light atomic nuclei to produce a heavier atomic nucleus, releasing energy. Fusion powers hydrogen bombs and makes stars shine, including the Sun. *See* proton-proton chain, and thermonuclear fusion.

G

GALLEX: Acronym for GALLium EXperiment, operated between 1991 and 1997 within the Gran Sasso Underground Laboratories below a peak in the Apennine mountains of Italy, using 30 tons of gallium and 1,000 tones of gallium chloride solution. The experiment measures solar neutrinos of low energy from the proton-proton reaction. From 1998 to 2002 measurements were carried out at the same location by the Gallium Neutrino Obsrevatory, abbreviated GNO.

Gamma ray: *See* gamma-ray radiation.

Gamma-ray radiation: The most energetic form of electromagnetic radiation, with photon energies in excess of 100 keV or 0.1 MeV. The wavelength of gamma-ray radiation is less than 0.01 nanometer or 10^{-11} meters, the shortest possible. Because the Earth's atmosphere absorbs the radiation at that end of the electromagnetic spectrum, gamma ray studies of cosmic objects are conducted from satellites such as the ***Ramaty High Energy Solar Spectroscopic Imager,*** abbreviated ***RHESSI.*** See ***Ramaty High Enrgy Solar Spectroscopic Imager,*** and electromagnetic spectrum.

Gas pressure: The outward pressure caused by the motions of gas particles and increasing with their temperature. Gas pressure supports main-sequence stars like the Sun against the inward force of their immense gravity.

Gauss: The c.g.s., or centimeter-gram-second, unit of magnetic field strength, named after the German physicist and astronomer **Karl Friedrich Gauss (1777–1855),** who showed in 1838 that the Earth's dipolar magnetic field must originate inside the planet. The SI, or Systeme International, unit of magnetic field strength is the Tesla, where 1 Tesla = 10,000 Gauss = 10^4 Gauss. The Tesla is named after the Croatian-born American electrical engineer **Nikola Tesla (1856–1943),** who was a pioneer in the use of alternating-current electricity. *See* Tesla.

Geomagnetic field: The magnetic field in and around the Earth. The intensity of the magnetic field at the Earth's surface is approximately 3.2×10^{-5} Tesla, or 0.32 Gauss, at the equator and 6.2×10^{-5} Tesla, or 0.62 Gauss, at the North Pole. To a first approximation, the Earth's magnetic field is like that of a bar magnet, or a dipole, currently displaced about 500 kilometers from the center of the Earth towards the Pacific Ocean and tilted at 11 degrees to the rotation axis.

Geomagnetic storm: A rapid, worldwide disturbance in the Earth's magnetic field, typically lasting a few hours, caused by the arrival in the vicinity of the Earth of a coronal mass ejection. A substorm is a magnetic disturbance observed in Polar Regions only, and is associated with changes in direction of the interplanetary magnetic field as it encounters the Earth. Auroral activity and disruption of radio communications are common during intense geomagnetic storms. *See* coronal mass ejection, and magnetic storm.

Geosynchronous orbit: The orbit of a satellite that travels above the Earth's equator from west to east at an altitude where the satellite's orbital velocity is equal to the rotational velocity of the Earth. In this orbit, a satellite remains nearly stationary above a particular point on the planet. Such an orbit has an altitude of about 35,900 kilometers.

Global Oscillation Network Group: Abbreviated GONG, a network of six identical solar telescopes located around the world and operated by the United States National Solar Observatory. The telescopes follow the Sun as the Earth rotates, providing almost continuous monitoring of the Sun's five-minute oscillations, or pulsations, and the associated sound waves that penetrate the Sun's interior. Beginning operations in 1995, the Global Oscillation Network Group has provided valuable information about the composition, motions, structure and temperature of the Sun's interior using the technique of helioseismology. *See* five-minute oscillations, helioseismology, and National Solar Observatory.

Global warming: A potentially dangerous increase in the temperature of the Earth due to an increase of heat-trapping gases, like carbon dioxide, in the atmosphere, resulting from human activity such as the burning of the fossil fuels coal, gas and oil. In 1896 the Swedish chemist **Svante Arrhenius (1859–1927)** showed that a doubling of the Earth's atmospheric carbon dioxide produces a global warming of 5 to 6 degrees Centigrade (9 to 11 degrees Fahrenheit), comparable to modern estimates. Measurements begun in 1958 by the American scientist **Charles Keeling (1928–2005)** have demonstrated an exponential rise in the Earth's atmospheric carbon dioxide over the subsequent two decades, and that by burning coal, gas and oil humans have increased the amount of carbon dioxide in the Earth's atmosphere by 30 percent since the industrial revolution. *See* greenhouse effect.

GNO: Aconym for Gallium Neutrino Observatory. *See* GALLEX.

GONG: Acronym for Global Oscillation Network Group. *See* Global Oscillation Network Group.

Granulation: A mottled, cellular pattern visible at high spatial resolution in the white light of the photosphere. The solar granulation exhibits a non-stationary, overturning motion, a visible manifestation of hot gases rising from the Sun's interior by convection. *See* granule, and supergranulation.

Granule: One of about a million bright regions, or cells, that cover the visible solar disk at any instant, and that comprise the granulation detected in high-resolution, white-light observations of the photosphere. The bright center of each granule, or convection cell, is the highest point of a rising column of hot gas. The dark edges of each granule are the cooled gas, which is descending because it is denser than the hotter gas. The mean angular distance between the bright centers of adjacent granules is about 2.0 seconds of arc, corresponding to about 1500 kilometers at the photosphere. Individual granules are often polygonal-shaped regions, and they appear and disappear on time scales of about ten minutes. *See* granulation.

Great circle: The line traced out on the Earth by a plane passing through the center of the globe, dividing it into two equal hemispheres. The name comes form the fact that no greater circles can be drawn on a sphere. A great circle halfway between the North and South Poles is called the equator, because it is equally distant between both poles.

Greenhouse effect: The warming of a planet's surface by heat trapped by gases in its atmosphere, owing to their infrared opacity. The French mathematician **Joseph (Jean Baptiste) Fourier (1768–1830)** first proposed the mechanism in 1827, while wondering how the Sun's heat could be retained to keep the Earth hot. Incoming sunlight is absorbed at the surface and re-radiated at longer wavelengths, as infrared heat radiation. As the Irish physicist **John Tyndall (1820–1893)** showed in 1861, significant heat is absorbed in the Earth's atmosphere by carbon dioxide and water vapor. These greenhouse gases are not transparent to the infrared, and the heat is reflected back to the ground. The natural greenhouse effect warms the surface of the Earth by about 33 degrees Celsius and keeps the oceans from freezing. How much heating the greenhouse effect causes depends on how opaque the atmosphere is to infrared radiation. On Venus, the dense carbon dioxide atmosphere has raised the surface temperature to around 750 kelvin, hot enough to melt lead. Concern has mounted that global warming of the Earth will result from increased concentrations of carbon dioxide and other so-called "unnatural" greenhouse gases released by human activity, particularly the burning of fossil fuels such as coal and oil. Measurements begun in 1958 by the American scientist **Charles Keeling (1928–2005)** have demonstrated an exponential rise in the amount of carbon dioxide in the Earth's atmosphere over the subsequent two decades. *See* global warming.

Green line: *See* coronal green line.

H

Hale's law: The leading, or westernmost spots of any sunspot group in one hemisphere of the Sun have the same magnetic polarity, while the following, or easternmost, spots have the opposite magnetic polarity. The direction of any bi-polar group in the southern hemisphere is the reverse of that in the northern hemisphere. All of the spots' magnetic polarities reverse each 11-year solar activity cycle.

H-alpha: *See* hydrogen-alpha line.

H and K lines: The strongest lines in the visible spectrum of ionized calcium, Ca II, lying in the violet at wavelengths of 393.4 and 396.8 nanometers. They are conspicuous features in the spectra of many stars, including the Sun. The designations H and K were given by Fraunhofer and are still commonly used. *See* Fraunhofer lines and spectroheliograph.

Hard X-rays: Electromagnetic radiation with photon energies of between 10 keV and 100 keV and wavelengths between about 10^{-10} and 10^{-11} meters.

Heavy water: A form of water in which the abundant, light, hydrogen atoms, H, are replaced by less abundant, deuterium atoms, a heavy form of hydrogen denoted by D, to form D_2O; here O denotes an atom of oxygen.

Helicity: The helicity of an object is a measure of the "twist it has, such as the degree of coiling of a magnetic field.

Heliographic latitude: The angle in degrees from the solar equator to the object as measured along a great circle passing through the poles of the Sun.

Heliopause: The place where the solar wind pressure balances other pressures found in interstellar space. The heliopause is located about 100 astronomical units from the Sun or at about 100 times the distance between the Earth and the Sun.

Helioseismology: The study of the interior of the Sun by the analysis of sound waves that propagate through the solar interior and manifest themselves by oscillations at the photosphere. These oscillations have periods of around five minutes and are observed spectroscopically as Doppler shifts in the absorption line spectrum. The technique of helioseismology has been used to infer the composition, motions, structure and temperature of the solar interior. Helioseismology is a hybrid name combining the Greek words *Helios* for the "Sun" and *seismos* for "tremor or quake." *See* five-minute oscillations, Global Oscillation Network Group, and ***SOlar and Heliospheric Observatory.***

Heliosphere: A vast region, cavity, or bubble carved out of interstellar space by the solar wind, and extending to about 100 astronomical units from the Sun or to about 100 times the mean distance between the Earth and the Sun. The heliosphere is the region of interstellar space surrounding the Sun where the Sun's magnetic field and the charged particles of the solar wind control plasma processes. The heliosphere is immersed in the local interstellar medium, and defines the extent of the Sun's influence. The heliosphere contains our Solar System, and extends well beyond the planets to the heliopause, an outer boundary that marks the place where the solar wind pressure balances other pressures found in interstellar space. *See* heliopause.

Heliostat: A moveable flat mirror used to reflect sunlight into a fixed solar telescope.

Helium: After hydrogen, the second most abundant element in the Sun and the Universe. The nucleus of a helium atom is called an alpha particle; it contains two protons and two neutrons. Helium was first observed as a spectral line during the solar eclipse on 18 August 1868, by the French astronomer **Jules Janssen (1824–1907)** and the British astronomer **Norman Lockyer (1836–1920),** and named helium by Lockyer after the Greek Sun god *Helios.* Helium was not found on Earth until 1895, when the Scottish chemist **William Ramsay (1852–1916)** discovered it as gaseous emission from a heated uranium mineral. Most of the helium in the Universe was made in the immediate aftermath of the Big Bang that gave rise to the expanding Universe, but helium is also synthesized from hydrogen in the cores of main-sequence stars. Helium is one of the rare, inert noble gases. Helium has two isotopes, the more common helium-4, containing two protons and two neutrons in its nucleus, and the rare helium-3 with two protons and one neutron.

Helium burning: The release of energy by fusing helium into carbon within a star.

Helmet streamer: Named after spiked helmets once common in Europe, helmet streamers form in the low corona over the magnetic inversion, or neutral, lines in large active regions, with a long-lived prominence commonly embedded in the base of the streamer. The footpoints of the streamer are in regions of opposite polarity so the streamer itself straddles the prominence. Higher up, the magnetic field is drawn out into interplanetary space within long, narrow stalks. Gas flowing out along these open magnetic fields might help to create the slow-speed solar wind. *See* coronal streamer.

Hertz: A unit of frequency equal to one cycle per second, abbreviated Hz. One kilohertz, abbreviated kHz, is one thousand Hz, one megahertz, abbreviated MHz, is a million Hz, and one gigahertz, abbreviated GHz, is one thousand million Hz.

HESSI*:** Acronym for the ***High Energy Solar Spectroscopic Imager. See ***RHESSI.***

High-speed stream. A stream within the solar wind having speeds of up to 800 kilometers per second, thought to originate mainly from coronal holes.

Homestake: The South Dakota gold mine where the chlorine neutrino detector is located. *See* chlorine experiment.

Hydrogen: The lightest, simplest and most abundant element in the Sun and in the Universe, and the first element in the periodic table, consisting of one proton and one electron. All of the hydrogen in the Universe was produced during the earliest moments of the Big Bang. Hydrogen is detected in the visible-light spectra of the Sun and other stars at wavelengths specified by the Balmer series, and at ultraviolet wavelengths by the Lyman series. Interstellar regions of neutral, unionized hydrogen atoms are called H I regions, and they emit radio radiation at 21-centimeters wavelength; bright stars ionize nearby interstellar hydrogen, producing H II regions. A hydrogen molecule consists of two hydrogen atoms, and molecular hydrogen is the most abundant molecule in interstellar space. Most of the mass of a hydrogen atom is in its nucleus, the proton, whose mass is 1.6726×10^{-27} kilograms or 1836 times the mass of an electron. A rare, heavy isotope of atomic hydrogen, known as deuterium, contains one proton and one neutron in its nucleus. Hydrogen is so light that it is found in only small amounts in Earth's atmosphere, but it is the main constituent of the massive giant planets Jupiter and Saturn. *See* alpha particle, Balmer series, and Fraunhofer lines.

Hydrogen-alpha line: The spectral line of neutral hydrogen in the red part of the visible spectrum, denoted by Hα. Light emitted or absorbed at a wavelength of 656.3 nanometers during an atomic transition in hydrogen, the lowest energy transition in its Balmer series. It is the dominant emission from the solar chromosphere. The wavelength also corresponds with a dark line produced by hydrogen absorption in the photosphere, designated with the letter C in 1814 by the German physicist and optician **Joseph von Fraunhofer (1787–1826)**. *See* Balmer series, chromosphere, filament, Fraunhofer lines, plage, prominence, spectral line, spectroheliograph, and spicule.

Hydrogen burning: The thermonuclear fusion of hydrogen nuclei into helium nuclei, and the process by which all main-sequence stars generate energy, including the Sun. Stars spend most of their lifetime burning hydrogen, and such stars are the most common kind of star. A star with the mass of the Sun spends around ten billion years in this stage; more massive stars spend less time. *See* proton-proton chain.

Hydromagnetic wave: A wave in which both the plasma and magnetic field oscillate.

Hz: Abbreviation for Hertz, a unit of frequency equivalent to one cycle per second. *See* Hertz.

I

Ice Age: A period of cool, dry climate on Earth causing a long-term buildup of extensive ice sheets far from the poles. The major ice ages last for about 100,000 years; they are separated by warmer interglacial periods that last roughly 10,000 years. The major ice ages may be triggered by rhythmic variations in the global distribution of sunlight with periods of 23, 41 and 100 thousand years. *See* Milankovitch cycles.

Impulsive flare: The most common type of solar flare, lasting minutes to hours. *See* flare, gradual flare and solar flare.

Interferometry: The use of an interferometer to combine the radiation received from a cosmic object by two or more telescopes.

Interplanetary magnetic field: The Sun's magnetic field carried into interplanetary space by the expanding solar wind. It has a magnetic field strength of about 6×10^{-9} Tesla, or 6 nanoTesla and 6×10^{-5} Gauss, at the Earth's orbital distance of about 1 astronomical unit from the Sun. The interplanetary magnetic field is carried radially outward by the solar wind, but since the magnetic field originates at the Sun, where the magnetic footpoints are located, solar rotation twists the interplanetary magnetic field into a spiral structure with the shape of an Archimedean spiral.

Invisible radiation: Those kinds of radiation to which the human eye is not sensitive; for example, radio waves, X-rays and gamma rays. The human eye is sensitive to visible light, also known as optical radiation since optics are used to detect it. *See* optical radiation and visible light.

Ion: An atom that has lost one or more electrons, or in less common instances gained an electron. An ion has a net electrical charge. By contrast, a neutral atom has an equal number of negatively charged electrons and positively charged protons, giving the atom a zero net electrical charge.

Ionization: The process in which a neutral atom or molecule is given a net electrical charge. The atomic process in which ions are produced by removing an electron from an atom, ion or molecule, typically by collisions with atoms or electrons, known as collisional ionization, or by interaction with electromagnetic radiation, called photoionization.

Ionized: State of an atom having fewer, or more, electrons than the number of protons, leaving it with an electrical charge.

Ionosphere: The upper region of a planet's atmosphere in which there are free electrons and ions produced when solar ultraviolet and X-ray radiation ionizes the constituents of the atmosphere. Most of the Earth's ionosphere lies between heights of about 50 and 300 kilometers above the ground, though the extent varies considerably with time, season and solar activity. The D region, between 50 and 90 kilometers, has low electron density. The E and F regions, at about 100 and 200 to 300 kilometers, form the main part of the ionosphere. The reflecting power of the ionosphere makes long-range broadcasting and telecommunication possible at radio frequencies up to about 30 MHz, or wavelengths longer than 10 meters.

Irradiance: The solar irradiance is the amount of solar energy received at the Earth outside its atmosphere per unit area per unit time, with units of watts per square meter. *See* solar constant.

K

K or °K: Abbreviation for degrees on the Kelvin scale of temperature. *See* Kelvin.

Kamiokande: A massive underground neutrino detector in Japan filled with water, replaced by the Super-Kamiokande detector. *See* Super-Kamiokande.

Kelvin: A unit for measuring absolute temperature abbreviated K or °K for degrees on the Kelvin scale, and named for the Irish physicist **William Thomson (1824–1907), Lord Kelvin of Largs.** On the Kelvin scale, the coldest possible temperature is 0 degrees kelvin, where atomic or molecular motion stops. Water freezes at 273 kelvin, room temperature is about 295 kelvin, and the boiling point of water is 373 kelvin. To convert a temperature in kelvin to the equivalent temperature in degrees Celsius, subtract 273.15, or $°C = °K - 273.15$, and the equivalent temperature in degrees Fahrenheit is $°F = (9\,°K/5) - 459.4$. *See* absolute temperature, Celsius, and Fahrenheit.

keV: Abbreviation for kilo electron volts, or one thousand electron volts. A unit of energy with $1\text{ keV} = 1.602\,191\,7 \times 10^{-16}$ Joules. The wavelength of radiation with a photon energy in keV is 1.24×10^{-9} meters / energy in keV. *See* electron volt.

Kilometer: A unit of distance, abbreviated km. It is equal to one thousand meters and to 0.6214 miles.

Kilometer per second: A unit of speed equal to 2237 miles per hour.

Kinetic energy: The energy that an object possesses as a result of its motion.

Kirchhoff's law: The ratio of the emission and absorption coefficients of a blackbody is equal to its brightness; a relation named for the German physicist **Gustav Kirchhoff (1824–1887)** who announced it in 1859.

L

Latitude: Distance north or south of the equator of the Earth or the Sun along a great circle connecting the poles. A great circle divides the sphere of the Earth or Sun in half, and the name comes from the fact that no greater circle can be drawn on the sphere.

Light: The kind of radiation to which the human eye is sensitive. *See* optical radiation, velocity of light, and visible spectrum.

Light element: An element of very low atomic number, such as hydrogen, helium and lithium, which have atomic numbers of one, two and three.

Limb: The apparent edge of a celestial object which is visible as a disk, such as the Sun or the Moon. Astronomers refer to the left edge of the solar disk as the Sun's east limb and to the right edge as its west limb.

LMSAL: Acronym for Lockheed Martin Solar and Astrophysics Laboratory.

Luminosity: The amount of energy radiated per unit time by a glowing object, often measured in units of watts. This intrinsic brightness of an object is also called the absolute luminosity, and it differs from how bright the object looks, which is known as the apparent luminosity. One Joule per second is equal to one watt of power, and to ten million ergs per second. The absolute luminosity, L, of the thermal emission from a blackbody is given by the Stefan-Boltzmann law in which $L = 4\pi\sigma R^2 T_{eff}^4$, where $\pi = 3.14159$, the Stefan-Boltzmann constant, $\sigma = 5.67051 \times 10^{-8}$ Joule m^{-2} $°K^{-4}$ s^{-1}, the radius is R and the effective temperature is T_{eff}. The brightness of radiation falls off with the inverse square of the distance, D, so the apparent luminosity $l = L/D^2$, and an object of a given absolute luminosity looks fainter when it is located further away.

M

Magnetic dynamo: *See* dynamo.

Magnetic field: A magnetic force field around the Sun, a planet, and any other magnetized body, generated by electrical currents. The Sun's large-scale magnetic field, like that of Earth, exhibits a north and south pole linked by lines of magnetic force, but the Sun also contains numerous dipolar sunspots linked by magnetic loops.

Magnetic field lines: Imaginary lines that indicate the strength and direction of a magnetic field. The lines are drawn closer together where the field is stronger. Charged particles move freely along magnetic field lines, but cannot move across them.

Magnetic network: *See* calcium network, chromospheric network, and network.

Magnetic polarity: *See* polarity.

Magnetic reconnection: A change in the topology of the magnetic field where the magnetic field lines reorient themselves via new connections. During magnetic reconnection, some magnetic field lines are broken and then rejoined in a new configuration. This can occur when two oppositely directed magnetic fields move toward each other and reconnect at the place that they touch. Magnetic reconnection is an important mechanism for releasing magnetic energy to heat the Sun's corona and to power explosive phenomenon on the Sun, such as coronal mass ejections and solar flares.

Magnetic storm: A disturbance in the Earth's magnetic field observed all over the Earth due to collision by a coronal mass ejection. A substorm is a magnetic disturbance observed in polar regions only.

Magnetism: One aspect of electromagnetism, a fundamental force of nature, whereby a magnetized object can force a charged particle into a new direction of motion and attract or repel other magnetized objects.

Magnetogram: A computer image, picture or map of the strength, direction and distribution of magnetic fields across the solar photosphere, based on Zeeman-effect measurements.

Magnetograph: An instrument used to map the strength, direction and distribution of magnetic fields across the solar photosphere using the Zeeman effect. Normally, the longitudinal, or line-of-sight, magnetic field is measured, but the transverse component is also measured with a vector magnetograph. A magnetogram is a map of the measured magnetic field.

Magnetohydrodynamics: The study of the behavior of a plasma in a magnetic field, abbreviated as MHD. A description of a plasma as a single conducting fluid. *See* Alfvén velocity, Alfvén wave, and plasma.

Magnetopause: The outer boundary layer of a planetary magnetosphere, where it joins the solar wind. The magnetopause separates the magnetosphere from the solar wind. At the magnetopause, the magnetic pressure of the planet's internal magnetic field deflects the flow of the solar wind around the magnetosphere.

Magnetosphere: The region of space surrounding a planet in which the planet's magnetic field predominates over the solar wind, and controls the motions of charged particles in it. The magnetosphere is shaped by interactions between a planet's magnetic field and the solar wind. The magnetosphere shields the planet from the solar wind, preventing or impeding the direct entry of the solar wind particles into the magnetic cavity. *See* bow shock, magnetopause, magnetotail, and Van Allen belts.

Magnetotail: The portion of a planet's magnetosphere formed on the planet's dark night side by the pulling action of the solar wind. The magnetotail is an elongated extension of the planet's magnetic fields on the side of the planet opposite to the Sun. Magnetotails can extend hundreds of planetary radii.

Mass-energy equivalence: Mass, denoted by m, can be transformed into energy, E, and every energy has a mass, according to the formula $E = mc^2$, where c is the velocity of light, first derived in its exact form by the German physicist **Albert Einstein (1879–1955)** in 1907.

Maunder Minimum: The 70-year period between 1645 and 1715 when few sunspots were observed. It is named after the English astronomer **E. Walter Maunder (1851–1928),** who in 1922 provided a full account of the missing sunspots, previously noticed by the German astronomer **Gustav F. W. Spörer (1822–1895)** in 1897.

Methane: A molecule denoted by the symbol CH_4, where C is an atom of carbon and H is a hydrogen atom. We cook and heat some homes with gaseous methane.

MeV: A unit of energy equal to one million electron volts and $1.602\ 177 \times 10^{-13}$ Joule. This unit is used to describe the total energy of high-velocity particles or energetic radiation photons.

MHD: Acronym for magnetohydrodynamics.

Microflares: Also called nanoflares, ubiquitous small brightenings on the Sun, each lasting a few minutes, which can be observed at extreme ultraviolet, radio, ultraviolet and X-ray wavelengths. Microflares are formed in the chromosphere or corona, and they may contribute to coronal heating. *See* nanoflares.

Milankovitch cycles: Three rhythmic fluctuations in the wobble and tilt of the Earth's rotational axis and the shape of the Earth's orbit that set the major ice ages in motion by altering the seasonal distribution of the Sun's light and heat on Earth. They are the 23,000-year precessional wobble of the Earth's rotational axis, the 41,000-year variation in the Earth's axial tilt, and the 100,000-year change in the eccentricity or shape of the Earth's orbit. These cycles are named after the Yugoslavian astronomer **Milutin Milankovitch (1879–1958)** who developed the relevant mathematical theory between 1920 and 1941. *See* ice age.

Mile: A unit of distance equal to 1.609 34 kilometers and 5280 feet.

Million: One thousand thousand, written as 1,000,000 or 10^6.

Minute of arc: A unit of angular measure equal to 1/60 degree and 60 seconds of arc.

Molecule: The smallest unit of a chemical compound. A molecule is composed of two or more atoms, bound together by interaction of their electrons.

Morton wave: A wave-like disturbance in the chromosphere initiated by a solar flare. A Morton wave is detected as a circular hydrogen-alpha brightening that expands away from a flare and across the solar disk to distances of about a billion meters at velocities of about a million meters per second. It is named after Gail E. Moreton who first observed them in 1960.

MSW effect: The transformation of a neutrino of one type or flavor into a neutrino of another kind while traveling through matter, named after **Lincoln Wolfenstein (1923–)** who originated the theory in 1978 and **Stanislav Mikheyev (1940–)** and **Alexis Y. Smirnov (1951–)** who further developed it about a decade later. The MSW effect could explain the solar neutrino problem if some of the electron neutrinos produced in the core of the Sun oscillate into muon or tau neutrinos on their way out of the Sun, thereby becoming invisible to most neutrino detectors. *See* neutrino oscillation, and solar neutrino problem.

N

Nanoflares: Ubiquitous low-level flares that might heat the corona, also called microflares. They may be detected at extreme ultraviolet, radio, ultraviolet and X-ray wavelengths. A billion of them would be required to release the same amount of energy as a normal flare. *See* microflares.

NanoTesla: A unit of magnetic field strength, abbreviated nT. It is equal to 10^{-9} Tesla, to 10^{-5} Gauss, and to 1 gamma. *See* Gauss, and Tesla.

NASA: Acronym for the National Aeronautics and Space Administration, United States. *See* National Aeronautics and Space Administration.

National Aeronautics and Space Administration: Abbreviated NASA, the United States government space agency formed in 1958. It is responsible for civilian manned and robotic activities in space, including launch vehicles, scientific satellites, and space probes. NASA Headquarters are in Washington, DC.

National Optical Astronomy Observatory: Abbreviated NOAO, it was formed in 1982 to consolidate under one director several ground-based astronomical optical observatories.

National Radio Astronomy Observatory: Abbreviated NRAO, with its headquarters at Green Bank, West Virginia, the National Radio Astronomy Observatory operates the Very Large Array in New Mexico, and other powerful, advanced radio telescope facilities in the western hemisphere.

National Solar Observatory: Abbreviated NSO, the United States National Solar Observatory includes three facilities dedicated to observing the Sun with high resolution at optical, or visible-light, wavelengths—the Sacramento Peak Observatory in New Mexico, with its 0.79-meter (30-inch) Richard B. Dunn Vacuum Tower Telescope, the Kitt Peak National Observatory's 1.5-meter (60-inch) McMath-Pierce Solar Telescope, also in Arizona, and the Global Oscillation Network Group of six solar telescopes that provide nearly continuous worldwide observations of the Sun *See* Global Oscillation Network Group, Kitt Peak National Observatory, and National Optical Astronomy Observatory.

Network: Chromosphere and photosphere features arranged in a cellular structure, associated with supergranular cells and magnetic fields. *See* calcium network and supergranulation cells.

Neutral line or region: A line that separates longitudinal magnetic fields of opposite magnetic polarity. A region where the longitudinal magnetic field strength approaches zero. Generally, neutral regions occur between regions of opposite polarity.

Neutrino: A subatomic particle with no electric charge and very little mass that responds to the weak nuclear force but not to the strong nuclear and electromagnetic forces. In 1930 the Austrian physicist **Wolfgang Pauli (1900–1958)** proposed an uncharged and unseen particle to remove energy during radioactive beta decay, and the Italian physicist **Enrico Fermi (1901–1954)** developed the idea in 1934, giving the putative particle its name *neutrino,* Italian for "little neutral one." Neutrinos travel at very near the velocity of light and interact very weakly with other matter. They are emitted during radioactivity and generated during thermonuclear reactions in the Sun and other stars. There are three types, or flavors of neutrinos, named the electron, muon and tau neutrinos after the particles they interact with. Vast quantities of electron neutrinos are created as the result of nuclear reactions in the Sun's core. Other types of neutrinos are produced in man-made particle accelerators, nuclear reactors and by cosmic rays entering the Earth's atmosphere.

Neutrino oscillation: The change of one type or flavor of neutrino to another while traveling through matter or a vacuum. Neutrinos come in three flavors, the electron, muon and tau neutrinos. *See* MSW effect, and solar neutrino problem.

Neutron: A subatomic particle with no electric charge found in all atomic nuclei except that of hydrogen. The English physicist **James Chadwick (1891–1974)** discovered the neutron in 1932 when bombarding atoms with energetic particles. The neutron has slightly more mass than a proton. The mass of the neutron is 1.008 665 atomic mass units and 1.6748×10^{-27} kilograms. The neutron is 1839 times heavier than an electron. When set free from an atomic nucleus, or created outside one, a neutron decays into a proton, an electron and an anti-electron neutrino with a neutron half-life of only 614 seconds or 10.25 minutes, and a mean life of just 887 seconds.

NOAO: Acronym for the National Optical Astronomy Observatory, United States. *See* National Optical Astronomy Observatory.

Nobel Prize: An international award given yearly since 1901 for achievements in physics, chemistry, physiology or medicine, literature and peace. Significant accomplishments in astronomy and astrophysics have been awarded the Nobel Prize in Physics. The awards are bestowed by the Nobel Foundation, which was endowed on 27 November 1895 in the last will of the Swedish industrial chemist and philanthropist **Alfred Nobel (1833–1896),** who invented dynamite.

Non-thermal particle: A particle that is not part of a thermal gas. A single temperature cannot describe these particles. *See* thermal gas.

Non-thermal radiation: The electromagnetic radiation produced by a non-thermal electron traveling at a speed close to that of light in the presence of a magnetic field. Such radiation is called synchrotron radiation after the man-made particle accelerator where it was first seen. More generally, the term non-thermal radiation denotes any electromagnetic radiation from an astronomical body that is not thermal in origin.

Northern lights: A popular name for an aurora when observed from northern latitudes.

NRAO: Acronym for the National Radio Astronomy Observatory, United States. *See* National Radio Astronomy Observatory.

NRL: Acronym for the Naval Research Laboratory, United States.

NSF: Acronym for the National Science Foundation, United States.

Nuclear energy: The energy obtained by nuclear reactions, the source of the Sun's luminosity.

Nuclear fission: A reaction in which a heavy atomic nucleus is broken apart into two or more lighter nuclei, releasing energy. This process powers man-made nuclear reactors on Earth. *See* fission.

Nuclear force: The force that binds protons and neutrons within atomic nuclei, and which is effective only at distances less than 10^{-15} meters.

Nuclear fusion: The process by which two or more light atomic nuclei fuse together to make a heavier atomic nucleus, releasing energy. This process occurs at the high temperatures and pressures found near the centers of stars, and it makes most stars shine, including the Sun.

Nucleus: The central, massive part of an atom. The atomic nucleus is composed of protons and (except for hydrogen) neutrons bound together by the strong nuclear force, and containing nearly all of an atom's mass. An atom's negatively charged electrons surround the nucleus, which has a positive electrical charge. The central region of a galaxy is also known as the nucleus of the galaxy.

O

Opacity: A measure of the ability of a gaseous atmosphere to absorb radiation and become opaque to it. A transparent gas has little or no opacity. Opacity also denotes the mean absorption of radiation, which is used with a diffusion equation to calculate the photon luminosity flow in the Sun or other stars.

Optical astronomy: The study of visible light from objects in space. *See* visible light.

Optical radiation: Electromagnetic radiation that is visible to the human eye, with wavelengths of approximately 385 to 700 nanometers. *See* optical spectrum, optics, photosphere, and visible light.

Optical spectrum: Spectrum of a source that spans the visible wavelength range, approximately from 385 to 700 nanometers. *See* visible light.

Optics: The science that describes the lenses and mirrors that are used to focus and collect visible light.

Orbiting Solar Observatory: Abbreviated ***OSO,*** a series of eight orbiting observatories launched between 1962 and 1971 by NASA to study the Sun in the ultraviolet and X-ray wavelengths from above the atmosphere.

Ozone: A form of molecular oxygen containing three atoms of oxygen instead of the normal two, and hence denoted O_3, where O is an oxygen atom. It is created by the action of ultraviolet sunlight on the Earth's atmosphere, and this ozone layer shields the Earth's surface from deadly ultraviolet radiation from the Sun. *See* ozone hole, and ozone layer.

Ozone depletion: *See* ozone hole.

Ozone hole: An exceptionally low concentration of ozone in the stratosphere. The ozone hole forms above the South Pole in the local winter. It was conclusively demonstrated in 1985, as the result of 27 years of continuous measurements of the ozone content above Halley Bay, Antarctica, begun by the British scientist **Gordon Miller Bourne "G. M. B." Dobson (1889–1976)** in 1957. Two American chemists, **Mario J. Molina (1943–)** and **F. Sherwood Rowland (1927–)** showed in 1974 that the chorine in man-made chemicals called chlorofluorocarbons, or CFCs for short, would destroy enormous amounts of ozone in the atmosphere. The CFCs were limited and eventually banned by in an international treaty known as the *Montreal Protocol,* first signed in 1987 and strengthened in later years. *See* ozone layer.

Ozone layer: A region in the lower part of Earth's stratosphere, located at about 20 to 60 kilometers above sea level, where the greatest concentration of ozone appears. The ozone layer shields the Earth's surface from the Sun's energetic ultraviolet rays.

P

Pair annihilation: Mutual destruction of an electron and positron with the formation of energetic gamma rays that have photon energy of 511 keV.

Particle: Fundamental unit of matter and energy, including subatomic particles that are smaller than the atom.

Penumbra: The lighter periphery of a sunspot seen in white light, surrounding the darker umbra. Also the outer part of a shadow cast during an eclipse, where only a partial eclipse is seen.

Photon: A corpuscle, particle, or quantum, of light or other electromagnetic radiation. A photon is a discrete unit or quantity of electromagnetic energy associated with waves of electromagnetic radiation. Photons have no electric charge and travel at the speed of light. Short wavelength, or high frequency, photons carry more energy than long wavelength, or low frequency, photons. *See* photon energy.

Photon energy: The energy of radiation of a particular frequency or wavelength. The photon energy is equal to the product of Planck's constant, 6.626×10^{-34} J s, and the frequency of the radiation. Short wavelength, or high frequency, photons have more energy than long wavelength, or low frequency, photons.

Photosphere: That part of the Sun from which visible light originates. The photosphere emits white light, or all the colors combined. It is the intensely bright, visible portion of the Sun, and the place where most of the Sun's energy escapes into space. The lowest layer of the Sun's atmosphere viewed in white light. The photosphere of any star is the region that gives rise to its visible continuum radiation. About 500 kilometers thick, the solar photosphere is a zone where the gaseous layers change from being completely opaque to radiation to being transparent. The effective temperature of the Sun's photosphere is 5780 kelvin. The continuum radiation of the photosphere is absorbed at certain wavelengths by slightly cooler gas just above it, producing the dark Fraunhofer lines. Sunspots, faculae, granules and the supergranulation are observed in the photosphere, and magnetograms describe the magnetic field in the photosphere. The in and out heaving motions, or oscillations, of the photosphere are used to decipher the internal dynamics and structure of the Sun. The transparent chromosphere lies just above the photosphere, and the transparent corona is found above the chromosphere. *See* chromosphere, corona, Fraunhofer line, granule, helioseismology, magnetogram, sunspot, supergranulation, and white light.

Photosynthesis: The chemical process whereby green plants use sunlight to manufacture carbohydrates from carbon dioxide and water and release oxygen as a by-product.

Plage: From the French word for "beach", a plage is a bright, dense region in the chromosphere found above sunspots or other active areas of the solar photosphere. It always accompanies and outlives sunspot groups. Plages appear much brighter in the monochromatic light of the hydrogen-alpha line or the calcium H and K lines than the surrounding parts of the chromosphere. Plages are associated with faculae, which occur just below them in the photosphere, and are found in regions of enhanced magnetic field. *See* calcium H and K lines, chromosphere, faculae, H and K lines, and hydrogen-alpha line.

Plasma: A completely ionized gas, consisting of electrons that have been pulled free of atoms, and ions, in which the temperature is too high for neutral, unionized atoms to exist. The high temperatures result in atoms losing their normal complement of electrons to leave positively charged ions behind, and it is too hot for the free electrons and ions to join together and form permanent atoms. The interior of the Sun and other main-sequence stars is a plasma consisting mainly of electrons and protons, the nuclei of hydrogen atoms. Plasma has been called the fourth state of matter, in addition to solid, liquid, and gas. Since the total negative electrical charge of the free electrons is equal to the total positive charge of the ions, plasma is electrically neutral over a sufficiently large volume. Most of the matter in the Universe is in the plasma state.

Polarity: The directionality of a magnet or magnetic field, being north- or south-seeking. According to one convention, magnetic lines of force emerge from regions of positive north polarity and re-enter regions of negative south polarity. However, compass needles point towards the Earth's magnetic North Pole.

Polarization: When radiation is polarized, its electromagnetic vibrations are not randomly oriented, but instead have a preferred direction, oscillating in a plane. Synchrotron radiation is polarized.

Pore: Small, short-lived dark area in the photosphere out of which a sunspot may develop.

Positron: A positively charged anti-particle of the electron, named a positron for a "positive electron". The positron is a subatomic particle with the mass of the electron but an equal positive electric charge. *See* antimatter , and beta decay.

Post-flare loop: An arcade of loops or a loop prominence system, often seen after a major two-ribbon flare, which bridges the ribbons.

Prominence: A region of high-density, cool, 10,000-kelvin gas embedded in the lower part of the low-density, hot, million-degree solar corona. A prominence is apparently suspended above the photosphere by magnetic fields. It can be seen at the limb of the Sun during a solar eclipse or with a coronagraph. A prominence is a filament viewed on the limb of the Sun in the light of the hydrogen-alpha line. A quiescent prominence occurs away from active regions and can last for weeks or many months. An active prominence is a short-lived, high-speed eruption associated with active regions, sunspots and flares. An eruptive prominence is often associated with a coronal mass ejection. *See* active region, coronagraph, coronal mass ejection, filament, and hydrogen-alpha line.

Proton: A subatomic particle with positive electric charge located in the nucleus of every atom or set free from it. In 1920 the New-Zealand born British physicist **Ernest Rutherford (1871–1937)** announced that the massive nuclei of all atoms are composed of the positively charged hydrogen nuclei, which he named protons. The nucleus of a hydrogen atom is a proton. The atomic number, denoted by Z, specifies the number of protons in the nucleus of any atom. The proton has a mass of 1.672623×10^{-27} kilograms, and a mass-energy equivalent to 938.3 MeV. A proton is 1836 times more massive than an electron. *See* atomic number, and nucleon.

Proton-proton chain: Abbreviated p-p chain, a sequence of thermonuclear reactions in which hydrogen nuclei, or protons, are transformed into helium nuclei. The German-born American physicist **Hans Bethe (1906–2005)** proposed the details of the proton-proton chain of thermonuclear reactions in 1938–39. It is the main source of energy in the Sun and of all main-sequence stars with a mass less than 1.5 times the Sun's mass. In the first stage of the proton-proton chain, two protons combine to form a deuterium nucleus, releasing a positron, a neutrino and radiation. In the second stage the deuterium nucleus combines with a proton to form an isotope of helium, denoted ^{3}He, again releasing radiation. In the last stage, two ^{3}He nuclei combine to form the normal helium nucleus, denoted ^{4}He, releasing two protons and radiation. Overall, four protons are converted into one helium nucleus, with the mass difference released as energy. *See* hydrogen burning.

Proton-proton reaction: *See* proton-proton chain.

Q

Quantum mechanics: The theory of atomic and subatomic systems based on the notion of quantized energy in radiation, the photons, and in the angular momentum and energy levels of electrons in atoms, arising from the failure of classical concepts of waves or particles individually to describe such systems.

Quantum tunneling: *See* tunnel effect.

Quiescent prominence: A prominence occurring away from active regions that can last for weeks or months. *See* active region and prominence.

Quiet Sun: The Sun when it is at the minimum level of activity in the solar cycle.

R

Radial velocity: The speed at which an object is apparently moving either toward or away from an observer along the line of sight. The Doppler effect measures the radial velocity, and the larger the Doppler shift in wavelength the greater the radial velocity. *See* Doppler effect, Doppler shift, and redshift.

Radian: A dimensionless unit of angular measure equal to $2.062\ 65 \times 10^{5}$ seconds of arc. There are 2π radians in a full circle of 360 degrees, where $\pi = 3.141\ 59$.

Radiation: A process that carries energy through space. *See* electromagnetic radiation.

Radiation belt: A ring-shaped region around a planet in which electrically charged particles – electrons, protons, and other ions – are trapped, following spiral trajectories around the direction of the magnetic field of the planet. The main radiation belts surrounding the Earth are known as the Van Allen belts, containing electrons or protons that are mainly from the solar wind. The Earth also has an inner radiation belt containing ions of material from interstellar space. *See* South Atlantic Anomaly and Van Allen belts.

Radiative zone: An interior layer of the Sun, lying between the energy-generating core and the convective zone, where energy travels outward by radiation.

Radio: *See* radio radiation.

Radioactive decay: *See* radioactivity.

Radioactivity: The spontaneous decay of certain rare, unstable, heavy nuclei into more stable light nuclei, with the release of energy and the emission of particles, including electrons, helium nuclei and neutrinos. The process by which certain kinds of atomic nuclei naturally decompose or decay, with the spontaneous emission of subatomic particles and gamma rays, and governed by the weak nuclear force.

Radio radiation: The part of the electromagnetic spectrum whose radiation has the longest wavelengths and smallest frequencies of all types, with wavelengths ranging from about 0.001 meter to 30 meters and frequencies ranging between 10 MHz and 300 GHz.

Ramaty High Energy Solar Spectroscopic Imager: NASA launched the ***Ramaty High Energy Solar Spectroscopic Imager,*** abbreviated ***RHESSI,*** on 5 February 2002, with an anticipated two-year lifetime. It obtained high-resolution imaging spectroscopy of solar flares at hard X-ray and gamma-ray wavelengths, in order to study solar flare particle acceleration and flare energy release. It determined the frequency, location and evolution of impulsive flare energy release in the corona and located the sites of particle acceleration and energy deposition at all phases of solar flares. The spacecraft was named after **Reuven Ramaty (1937–2001),** a pioneer in the fields of solar physics, gamma-ray astronomy, nuclear physics and cosmic rays.

Redshift: An increase in wavelength of electromagnetic radiation; at visible-light, or optical, wavelength. The shift is toward longer, redder wavelengths. The redshift of a star, galaxy, or other cosmic object is attributed to the Doppler shift of the radiation from a source that is moving away from the observer along the line of sight, and the larger the redshift, the faster the star is moving away from us. *See* Doppler effect, Doppler shift, and radial velocity.

Resolution: The degree to which the fine details in an image are separated and detected or seen. Also known as angular resolution and resolving power, the smallest angular size distinguishable by a telescope, which determines the ability of a telescope to discern fine detail of a celestial object. The angular resolution, denoted by θ, in radians is given by the ratio of the wavelength, λ, to the diameter, D, of the primary mirror or reflector, with $\theta = \lambda/D$ radians. To convert to seconds of arc, or arc seconds, 1 radian $=$ $2.062\,65 \times 10^5$ seconds of arc. The human eye has an angular resolution, or resolving power, of about 1 minute of arc. The resolving power of an optical telescope, operating at visible-light wavelengths, is found by dividing 110 by the aperture in millimeters, or by dividing 4.56 by the aperture in inches; the aperture is the diameter of the objective lens or the primary mirror. However, the resolution of such an optical telescope is limited to about 1 second of arc due to atmospheric turbulence, and this means that a telescope with a mirror or lens larger than about 0.12 meters in diameter has an effective angular resolution of 1 second of arc when observing from the ground, but the full angular resolution is possible if the telescope is in space. The atmosphere does not limit the resolution of a radio telescope. *See* scintillation, and seeing.

Resolving power: *See* resolution.

RHESSI: Acronym for ***Ramaty High Energy Spectroscopic Imager. See Ramaty High Energy Spectroscopic Imager.***

Rotation: The spin of an object about its own axis. The Earth rotates once a day.

S

Scintillation: The twinkling of the stars caused by wind-blown clouds or other non-uniform density distributions in our atmosphere, and the fluctuations of distant radio sources caused by the solar wind or winds in interstellar space. The stars twinkle, causing an apparent change in their color and brightness, but the Moon and planets like Venus do not scintillate because of their larger angular extent. Scintillation in the

Earth's atmosphere usually limits the angular resolution of even the best ground-based optical, or visible-light, telescope to about one second of arc, but this does not affect the resolution of a radio telecope. *See* resolution, and seeing.

Season: A periodic change in the weather conditions on a planet caused by the tilt of its rotational axis and its orbit around the Sun. The annual orbit of the Earth around the Sun produces summer and winter when the relevant hemisphere is pointed toward or way from the Sun. On Earth we traditionally divide the year into four seasons, spring, summer, autumn and winter, each lasting about three months.

Second of arc: A unit of angular measure of which there are 60 in 1 minute of arc and therefore 3600 in 1 degree. One second of arc is equal to 725 kilometers on the visible disk of the Sun. A second of arc is denoted by the symbol ″, and it is also called an arc second. There are 2.06265×10^5 seconds of arc in one radian, and 2π radians in a full circle of 360 degrees, where the constant $\pi = 3.14159$.

Seeing: Fluctuations in a visible-light image due to refractive inhomogenieties in the Earth's atmosphere. Seeing is caused by the random turbulent motion in the Earth's atmosphere. In conditions of good seeing, images are sharp and steady; in poor seeing, they are extended and blurred and appear to be in constant motion. Seeing usually limits the angular resolution of ground-based optical telescopes to about one second of arc. *See* scintillation.

Shock wave: A sudden discontinuous change in density and pressure propagating in a gas or plasma at supersonic velocity. There is also a change in particle flow speed and in the magnetic and electric field strengths associated with a shock.

***Skylab*:** An American space station, launched into Earth orbit in May 1973. Three crews, each of three men, were sent to the station for periods of several weeks between 1973 and 1974. The station burnt up on re-entering the atmosphere in 1979.

SMM*:** Acronym for the ***Solar Maximum Mission.

SNO: Acronym for the Sudbury Neutrino Observatory, Canada. *See* Sudbury Neutrino Observatory.

SNU: Abbreviation for the solar neutrino unit equal to 10^{-36} neutrino reactions per target atom per second.

Soft X-rays: Electromagnetic radiation with photon energies of 1 to 10 keV and wavelengths between about 10^{-9} and 10^{-10} meters.

SOHO*:** Acronym for the ***SOlar and Heliospheric Observatory. *See* ***SOlar and Heliospheric Observatory.***

***Solar-A*:** A mission of the Japanese Institute of Space and Astronautical Science, abbreviated ISAS, launched on 30 August 1991 to study the Sun at X-ray wavelengths. *Solar-A* was renamed the *Yohkoh* or "sunbeam" mission after its sucessful launch, and it operated until 2001. See *Yohkoh.*

Solar activity cycle: A cyclical variation in solar activity with a period of about 11 years between maxima, or minima, of solar activity. Waxing and waning of various forms of solar activity, such as sunspots, flares, and coronal mass ejections, characterize the solar activity cycle. Over the course of an 11-year cycle, sunspots vary both in number and latitude. The complete cycle of the solar magnetic field is 22 years. A dynamo driven by differential rotation and convection may maintain the solar cycle. Activity cycles similar to the solar cycle are apparently typical of stars with convection zones. *See* butterfly diagram, Hale law, solar maximum, solar minimum, Spörer's law, and sunspot cycle.

SOlar and Heliospheric Observatory: NASA launched the ***SOlar and Heliospheric Observatory,*** abbreviated ***SOHO,*** on 2 December 1995. It reached its permanent position on 14 February 1996, with operations continuing into 2006. ***SOHO*** orbits the Sun at the first, L_1, Lagrangian point where the gravitational forces of the Earth and Sun are equal. It is a cooperative project between ESA and NASA to study the Sun from its deep core, through its outer atmosphere and the solar wind out to and beyond the Earth's orbit. It has obtained important new information about helioseismology, the structure and dynamics of the solar interior, the heating mechanism of the Sun's million-degree outer atmosphere, the solar corona, and the origin and acceleration of the solar wind.

Solar atmosphere: The outer layers of the Sun, from the photosphere through the chromosphere, transition region and corona. The term atmosphere is used to describe the outermost gaseous layers of the Sun because they are relatively transparent at visible wavelengths.

Solar-B: A mission of the Japanese space agency, the Institute of Space and Astronautical Science abbreviated ISAS, proposed as a follow-on to the Japan/United States/United Kingdom ***Yohkoh,*** or ***Solar-A,*** collaboration. ISAS has scheduled the launch of ***Solar-B*** for September 2006. The spacecraft consists of a coordinated set of optical, extreme-ultraviolet and X-ray instruments to measure the detailed density, temperature and velocity structures in the photosphere, transition region and low corona with high spatial, spectral and temporal resolution, resulting in new information about the Sun's varying magnetic fields and their relationship to solar eruptions and atmosphere expansion.

Solar constant: The total amount of solar energy, integrated over all wavelengths, received per unit time and unit area at the mean Sun-Earth distance outside the Earth's atmosphere. Its value is 1366.2 Joule per second per square meter, which is equivalent to 1366.2 watt per square meter, with an uncertainty of $\pm$ 1.0 in the same units. The solar constant is not constant, but instead varies in tandem with the 11-year solar activity cycle, by about 0.1 percent from maximum to minimum. Temporary dips of up to 0.3 percent and a few days duration are due to the presence of large sunspots on the visible solar disk. The large number of sunspots near the peak in the 11-year cycle is accompanied by a rise in magnetic activity, which creates an increase in the luminous output of plages that exceeds the cooling effects of sunspots. *See* plage, solar activity cycle, and sunspot.

Solar core: The region at the center of the Sun where nuclear reactions release vast quantities of energy.

Solar corona: *See* corona.

Solar cycle: The approximately 11-year variation in solar activity, as well as the number and position of sunspots. Taking Hale's law of sunspot magnetic polarity into account, it leads to a 22-year magnetic cycle. *See* Hale's law, solar activity cycle, and sunspot cycle.

Solar dynamo: *See* dynamo.

Solar eclipse: A blockage of light from the Sun when the Moon is positioned precisely between the Sun and the Earth. A total solar eclipse is seen at places where the umbra of the Moon's shadow cone falls on and moves over the Earth's surface. Although the total duration of a solar eclipse can be as much as four hours, the Sun is completely covered by the Moon, at totality, for at most 7.5 minutes. During the brief moments of a total solar eclipse, darkness falls, and the outer parts of the Sun, the chromosphere and the corona, are seen. At any given point of Earth's surface, a total solar eclipse occurs, on the average, once every 360 years. *See* Baily's beads, chromosphere, corona, and prominence.

Solar energetic particle: Abbreviated by SEP, charged particles, mainly protons and electrons, accelerated to energies greater than 1 MeV by explosive processes on the Sun and detected by spacecraft and on Earth.

Solar flare: A sudden and violent release of matter and energy within a solar active region in the form of electromagnetic radiation, energetic particles, wave motions and shock waves, lasting minutes to hours. A sudden brightening in an active region observed in chromospheric and coronal emissions that typically lasts tens of minutes. The frequency and intensity of solar flares increase near the maximum of the 11-year solar activity cycle. Solar flares accelerate charged particles into interplanetary space. The impulsive or flash phase of flares usually lasts for a few minutes, during which matter can reach temperatures of hundreds of millions of degrees. The flare subsequently fades during the gradual or decay phase lasting about an hour. Most of the radiation is emitted as X-rays but flares are also observed at visible hydrogen-alpha wavelengths and radio wavelengths. They are probably caused by the sudden release of large amounts (up to 10^{25} Joule) of magnetic energy in a relatively small volume in the solar corona. *See* active region, flare, and solar activity cycle.

Solar limb: The apparent edge of the Sun as it is seen in the sky.

Solar mass: The amount of mass in the Sun, equal to 1.989×10^{30} kilograms.

Solar maximum: The peak of the sunspot cycle when the numbers of sunspots is greatest, and the output of particles and radiation is maximized. The highest level of solar activity, sometimes defined as the month(s) during the solar activity cycle when the 12-month mean of the monthly average of sunspot numbers reaches a maximum, such as July 1989.

Solar Maximum Mission: NASA launched the ***Solar Maximum Mission,*** abbreviated ***SMM,*** on 14 February 1980, with an in-orbit repair from the ***Space Shuttle Challenger*** on 6

April 1984 and a mission end on 17 November 1989. ***SMM*** obtained important new insights to powerful eruptions on the Sun known as solar flares and coronal mass ejections, during a maximum in the 11-year solar activity cycle.

Solar minimum: The beginning or end of a sunspot cycle, marked by the near absence of sunspots, and a relatively low output of radiation and energetic particles. The lowest level of solar activity, sometimes defined as the month(s) during the solar activity cycle when the 12-month mean of the monthly average of sunspot numbers reaches a minimum, such as September 1986 or November 1996.

Solar neutrino problem: Massive, subterranean neutrino detectors find only one-third to one-half the number of solar neutrinos that theoretical calculations predict. The solar neutrino problem has been explained by a transformation of solar neutrinos into an undetectable form on their way out of the Sun. *See* MSW effect, neutrino oscillation, and Sudbury Neutrino Observatory.

Solar neutrino unit: Abbreviated SNU, a unit of solar neutrino capture rate by subterranean detectors, with 1 SNU = 10^{-36} solar neutrino captures per second per target atom.

Solar probe*:** A future NASA spacecraft, the first to fly through the atmosphere of the Sun, taking *in situ* measurements down to 3 solar radii above the solar surface where the radiation temperatures exceed 2 million kelvin. ***Solar Probe will study the heating of the solar corona and the origin and acceleration of the solar wind.

Solar TErrestrial RElations Observatory: NASA plans to launch the two identical spacecraft of the ***Solar TErrestrial RElations Observatory,*** abbreviated ***STEREO,*** in 2006. One spacecraft will lead Earth in its orbit, and one will be lagging; the combined images and radio tracking from the two spacecraft will provide a three dimensional view of coronal mass ejections from their onset at the Sun to the orbit of the Earth, improving our understanding of these solar explosions and aiding space weather forecast capabilities.

Solar wind: A steady flow of energetic charged particles, mainly electrons and protons, and entrained magnetic fields moving out from the Sun in all directions into interplanetary space at supersonic speed. In 1951 the German astronomer **Ludwig Biermann (1907–1986)** proposed that the straight ion tails of comets, which always point away from the Sun, are accelerated by a continuous, electrified flow of charged particles from the Sun. In 1958 the American physicist **Eugene N. Parker (1927–)** showed that the million-degree electrons and protons in the corona will overcome the Sun's gravity and accelerate to supersonic speeds, naming the resultant radial outflow the solar wind. He also proposed that the solar wind carries the Sun's magnetic field with it, forming a spiral magnetic pattern in interplanetary space as the Sun rotates. Measurements from the Soviet ***Luna 2*** spacecraft in 1959 confirmed the existence of the solar wind by direct measurements of its protons, and in 1962–63 the American ***Mariner 2*** spacecraft determined its density, about 5 million electrons and 5 million protons per cubic meter near Earth, and measured its speed. The solar wind has a fast component, or high-speed stream, with a velocity of about 800 kilometers per second, and a slow-speed component moving at about half this speed. In 2000, comparisons of observations with the ***Ulysses*** spacecraft, which passed over the Sun's poles, with those of the ***Yohkoh*** and ***SOHO*** solar spacecraft demonstrated that much, if not all, of the high-speed solar wind comes from open magnetic fields in polar coronal holes, at least during the minimum in the Sun's 11-year activity cycle, and that the slow wind is confined to low latitudes near an equatorial streamer belt. The solar wind carries away about 10^{-13} of the Sun's mass per year. Although the solar wind is diverted around the Earth's magnetosphere, some of its particles enter the magnetosphere. *See* aurora, corona, coronal hole, coronal streamer, heliosphere, high-speed stream, magnetosphere, and Van Allen belts.

Sound speed: *See* velocity of sound.

Sound waves: *See* acoustic waves.

South Atlantic Anomaly: A region over the South Atlantic Ocean where the lower Van Allen belt of energetic electrically charged particles is particularly close to the Earth's surface, presenting a hazard for artificial satellites.

Space weather: Varying conditions in interplanetary space and the Earth's magnetosphere controlled by the gusty, variable solar wind, with Sun storms caused by solar flares or coronal mass ejections.

Spectral line: A feature observed in the spectra of stars and other luminous objects at a specific frequency or wavelength, either in bright emission or in dark absorption. A spectral line looks like a line in a display of radiation intensity as a function of wavelength or frequency. Atoms produce this spectral feature as they absorb or emit light, and it can be used to determine the chemical ingredients of the radiating source. Spectral lines are also used to infer the radial velocity and magnetic field of the radiator. *See* absorption line, Doppler effect, emission line, radial velocity, spectrum, and Zeeman effect.

Spectrograph: Also known as a spectrometer, a spectrograph spreads light or other electromagnetic radiation into its component wavelengths, collectively known as a spectrum, and records the result electronically or photographically. Spectra are now often recorded with Charge Coupled Devices, or CCDs, from which a computer can analyze information in digital form. High-dispersion instruments spread the spectral lines widely so that a particular wavelength can be studied in detail. An example of such a high-dispersion device is the spectroheliograph.

Spectroheliogram: A monochromatic image of the Sun produced by means of a spectroheliograph or by the use of a narrow-band filter. See calcium H and K lines, H and K lines, hydrogen alpha line, and spectroheliograph.

Spectroheliograph: A type of spectrograph used to image the Sun in the light of one particular wavelength only, invented independently in 1891 by the French astronomer **Henri Deslandres (1853–1948)** and by the American astronomer **George Ellery Hale (1868–1938).**

Spectrometer: *See* spectrograph.

Spectroscope: An instrument that spreads electromagnetic radiation into its various wavelengths, especially the component colors of visible light. The display, known as a spectrum, can be used to determine the ingredients, magnetic fields and motions of the cosmic source of the radiation. *See* spectral line.

Spectroscopy: The measurement and study of a spectrum, including the wavelength and intensity of emission and absorption lines, to determine the chemical composition, motion or magnetic field of the radiating source. *See* spectrum.

Spectrum: The distribution of the intensity of electromagnetic radiation with wavelength. Electromagnetic radiation, arranged in order of wavelength from long wavelength, low frequency, radio emissions to short wavelength, high frequency, gamma rays; also, a narrower band of wavelengths, called the visible spectrum, as when sunlight is dispersed by a prism or rainbow into its component colors. A continuous spectrum is an unbroken distribution of radiation over a broad range of wavelengths, such as the separation of white light into its component colors from red to violet. Continuous spectra are often punctuated with emission or absorption lines, called line spectra, which can be examined to reveal the composition, motion and magnetic field of the radiating source. *See* absorption line, Doppler effect, Doppler shift, emission line, Fraunhofer lines, spectral line, spectroheliograph, and Zeeman effect

Speed of light: *See* velocity of light.

Speed of sound: *See* velocity of sound.

Spicule: Narrow, predominantly radial, spike-like structures extending from the solar chromosphere into the corona, observed in hydrogen-alpha lines and concentrated at the cell boundaries of the supergranulation. A spicule lasts about 5 minutes and has a velocity of about 25 kilometers per second. They are about 1 kilometer thick and more than 10,000 kilometers long. *See* chromosphere, hydrogen-alpha line, and supergranulation.

Spörer minimum: A period of low sunspot activity in the fifteenth century (about AD 1420–1570), named after the German astronomer **Gustav Friedrich Wilhelm Spörer (1822–1895)** who also called attention to the Maunder minimum as early as 1887. *See* Maunder minimum.

Spörer's law: The appearance of sunspots at lower solar latitudes over the course of the 11-year solar activity cycle, drifting from mid-latitudes of 30 to 40 degrees north and south towards the equator as the cycle progresses. The sunspot migration to lower average latitudes was first discovered in 1869 by the English astronomer **Richard Carrington (1826–1875),** over an incomplete part of one solar cycle, and investigated in greater detail by the German astronomer **Gutav Friedrich Wilhelm Spörer (1822–1895).** *See* solar activity cycle, sunspot, and sunspot cycle.

Standard Solar Model: A theoretical model of the evolution and internal properties of the Sun, based on physical laws and constrained by assumed initial composition and age of the Sun as well as its observed mass, radius and luminosity. This mathematical description of the solar interior specifies the variation with radius of density, temperature, luminosity and pressure. Theoretical neutrino fluxes are also calculated from the model. The Standard Solar Model is consistent with helioseismology measurements of the temperatures inside the Sun. *See* helioseismology and solar neutrino problem.

STEREO: Acronym for ***Solar TErrestrial RElations Observatory***. See ***Solar TErrestrial RElations Observatory***.

Stratosphere: The layer of a planet's atmosphere in which the temperature remains roughly constant with altitude. The stratosphere lies between heights of about 15 and 50 kilometers, immediately above the troposphere and below the ionosphere.

Streamer: *See* coronal streamer, and helmet streamer.

Subatomic: Smaller in size than an atom.

Subatomic particle: *See* particle, and subatomic.

Sudbury Neutrino Observatory: Abbreviated SNO, a massive underground neutrino detector in Canada filled with heavy water. Observations from SNO have confirmed that solar neutrinos change form on the way from the Sun to the Earth, and that the amount of neutrinos emitted by the Sun is in accordance with theoretical expectations. *See* MSW effect and solar neutrino problem.

Sun: The central star of the Solar System, around which all the planets, asteroids and comets revolve in their orbits, formed together with the planets 4.6 billion years ago. The Sun is a yellow dwarf star on the main sequence, of spectral type G2 with an effective temperature of 5780 kelvin. The absoulte visual magnitude of the Sun is +4.82, and its apparent visual magnitude is −26.72. The Sun has a radius of 0.7 million kilometers, or 109 times the Earth's radius, a mass of 1.989×10^{30} kilograms, or 333 000 times the mass of the Earth, and a mean mass density of 1409 kilograms per cubic meter. The Sun has an absolute luminosity of 3.85×10^{26} Joule per second, and is much more luminous than the average star, outshining 95 percent of the stars in the Milky Way. The total amount of solar energy received per unit time and unit area at the mean Earth-Sun distance, or at one astronomical unit, is called the solar constant and amounts to 1366 Joule per second per square meter. The mean distance of the Sun from the Earth is one astronomical unit, which is 149.598 million kilometers. The visible-light spectrum of the photosphere exhibits dark absorption lines, also known as Fraunhofer lines, which can be examined to determine the ingredients of the Sun. Its principal chemical constituents are hydrogen and helium, which account for 92.1 and 7.8 percent of the atoms, respectively, and for respective mass fractions of 70.68 and 27.43 percent. The Sun's energy source is the nuclear fusion of hydrogen into helium by the proton-proton chain, taking place near the center of the Sun, which has a temperature of 15.6 million kelvin. Overlying this core is the radiative zone where the high-energy radiation, produced in the core fusion reactions, collides with electrons and ions to be re-radiated in the form of less energetic radiation. Outside the radiative zone is a convective zone in which currents of gas flow upward to release energy at the photosphere before flowing downward to be reheated. The visible disk, or photosphere, from which the light we see comes, is some hundreds of kilometers thick, and white-light images of the photosphere reveal the granules and supergranulation that mark the top of convection cells. It takes about 170 thousand years for radiation to work its way out from the Sun's core to the bottom of the convective zone, and only about 10 days for the heated material to carry energy through the convective zone to the photosphere. Sunlight travels from the photosphere to the Earth in 499 seconds. The internal structure of the Sun can be determined by observing the five-minute oscillations in the photosphere, using the technique of helioseismology. *See* absorption line, chromosphere, convective zone, corona, coronagraph, coronal hole, coronal loop, coronal mass ejection, flare, Fraunhofer line, GONG, granulation, helioseismology, photosphere, proton-proton chain, solar activity cycle, ***SOlar and Heliospheric Observatory,*** solar constant, solar eclipse, solar wind, spectroheliograph, sunspot, sunspot cycle, supergranulation, ***Transition Region And Coronal Explorer, Ulysses, Yohkoh,*** and Zeeman effect.

Sunspot: A dark, temporary concentration of strong magnetic fields in the Sun's photosphere. A sunspot is cooler than its surroundings and therefore appears darker. A typical spot has a central umbra surrounded by a penumbra, although either feature can exist without the other. In the umbra, the effective temperature can be about 4000 kelvin compared with 5780 kelvin in the surrounding photosphere. Sunspots are associated with strong magnetic fields of 0.2 to 0.4 Tesla, and vary in size from about 1,000 to 50,000 kilometers across. They occasionally grow to about 200,000 kilometers in size, becoming visible to the unaided eye. Their duration varies from a few hours to a few weeks, or months for the very biggest. The number and location of sunspots depend on the 11-year solar activity cycle or sunspot cycle. They usually occur in pairs or groups of opposite magnetic polarity that move in unison across the face of the Sun as it rotates. The leading, or preceding, spot is called the P spot; the following one is termed the F spot. *See* Hale's law, penumbra, solar activity cycle, Spörer's law, sunspot cycle, umbra, and Zeeman effect.

Sunspot belts: The heliographic latitude zones where sunspots are found, moving from mid-latitudes to the solar equator during the 11-year sunspot cycle. *See* solar activity cycle and sunspot cycle.

Sunspot cycle: The recurring, eleven-year rise and fall in the number and position of sunspots discovered in 1843 by the German astronomer **Samuel Heinrich Schwabe (1789–1875).** The conventional onset for the start of a sunspot cycle is the time when the smoothed number of sunspots, the 12-month moving average, has decreased to its minimum value. At the commencement of a new cycle sunspots erupt around latitudes of 35 to 45 degrees north and south. Over the course of the cycle, subsequent spots emerge closer to the equator, continuing to appear in belts on each side of the equator and finishing at around 7 degrees north and south. This pattern can be demonstrated graphically as a butterfly diagram. All forms of solar activity vary in step with the eleven-year sunspot cycle, including the number and frequency of coronal mass ejections and solar flares and the X-ray intensity of the Sun. *See* coronal mass ejection, flare, solar activity cycle, solar flares, Spörer's law, and X-rays.

Supergranulation: A system of convective cells, called supergranules, with diameters of about 30,000 kilometers and lifetimes of 1 to 2 days, which cover the visible disk of the Sun, the photosphere. The supergranulation was discovered in 1960 by the American physicist **Robert Leighton (1919–1997)** and his colleagues using an instrument developed by them, which produced an image, called a Dopplergram, of motions in the photosphere. The gas flows horizontally from the center to the edge of each supergranule, carrying the photosphere magnetic field to the edges of the supergranules and forming polygonal-shaped structures called the magnetic network that is also detected as bright calcium emission in the overlying chromosphere. The spicules are also concentrated at the boundaries of the supergranule cells. *See* chromospheric network, and spicule.

Super-Kamiokande: A massive underground neutrino detector in Japan filled with pure water, replacing the Kamiokande detector. *See* Kamiokande.

Supersonic: Moving at a speed greater than that of sound, or with a velocity that exceeds the velocity of sound. In air on Earth, the velocity of sound is about 340 meters per second, but the velocity of sound depends on both the temperature and composition of the gas. *See* velocity of sound.

Synchrotron radiation: Electromagnetic radiation emitted by an electron traveling almost at the speed of light in the presence of a magnetic field. Such radiation is linearly polarized. The name arises because it was first observed in 1948 when electrons were accelerated in the General Electric synchrotron particle accelerator. The English physicist **George A. Schott (1868–1937)** discussed such a radiation process by electrons in a paper published in 1907 and in his 1912 book on *Electromagnetic Radiation.* The acceleration of the electrons causes them to emit radiation that is strongly polarized and increases in intensity at longer wavelengths. The wavelength region in which the emission occurs depends on the energy of the electron – 1 MeV electrons radiate mostly in the radio region.

T

Temperature: A measure of the heat of an object and the average kinetic energy of the randomly moving particles in it.

Termination shock: A discontinuity in the solar wind flow in the outer heliosphere where the solar wind slows from supersonic to subsonic motion as it interacts with the interstellar plasma. It marks the outer edge of the solar system. *See* heliopause.

Tesla: The Systeme International, or SI, unit of magnetic flux density, named after **Nikola Tesla (1856–1943),** a Croatian-born American pioneer in the fields of alternating-current electricity, and electrical power generation and distribution. The Tesla is a measure of the strength of a magnetic field. The centimeter-grams-second, or c.g.s., unit of magnetic field strength, often used in astrophysics, is the Gauss, where 1 Tesla = 10,000 Gauss = 10^4 Gauss. *See* Gauss.

Thermal bremsstrahlung: Emission of radiation by energetic electrons encountering the field of a positive ion in a gas that is in thermal equilibrium. *See* bremsstrahlung and thermal equilibrium.

Thermal energy: Energy associated with the motions of the molecules, atoms, or ions.

Thermal equilibrium: A physical system in which all parts have exchanged heat and are characterized by the same temperature at all points. In such equilibrium, a single temperature characterizes the velocity distribution.

Thermal gas: A collection of particles that collide with each other and exchange energy frequently, giving a distribution of particle energies that can be characterized by a single temperature. *See* non-thermal particle.

Thermal particle: A particle that is part of a thermal gas. *See* non-thermal particle.

Thermal radiation: Electromagnetic radiation emitted by a gas in thermal equilibrium. Thermal radiation arises by virtue of an object's heat, or temperature. Energetic electrons that are not necessarily in thermodynamic equilibrium emit non-thermal radiation. *See* blackbody radiation, and non-thermal radiation.

Thermonuclear fusion: The fusion of atomic nuclei at high temperatures to form more massive nuclei with the simultaneous release of energy. Thermonuclear fusion is the power source at the core of the Sun. *See* fusion.

Thermosphere: The atmospheric region where the temperature rises due to heating in the ionosphere.

Torsional oscillations: Zones of alternating fast and slow rotation appearing in the photosphere and below, moving from poles to equator.

TRACE*:** Acronym for ***Transition Region And Coronal Explorer. See ***Transition Region And Coronal Explorer.***

Transition region: A thin region of the solar atmosphere, less than 100 kilometers thick, between the chromosphere and corona characterized by a large rise of temperature from about 10 thousand to a million degrees kelvin. The density decreases as the temperature increases in such a way to keep the gas pressure spatially constant.

Transition Region And Coronal Explorer: NASA launched its ***Transition Region And Coronal Explorer,*** abbreviated ***TRACE,*** spacecraft in April 1998. It provided images of the solar atmosphere at temperatures from 10 thousand to 10 million kelvin with high temporal resolution and second of arc angular resolution using observations at ultraviolet and extreme ultraviolet wavelengths. ***TRACE*** showed that the million-degree corona is comprised of thin magnetic loops that are naturally dynamic and continuously evolving.

Trillion: A million million, and a thousand billion or 10^{12}.

Troposphere: Lowest level of the Earth's atmosphere, where most of the weather takes place, from the ground to about 15 kilometers above the surface. The temperature of the troposphere decreases from 290 kelvin at the ground to 240 kelvin at its top. The troposphere is any region of a planetary atmosphere in which convection normally takes place. *See* convection.

Tunnel effect: A quantum mechanical effect that permits an energetic nuclear particle to escape the nucleus of a radioactive atom. It also enables two colliding protons in the Sun to overcome the electrical repulsion between them, resulting in their nuclear fusion. The tunnel effect involves a quantum leap over, or under, an otherwise insurmountable barrier.

Turbulence: The chaotic mass motions associated with convection.

U

Ultraviolet radiation: Abbreviated UV, electromagnetic radiation with a slightly higher frequency and somewhat shorter wavelength than visible blue light. The German physicist **Johann Wilhelm Ritter (1776–1810)** discovered ultraviolet light in 1801, when examining the spectrum of sunlight at wavelengths just a bit shorter than the wavelength of violet light. Ultraviolet radiation has wavelengths between about 10^{-8} and 3.5×10^{-7} meters, or from 10 to 350 nanometers, with the extreme ultraviolet lying in the short wavelength part of this range. The Earth's atmosphere absorbs most cosmic ultraviolet radiation, so thorough studies of the ultraviolet output of cosmic objects must be conducted from space. Such observations have not led to as many discoveries of totally new cosmic objects as the first detections of the radio and X-ray Universe. Observations of the extreme ultraviolet radiation from the Sun were pioneered by NASA's ***Orbiting Solar Observatories,*** launched from 1962 to 1971, and continued to the contemporary instruments aboard the ***SOlar and Heliospheric Observatory,*** or ***SOHO*** for short, and the ***Transition Region And Coronal Explorer,*** abbreviated ***TRACE.***

Ulysses Mission: The ***Ulysses*** spacecraft was a joint undertaking of ESA and NASA, launched by the ***Space Shuttle Discovery*** on 6 October 1990 to study the interplanetary medium and the solar wind at different solar latitudes. It provided the first opportunity for measurements to be made over the poles of the Sun, using the gravity assist technique to take it out of the plane of the Solar System. After an encounter with Jupiter in February 1992, the spacecraft moved back towards the Sun to pass over the solar South Pole in September 1994 and the North Pole in July 1995, and then again over the South Pole in November 2000 and the North Pole in December 2001. Comparisons of ***Ulysses*** observations with those from other spacecraft, such as ***Yohkoh*** and ***SOHO***, showed that much, if not all, of the high-speed solar wind comes from open magnetic fields in polar coronal holes, at least during the minimum of the Sun's 11-year cycle of magnetic activity.

Umbra: The dark inner core of a sunspot with a penumbra, or a sunspot lacking a penumbra, visible in white light. Also the inner part of the shadow cast during an eclipse, where a total eclipse is visible. *See* penumbra, solar eclipse, and sunspot.

V

Van Allen belts: Two concentric, ring-shaped regions of high-energy charged particles that girdle the Earth's equator within the Earth's magnetosphere, named after the American scientist **James A. Van Allen (1914–),** who discovered them with his colleagues in 1958–59, using the first two successful United States artificial satellites ***Explorer 1*** and ***3.*** The inner belt lies between 1.2 and 4.5 Earth radii, measured from the Earth's center, and the outer belt is located between 4.5 and 6.0 Earth radii. The inner Van Allen belt contains protons with energies greater than 10 MeV and electrons exceeding 0.5 MeV. The outer belt also contains protons and electrons, most of which have energies under 1.5 MeV. The outer belt contains mainly electrons from the solar wind, and the inner belt mainly protons from the solar wind. Another source of radiation-belt particles is neutrons produced when cosmic rays and energetic particles bombard Earth's atmosphere; some of these neutrons decay into protons and electrons that are trapped by the Earth's magnetic field. Within the inner belt is a radiation belt consisting of particles produced by interactions between the solar wind and heavier cosmic ray particles. Because the Earth's magnetic field is offset from the planet's center by about 500 kilometers, the inner belt dips down towards the Earth's surface in the region of the South Atlantic Ocean, off the coast of Brazil. *See* radiation belt, and South Atlantic Anomaly.

Velocity: A quantity that measures the rate of movement and the direction of movement of an object.

Velocity of light: The fastest speed that anything can move. The velocity of light, denoted by c, equals 299 792.458 kilometers per second.

Velocity of sound: Denoted by c_s, the velocity of sound is proportional to the square root of the gas temperature, T, and inversely proportional to the square root of the mean molecular weight, μ, or $c_s \propto (T/\mu)^{1/2}$.

Very Large Array: Abbreviated VLA, a radio interferometer near Socorro, New Mexico consisting of twenty-seven dishes of 25 meters diameter, moveable along the arms of a giant Y with dish separations of up to 34 thousand meters. The VLA operates at wavelengths between 0.018 and 0.90 meters.

Visible light: The form of electromagnetic radiation that can be seen by human eyes, occupying a narrow range of wavelengths in the electromagnetic spectrum. Visible light is also known as optical radiation since optics is used to detect it. Visible light extends roughly from violet wavelengths at 385 nanometers, or

3.85×10^{-7} meters, to red wavelengths at 700 nanometers, or 7.0×10^{-7} meters, and lies between the ultraviolet and infrared parts of the electromagnetic spectrum. *See* electromagnetic radiation, invisible radiation, optical astronomy, optical radiation, and optics.

Visible radiation: *See* visible light.

VLA: Acronym for the Very Large Array, United Stares. *See* Very Large Array.

***Voyager 1* and *2*:** Two almost identical planetary probes launched by the United States in 1977.

W

Water: A substance, denoted by H_2O, composed of atoms of hydrogen, H, and oxygen, O, two of the most abundant elements in the Universe. Water can exist on Earth as a liquid, gas or ice, and in its liquid form it is vital for life. Water molecules have been observed in the atmosphere of Mars and large amounts of frozen water are now found in the surface of Mars. Water molecules are also found in interstellar space.

Wavelength: The distance between successive crests or troughs of an electromagnetic or other wave. Wavelengths are inversely proportional to frequency. The longer the wavelength, the lower the frequency. The product of the wavelength and the frequency of electromagnetic radiation is equal to the velocity of light. *See* velocity of light.

Waves: Propagation of energy by means of coherent vibration.

White light: The visible portion of sunlight that includes all of its colors. Sunlight integrated over the visible portion of the spectrum, from 400 to 700 nanometers, so that all colors are blended to appear white to the eye.

White-light flare: An exceptionally intense and rare solar flare that becomes visible in white light, or in all the colors of the Sun combined.

X

X: The mass fraction of hydrogen. Observations of sunlight and meteorites indicate that $X = 0.706 \pm 0.025$ for the solar material outside its core. Owing to nuclear reactions, there is now less hydrogen in the Sun's core.

X-ray: *See* X-ray radiation.

X-ray radiation: An energetic form of electromagnetic radiation that has short wavelengths, between those of gamma rays and ultraviolet radiation. The German physicist **Wilhelm Konrad Röntgen (1845–1923)** discovered X-rays in 1895. *See* electromagnetic spectrum, hard X-rays, soft X-rays, and ***Yohkoh Mission.***

Y

Y: The mass fraction of helium. Observations of sunlight and meteorites indicate that $Y = 0.274 \pm 0.026$ for solar material outside the core. Owing to nuclear reactions, there is now more helium in the Sun's core.

Yohkoh Mission: The Japanese Institute of Space and Astronautical Science, abbreviated ISAS, launched its ***Solar-A*** spacecraft on 30 August 1991, renaming it ***Yohkoh,*** which means "sunbeam" in English. ***Yohkoh*** collected high-energy radiation from the Sun at soft X-ray and hard X-ray wavelengths, with high angular and spectral resolution, for more than a decade, providing new insights to the mechanisms of solar flare energy release and the heating of the million-degree solar corona. It was a collaborative effort of Japan, the United States and the United Kingdom. *See* hard X-rays and soft X-rays.

Z

Z: The mass fraction of elements heavier than hydrogen and helium. Observations of sunlight and meteorites indicate that $Z = 0.01886 \pm 0.0085$ for solar material.

Zeeman components: The linearly polarized, π, and circularly polarized, σ, components which comprise a line split in the presence of a strong magnetic field. *See* Zeeman effect, and Zeeman splitting.

Zeeman effect: A splitting of a spectral line into components by a strong magnetic field, named after the Dutch physicist **Pieter Zeeman (1865–1943),** who discovered the effect in 1897. If the components cannot be resolved, there is an apparent broadening or widening of the spectral line. The amount of splitting measures the strength of the magnetic field, and the direction of the magnetic field can be inferred from the polarization of the components. The Zeeman effect is thereby used to determine the direction, distribution and strength of the longitudinal magnetic fields in the photosphere.

It has also been used to measure the magnetic fields in other stars and in the interstellar medium. In the simplest case, denoted as normal Zeeman splitting, a line splits into three components. One component, the π component, is not displaced in wavelength or frequency, and is linearly polarized. Two components, the σ components, are shifted by equal amounts to lower and higher wavelengths or frequencies, the magnitude of the shift being proportional to the magnetic field strength. In the general case, the σ components are both circularly and linearly polarized. *See* magnetograph, Zeeman components, and Zeeman splitting.

Zeeman splitting: The splitting of atomic spectral lines in a magnetic field. *See* Zeeman components, and Zeeman effect.

DIRECTORY OF WEB SITES

ACTIVITY CYCLE OF THE SUN

http://www.sec.noaa.gov/SolarCycle/index.html

ECLIPSES OF THE SUN

http://www.MrEclipse.com

http://sunearth.gsfc.nasa.gov/eclipse/eclipse.html

http://www.totalsolareclipse.net

EDUCATIONAL

http://ase.tufts.edu/cosmos/

http://solar-center.stanford.edu/

http://helios.gsfc.nasa.gov/

http://istp.gsfc.nasa.gov/istp/outreach/

http://istp.gsfc.nasa.gov/exhibit/

http://www.lmsal.com/YPOP/

NASA

http://science.hq.nasa.gov/

http://sec.gsfc.nasa.gov

http://umbra.nascom.nasa.gov/solar_connections.html

OBSERVATORIES

http://www.hao.ucar.edu/

http://www.nso.edu/

http://www.solarphysics.kva.se/

SOLAR SPACE MISSIONS

RHESSI:

http://hesperia.gsfc.nasa.gov/hessi/

http://hessi.ssl.berkeley.edu/

SOHO:

http://sohowww.nascom.nasa.gov/

SOLAR-B:

http://science.nasa.gov/ssl/pad/solar/solar-b.stm

STEREO:

http://umbra.nascom.nasa.gov/stereo_facts.html

http://sd-www.jhuapl.edu/STEREO/

TRACE:

http://trace.lmsal.com/

ULYSSES:

http://ulysses.jpl.nasa.gov/

YOHKOH'S SOFT X-RAY TELESCOPE:

http://www.lmsal.com/SXT/

SPACE WEATHER:

http://solar.sec.noaa.gov/

http://www.spaceweather.com

http://www.windows.ucar.edu/spaceweather

http://www.esa-spaceweather.net/

FURTHER READING

AN ANNOTATED LIST OF BOOKS PUBLISHED BETWEEN 1995 AND 2005.

Ashwanden, Markus J.: **Physics of the Solar Corona.** Springer, New York 2004.

This technical book provides a thorough introduction to the subject of solar physics and the million-degree solar corona. It discusses observations of the solar corona at radio, X-ray and extreme-ultraviolet wavelengths. Basic plasma physics, plasma instabilities and plasma heating processes are included, within the contexts of coronal heating and radiation, solar flares and coronal mass ejections.

Bone, Neil: **The Aurora: Sun-Earth Interactions.** John Wiley and Sons, New York 1996.

A complete review of all aspects of the aurora, describing the mechanism of auroras and their observation, including the history of aurora investigations from the earliest ideas to scientific investigations with satellites. Solar activity, Earth's magnetosphere and solar-terrestrial interactions are also reviewed.

Carlowicz, Michael J. and Lopez, Ramon E.: **Storms from the Sun: The Emerging Science of Space Weather.** The Joseph Henry Press, Washington, D.C. 2002.

An excellent popular discussion of space weather and its impact on our society, which has become increasingly dependent on space-based technologies. It describes the effects of solar flares and coronal mass ejections on the Earth, including disruption of radio communications and damage to Earth-orbiting satellites. There are colorful descriptions of dramatic past events that illustrate these effects of space weather.

Golub, Leon and Pasachoff, Jay M.: **Nearest Star: The Surprising Science of Our Sun.** Harvard University Press, Cambridge, Massachusetts 2001.

A well-written description of our current understanding of the Sun and its effects on Earth, placed within both a historical and modern context. Fundamental questions are included, such as why does the Sun shine, is the Sun a variable star, and why is the corona so hot?

Golub, Leon, and Pasachoff, Jay M.: **The Solar Corona.** Cambridge University Press, New York 1997.

An advanced textbook that presents our recent understanding of coronal physics, written for graduate and advanced undergraduate students.

Jokipii, J. Randy, Sonett, Charles P., and Giampapa, Mark S. (editors): **Cosmic Winds and the Heliosphere.** The University of Arizona Press, Tucson 1997.

A collection of 28 technical articles that discuss the solar heliosphere, the solar wind, the physics of wind origins, winds from other stars, the physical properties of the solar wind, and interactions of winds with the surrounding medium.

Lang, Kenneth R.: **The Cambridge Encyclopedia of the Sun.** Cambridge University Press, New York 2001.

A fundamental, up-to-date reference source of information about the Sun. It includes the properties of the Sun as a star and its place in the Galaxy and the Universe, as well as the solar interior, solar flares and coronal mass ejections, the Sun's winds and solar-terrestrial interactions. Numerous tables of fundamental data also complement the text.

Lang, Kenneth R.: **The Sun From Space.** Springer-Verlag, New York 2000.

A comprehensive account of solar astrophysics and how our perception and knowledge of the Sun have gradually changed. Timelines and hundreds of seminal papers are provided for key discoveries during the past two centuries, but the emphasis is on the last decade which has seen three successful solar spacecraft missions: *SOHO, Ulysses* and *Yohkoh.* Together these have confirmed many aspects of the Sun and its output, and provided new clues to the numerous open questions that remain. This well-illustrated book is written in a clear and concise style, covering all levels from the amateur astronomer to the expert.

Lang, Kenneth R.: **Astrophysical Formulae. Volume I. Radiation, Gas Processes and High Energy Astrophysics. Volume II. Space, Time, Matter and Cosmology.** Springer-Verlag, New York 1999.

The third, enlarged edition of a comprehensive, widely used reference to the fundamental formulae employed in astronomy, astrophysics and general physics, including 4,000 formulae and 5,000 references to original papers. It includes all aspects of astronomy and astrophysics that are relevant to studies of the Sun.

Lang, Kenneth R.: **Sun, Earth and Sky.** Springer-Verlag, New York 1995.

The first edition of this book. It introduces the Sun, its physics and its impact on life here on Earth, written in a light and friendly style with apt metaphors, similes and analogies, poetry, art, history, and vignettes of scientists at work.

Schrijver, Carolus J. and Zwaan, Cornelius: **Solar and Stellar Magnetic Activity.** Cambridge University Press, New York 2000.

A comprehensive review and synthesis of our current understanding of the origin, evolution and effects of magnetic fields in the Sun and other cool stars. Solar topics include solar differential rotation and meridional flow, solar magnetic configurations and structure, the global solar magnetic field, the solar dynamo and the solar atmosphere.

Author Index

Abbott, Charles Greeley (1872–1973) 213
Adhémar, Joseph Alphonse (1797–1862) 227
Alfvén, Hannes (1908–1995) 118
Alighieri, Dante (1265–1321) 234
Anderson, Carl D. (1905–1991) 30, 233
Ångström, Anders Jonas (1814–1874) 238
Appleton, Edward (1892–1965) 206
Aristotle (384–322 BC) 82, 174
Arrhenius, Svante August (1859–1927) 200, 235, 248
Aston, Francis W. (1877–1945) 26

Babcock, Horace W. (1912–2003) 103, 234
Bahcall, John N. (1934–2005) 46, 53, 234
Bailey, Francis (1774–1844) 109, 234, 239
Balmer, Johann (1825–1898) 12, 239
Bartels, Julius (1899–1964) 183, 235
Bellay, Joachim du (1522–1560) 129, 234
Bethe, Hans A. (1906–2005) 30, 31, 53, 234, 257
Biermann, Ludwig Franz (1907–1986) 117, 125, 234, 261
Birkeland, Kristian (1867–1917) 167, 178, 181, 235
Bohr, Niels (1885–1962) 12, 239
Bourdillon, Francis William (1852–1921) 8, 233
Braun, Douglas (1961–) 163
Braun, Karl Ferdinand (1850–1918) 205
Browning, Robert (1812–1889) 177, 235
Buchli, James (1945–) 193, 235
Bunsen, Robert William (1811–1899) 14, 233
Bush, George W. (1946–) 224

Carrington, Richard C. (1826–1875) 82, 100, 140, 235, 262
Celsius, Anders (1701–1744) 181, 240
Chadwick, James (1891–1974) 11, 254
Chamberlin, Thomas Chrowder (1843–1928) 235
Chapman, Sydney (1888–1970) 125, 181
Cherenkov, Pavel A. (1904–1958) 50, 240
Chrichton, Michael (1942–) 225
Chupp, Edward L. (1927–) 148
Church, Frederic (1826–1900) 175
Compton, Arthur H. (1892–1962) 242
Cook, Captain James (1728–1779) 176
Cowan, Clyde L. (1919–1974) 43, 234
Critchfield, Charles (1910–1994) 30
Croll, James (1821–1890) 227
Crutzen, Paul J. (1933–) 211

Davis, Raymond Jr. (1914–) 47, 52, 240
De Bort, Leon Philippe Teisserenc (1855–1913) 205
De Mairan, Jacques D'Ortous (1678–1771) 216
Delaunay, Robert (1885–1941) 40, 192
Descartes, Rene (1596–1650) 177
Deslandres, Henri (1853–1948) 85, 90, 141, 262
Deubner, Franz-Ludwig (1934–) 66
Dirac, Paul Adrien Maurice "P.A.M." (1902–1984) 30, 233
Dobson, Gordon Miller Bourne "G.M.B." (1889–1976) 208, 236, 255
Doppler, Christian Johann (1803–1853) 17, 18, 243
Du Bellay, Joachim (1522–1560) 129, 234
Duvall, Thomas L. Jr. (1950–) 87

Eddington, Arthur Stanley (1882–1944) 26, 233
Eddy, John A. "Jack" (1931–) 216
Edlén, Bengt (1906–1993) 114, 241, 242
Einstein, Albert (1879–1955) 77, 253
Eliot, T. S. (1888–1965) 206
Emerson, Ralph Waldo (1803–1882) 4
Evershed, John (1864–1956) 87, 245

Fahrenheit, Gabriel Daniel (1686–1736) 246
Fermi, Enrico (1901–1954) 42, 254
Ferraro, Vincent (1907–1974) 181
Forbush, Scott (1904–1984) 165
Foukal, Peter V. (1945–) 214
Fourier, Joseph (Jean Baptiste) (1768–1830) 199, 248
Fowler, William A. "Willy" (1911–1995) 46
Fraunhofer, Joseph von (1787–1826) 11, 240, 247, 250
Frazier, Edward N. (1939–) 64, 234
Friedman, Herbert (1916–2000) 119
Friedrich, Caspar David (1774–1840) 3, 4
Friis-Christensen, Eigil (1944–) 218
Fritz, Herman (1830–1883) 178
Frost, Robert (1874–1963) 231, 236

Galilei, Galileo (1564–1642) 74, 81, 174
Gamow, George (1904–1968) 28

Gauss, Karl Friedrich (1777–1855) 86, 247
Giacconi, Riccaardo (1931–) 52
Gilbert, William (1544–1603) 166
Ginzburg, Vitalii L. (1916–) 114
Glashow, Sheldon (1932–) 54
Gold, Thomas (1920–2004) 167
Golding, William (1911–1993) 203
Gosling, John T. "Jack" (1938–) 235
Graham, George (1674–1751) 181
Gribov, Vladimir (1930–1997) 55
Grotrian, Walter (1890–1954) 114, 241, 242

Hadley, George (1685–1768) 197
Hale, George Ellery (1868–1938) 84, 85, 90, 141, 262
Harkness, William (1849–1900) 113
Hays, James D. (1933–) 229
Heaviside, Oliver (1850–1925) 206
Helmholtz, Hermann von (1821–1894) 23
Herschel, John (1792–1871) 181
Herschel, William (1738–1822) 8, 34
Hess, Victor (1883–1964) 32, 165, 242
Hodgson, Richard (1804–1872) 140, 235
Howard, Robert "Bob" (1932–) 76
Hoyle, Fred (1915–2001) 193, 235
Hudson, Hugh S. (1939–) 213
Humboldt, Baron Alexander von (1769–1859) 181

Imbrie, John (1925–) 229
Inhofe, James M. (1934–) 224

Janssen, Pierre Jules César "P.J.C." (1824–1907) 14, 88
Johnson, Manuel John (1805–1859) 234
Joyce, James (1882–1941) 54

Kahn, Franz Daniel "F. D." (1926–1998) 63
Keeling, Charles (1928–2005) 218, 248, 249
Kennelly, Arthur E. (1861–1939) 206
Kepler, Johannes (1571–1630) 123
Kirchhoff, Gustav Robert (1824–1887) 14, 233, 251
Koshiba, Masatoshi (1926–) 50, 52
Kosovichev, Alexander G. "Sasha" (1953–) 87

LaBonte, Barry (1950–) 76
Lane, Jonathan Homer (1819–1880) 26
Lassen, Knud (1921–) 218
Lean, Judith (1953–) 214
Leibacher, John (1941–) 65
Leighton, Robert (1919–1997) 35, 63, 234, 246, 264
Lin, Robert P. "Bob" (1942–) 150
Lindsey, Charles (1947–) 163
Lockyer, Joseph Norman (1826–1920) 15, 88, 249
Loomis, Elias (1811–1889) 177
Lorentz, Hendrik (1853–1928) 84, 245
Louis XIV "Sun King" (1638–1715) 216
Lovelock, James (1919–) 202
Ludwig, George H. (1927–) 235
Lyot, Bernard (1897–1952) 112, 113, 241

Marconi, Guglielmo (1874–1937) 205
Margulis, Lynn (1938–) 203
Martyn, David F. (1906–1970) 114
Maunder, E. Walter (1851–1928) 100, 183, 216, 217, 235, 253
McCandless, Bruce II (1937–) 187
McDonald, Arthur D. "Art" (1943–) 60, 234
McIlwain, Carl E. (1931–) 235
Melville, Herman (1819–1891) 29, 233
Mikheyev, Stanislav P. (1940–) 55, 253
Milankovitch, Milutin (1879–1959) 222, 253
Millikan, Robert (1868–1953) 242
Miró, Joan (1893–1983) 4, 106, 164
Molina, Mario (1943–) 210, 211, 236, 255
Monet, Claude (1840–1926) 4, 22, 80, 136

Nansen, Fridtjof (1861–1930) 176, 235
Nasmyth, James (1808–1890) 82
Nietzsche, Friedrich (1844–1900) 4, 233
Nobel, Alfred (1833–1896) 255
Noyes, Robert W. (1934–) 63, 234

Parker, Eugene (1927–) 125, 149, 234, 241, 261
Pauli, Wolfgang (1900–1958) 41, 42, 234, 254
Pawsey, Joseph L. (1908–1962) 114
Payne, Cecilia H. (1900–1979) 14
Perl, Martin L. (1927–) 54
Perrin, Jean Baptiste (1870–1942) 28, 233
Peterson, Donald H. (1933–) 186
Petscheck, Harry E. (1930–2005) 149
Pontecorvo, Bruno (1913–1993) 55

Queen Elizabeth I (1533–1603) 166

Ramaty, Reuven (1937–2001) 258
Ramsay, William (1852–1916) 15, 249
Reines, Frederick (1918–1998) 43, 44, 54, 234
Remek, Vladimir (1948–) 176, 235
Revelle, Roger (1909–1991) 220, 236
Ritter, Johann Wilhelm (1776–1810) 265
Roberts, Walter Orr (1915–1990) 92
Röntgen, Wilhelm Konrad (1845–1923) 267
Roscoe, Henry Enfield (1833-1915) 233
Rowland, F. Sherwood (1927–) 210, 211, 236, 255
Russell, Morgan (1886–1953) 192
Rutherford, Ernest (1871–1937) 11, 26, 257

Sabine, Colonel Edward (1788–1883) 181, 235
Salam, Abdus (1926–1996) 54
Schatzman, Evry (1920–) 117
Schott, George A. (1868–1937) 264
Schwabe, Samuel Heinrich (1789–1875) 98, 181, 234, 264

Schwarzschild, Martin (1912–1997) 117
Secchi, Pietro Angelo (1818–1878) 92
Shackleton, Nicholas J. "Nick" (1937–) 229
Simon, George W. (1934–) 63, 234
Smirnov, Alexei Y. (1951–) 55, 253
Spörer, Gustav (1822–1895) 100, 216, 217, 253, 262
Stein, Robert F. (1935–) 65
Stevens. Wallace (1879–1955) 177, 235
Stoney, George Johnstone (1826–1911) 245
Størmer, Carl (1874–1957) 171, 172
Strömgren, Bengt (1908–1987) 30
Suess, Hans (1909–1993) 220, 236
Sweet, Peter A. (1921–2005) 149

Tesla, Nikola (1856–1943) 86, 247, 265
Thomson, Joseph John "J. J." (1856–1940) 245
Thomson, William "Lord Kelvin" (1824–1907) 23, 233, 251
Totsuka, Yoji (1942–) 50, 234
Trieman, Sam Bard (1925–1999) 172
Turck-Chièze, Sylvaine (1951–) 46, 53
Turner, Joseph Mallord William (1775–1851) 3, 5
Tyndall, John (1820–1893) 199, 235, 248

Ulrich, Roger (1942–) 65
Updike, John (1932–) 41, 234
Urey, Harold Clayton (1893–1981) 243

Van Allen, James A. (1914–) 172, 173, 235, 266
Van Gogh, Vincent (1853–1890) 4, 62

Webber, Andrew Lloyd (1948–) 206
Weinberg, Steven (1933–) 54
Weizsacker, Carl Friedrich von (1912–) 30
Willson, Richard C. (1937–) 213
Wolf, Johann Rudolf (1816–1893) 217
Wolfenstein, Lincoln (1923–) 55, 253
Wollaston, William Hyde (1766–1828) 247

Young, Charles A. (1834–1908) 113

Zarathustra (about 1300 BC, Greek Zoroaster) 3
Zeeman, Pieter (1865–1943) 84, 267
Zhao, Junwei (1971–) 87

Subject Index

Absolute temperature 237
Absolute zero 237
Absorption lines 12, 237
Abundance, of elements in Sun 13, 74
ACE, see *Advanced Composition Explorer*
Acoustic waves 237
ACRIM 213
Active Cavity Radiometer Irradiance Monitor 213
Active region 95, 237
Active-region belts 99
Advanced Composition Explorer 132, 133
Age, Earth 25
—, Sun 23, 24, 36
Alfvén waves 118, 238
—, polar coronal holes 132
Alpha particle 238
—, helium nuclei 42
Altamaha river delta, Georgia 196
Annihilation, of electrons and positrons 32
Antielectron 238
Anti-matter 10
Anti-particle, of electron 30
Appleton layers, ionosphere 206
Astronaut, space weather hazard 185
Astronomical unit 24, 238
Atmosphere 203–206, 238
—, buildup of carbon dioxide 218, 219
—, carbon dioxide 218, 219
—, greenhouse gases 220
—, pressure 204
—, Sun 81
—, temperature 204
—, variable solar heating 215
Atoms 11, 238
AU 239
Aurora 174–180, 239
—, and coronal mass ejections 180, 182
—, and magnetic reconnection of magnetotail 180
—, and upper atmosphere 179
—, particles 174
Aurora australis 174, 176
Aurora borealis 174–177
Aurora oval 178, 179, 182, 239

Baily's beads 239
Balmer lines 12
Balmer series 239
Benard convection 34
Beta decay 41–43, 239
Beta particles 239
—, electrons 42
Biosphere 195
Bipolar pairs, of sunspots 94
Bohr atom 12, 239
Boron-8 neutrinos 45
Bow shock 168, 169, 239
—, stellar 135
Bremsstrahlung 239
—, solar flares 143, 144
Burst 239
Butterfly diagram 100, 240

Calcium H and K lines, chromosphere 89, 93
Calcium magnetic network 93
Canopy, magnetic 94
Carbon cycle 195
Carbon dioxide 240
—, atmosphere increase 218, 219
—, greenhouse gas 200
—, ice ages 229, 230
Cataracts, and ozone depletion 211
Celsius, degrees 25, 240
CFCs 211
CGRO, see *Compton Gamma Ray Observatory*
Charged current reaction 60
Charged particles 240
Cherenkov light 60
Cherenkov radiation 50, 60, 240
Chlorine experiment 240
Chlorofluorocarbons 211
Chromosphere 88, 240
—, calcium H and K 89, 93
—, filaments 91, 152–155, 246
—, hydrogen alpha emission 91
—, magnetic network 93
—, network 241
—, plage 91
—, temperature 24, 115
—, spicules 92
Climate 241
Clouds, and global warming 222
CME, *See* coronal mass ejection
CO_2: *See* carbon dioxide.

Colors 7
Comets, solar wind 123, 125
Communication satellites, disabled by space weather 188, 189
Composition, stars 14
—, Sun 13
Compton Gamma Ray Observatory 147, 148, 152
Continental drift 197
Continuous spectrum 241
Continuum 241
Convection 34, 241
Convective zone 32, 33, 241
—, depth 74
—, differential rotation 75
Copuscular radiation 125
Corona 107, 108, 241
—, dimming 157
—, expansion 125
—, green emission line 113
—, heating 116–119
—, magnetic fields 124
—, million-degree temperature 113–114
—, structure 111, 112
—, temperature 24
Coronagraph 112, 113, 241
Coronal dimming 157
Coronal green line 242
Coronal heating 116–119
—, magnetic reconnection 118
—, magnetic waves 118
—, nanoflares or microflares 118
Coronal holes 119–123, 242
—, fast heavy ions 131
—, origin of fast solar wind 128, 130
Coronal loops 119–123, 242
—, magnetic reconnection 149
Coronal mass ejection 112, 113, 138, 139, 156–158, 242
—, and auroras 180, 182
—, and space weather 184
—, geomagnetic storms 181, 183
—, halo 160
—, intense geomagnetic storms 181, 183
—, interplanetary shocks 182
—, magnetic reconnection 158
—, prediction 159–161
—, size 156
—, space weather 184
—, *STEREO* mission 162
Coronal streamer 242
—, origin of slow solar wind 128, 130
Coronal transient 112
Coronium 113, 242
Corpuscular radiation 242
Cosmic rays 165, 242
—, acceleration in supernovae 165
—, and eleven year cycle of solar activity 165
—, and Van Allen radiation belts 173
—, discovery 30, 165
Current sheet 243
Currents, aurora oval 179
Cusp, X-ray 160, 161

D layer 243
Density 243
Depth, of convective zone 74
Deserts 195
Deuterium 243
Deuteron 243
Differential rotation 243
—, and sunspots 82
—, photosphere 74
—, Sun 74
Dipole 243
Disparition brusque 154, 155
Distance, Sun 24
Doppler effect 17, 18, 243
Doppler shift 243
Dopplergram 63
Dynamo 101, 102, 244
—, solar 75, 76

E layer 244
Earth 244
—, age 25
—, atmosphere 203–206
—, bow shock 168, 169
—, dipole magnetic field 166, 167
—, fate 203
—, global warming 200, 217–227, 248
—, greenhouse effect 199–203
—, magnetotail 168
—, temperature 217, 218
—, Van Allen radiation belts 171–173
—, view from space 193
Earth Radiation Budget radiometer 213
Electromagnetic radiation 244
Electromagnetic spectrum 17, 18, 244
Electromagnetic waves 16
Electron 245
Electron neutrino 54, 245
Electron volt 19, 245
Electron-positron annihilation, solar flares 147
Electrons, from solar flares 142–146
Electroweak theory 54
Elements 245
—, synthesis inside stars 13
Eleven-year magnetic activity cycle 99–102
Emission lines, chromosphere 88
Energetic proton event 157
Energy, nuclear 28
—, of mass 31
—, photon 19
—, subatomic 28
Energy-generating core 32, 33
ERB 213
Erupting filament 152–155
Erupting prominence 110, 138, 139, 152–155, 158

eV, *See* electron volt
Evershed effect 87, 245
Exosphere 206
Extreme ultraviolet radiation 245

F layer 246
Faculae 91, 213, 245
Fahrenheit 246
Fahrenheit, degrees 25
Faint-young-Sun paradox 37, 202, 246
Fast solar wind 126, 128, 130
Fate, Earth 203
—, Sun 36–38
Fibrils 246
Filament 91, 246
Five-minute oscillations 63, 246
—, and solar interior 71
—, photosphere 63
Flare 246
—, *See also* solar flare
Flare loops 146
Flash spectrum 246
Footpoint 246
Forbidden lines 114, 246
Forecast, space weather 191
Fraunhofer absorption lines 11, 12
Fraunhofer lines 247
Free electron 247
Frequency 18, 247
Fusion 247

GALLEX 247
GALLEX, *see* Gallium experiment
Gallium 50, 51
Gallium experiment, solar neutrino detector 51, 52
Gallium Neutrino Observatory 51, 52
Gamma rays 19
—, image of solar flare 150, 151
—, solar flares 146–148
Gamma-ray radiation 247
Gas pressure 26, 247
General Theory of Relativity, and solar oblateness 77
Geomagnetic field 247
Geomagnetic storms 181–183, 247
—, and sunspot cycle 181
—, eleven-year solar activity cycle 181
—, intense 181, 183
—, recurrent 183
Glaciers, shrinking 221
Global Oscillation Network Group 248
Global warming 200, 217–227, 248
—, and clouds 222
—, and oceans 222
—, current 221, 224
—, future consequences 223
—, *Kyoto Protocol* 225
—, melting permafrost 221
Global warming, preventive legislation 225, 226
—, rising sea levels 221, 223
—, shrinking glaciers 221
—, signs 221, 222
—, threat 224
—, uncertain computations 222
GNO, see Gallium Neutrino Observatory
Gradual phase, solar flares 144, 145
Granulation 34, 35, 84, 248
Granule 34, 248
Gravitational contraction 23
Greenhouse effect 248
—, natural 199
—, young Earth 37, 202
Greenhouse gases 200, 220
—, ice ages 229, 230

H and K lines 249
Hadley cells 197
Hale's law of sunspot polarity 101, 249
Hard X-rays 249
Heating, chromosphere 117
—, corona 116–117
Heat-trapping gases 200, 220
—, ice ages 229, 230
Heaviside layer, ionosphere 206
Heavy water 58, 249
—, neutrino reactions 59
Heliopause 249
Heliopolis 3
Helioseismology 71, 249
Heliosheath 134, 135
Heliosphere 125, 126, 134, 249
Helium 249
—, discovery in Sun 14, 88, 89
—, discovery on Earth 15, 89
—, mass 26
Helium burning 250
Helium nucleus 11
Helmet steamers 110, 111, 250
High-speed stream 250
Homestake solar neutrino detector 47, 48, 52, 250
Hubble Space Telescope 135, 189
Hydrogen 250
—, most abundant element 14
Hydrogen-alpha flare ribbons 141
Hydrogen-alpha line 250
Hydrogen-alpha Sun 91
Hydrogen burning 250
Hydrogen nucleus 11

Ice ages 227, 250
—, and distribution of sunlight 227
—, astronomical cycles 227
—, carbon dioxide 229, 230
—, cause 227
—, greenhouse gases 229, 230
—, methane 229, 230

Ice caps, disappearance 231
Ice cores 229
Impulsive flare 250
Impulsive phase, solar flares 144, 145
Infrared radiation 8
Interferometry 250
Interplanetary magnetic field 251
Interstellar wind 135
Invisible radiation 19, 251
Ionization 251
Ionosphere 204–206, 251
—, and aurora 179
—, and radio communication 205
Irradiance 251

K or °K 251
Kamioka Liquid scintillator Anti-Neutrino Detector, KamLAND 57
Kamiokande solar neutrino detector 50, 52, 251
KamLAND 57
Kelvin 251
Kelvin, temperature scale 23
KeV 251
Kodaikanal Observatory, India 87
Kyoto Protocol 225

Latitude 251
Leptons 54
Light 252
Light element 252
Limb 252
Little Ice Age 216
LL Orionis 135
Luminosity, Sun 23, 24
Lunik 2 126
Luxor 3

Magnetic canopy 94
Magnetic carpet 118, 119
Magnetic field 252
—, amplification 101, 102
—, corona 124
—, Earth 166, 168
—, interplanetary space 127
—, photosphere 81–88
—, planets 166
—, solar dipolar 24
—, solar wind 127
—, sunspots 24
Magnetic field lines 95, 252
Magnetic loops 95–97
Magnetic network 88
—, chromosphere 93
Magnetic reconnection 252
—, coronal heating 118
—, coronal loops 149, 150
—, coronal mass ejection 158
—, Earth's magnetotail 170, 180
—, solar flares 148–152
Magnetic storm 252
Magentic waves 118
Magnetic waves, polar coronal holes 132
Magnetism 252
Magnetogram 88, 89, 101, 252
Magnetograph 252
Magnetopause 252
Magnetosphere 166, 168, 253
Magnetotail 168, 253
—, magnetic reconnection 170, 180
Mariner II 126
Mars 7
Mass, energy 31
—, helium 26
—, neutrinos 54
—, proton 26
—, Sun 24
Mass-Energy equivalence 31
Mass spectrograph 26
Maunder minimum 216, 217, 253
McMath solar telescope 9
Meridional flow 103
Meridional oscillation 76
Methane 220, 253
—, ice ages 229, 230
Microflares 253
Milankovitch cycles 227, 253
Model, solar flare 151, 152
Montreal Protocol 212
Motion, velocity 18
Mount Wilson Observatory, 60-foot tower telescope 86
MSW effect 55, 253
Muon neutrinos 54

Nanoflares 253
—, coronal heating 118
Negative hydrogen ion 81
Network, magnetic 88, 93, 254
Neutral current reaction 60
Neutral line or region 254
Neutrino 10, 41, 42, 254
—, electron 54
—, mass 54
—, muon 54
—, tau 54
Neutrino detector, Super-Kamiokande 53, 54
Neutrino detectors 47
Neutrino oscillation 54, 57, 60, 61, 254
Neutrinos 10
—, and cosmic rays 57
—, boron-8 45
—, discovery on Earth 43
—, flavors 53
—, from nuclear reactor 43
—, from particle accelerators 57
—, from supernova explosion SN 1987A 50
—, in atmosphere 56, 57
—, measured flux from Sun 49–52

Neutrinos, proton-proton reaction 44, 45
—, solar 41, 44–47
Neutron 11, 254
—, solar flares 147, 148
Neutron-capture line, gamma rays 147
New Guinea 197
Nitrous oxide 220
Non-thermal particle 255
Non-thermal radiation 255
—, solar flares 143, 144
Northern lights 174–177, 255
Nuclear de-excitation gamma rays 147
Nuclear energy 28, 255
Nuclear force 255
Nuclear fusion 29, 255
Nuclear reactions, solar flares 146–148
Nuclear reactor, neutrinos 43
Nucleosynthesis, of elements in stars 27
Nucleus 255
—, of atom 11

Oblateness, Sun 76, 77
Oceans, and global warming 222
—, origin 201
Oil, depletion 231
Optical radiation 255
Optical spectrum 255
Optics 255
Origin, blue sky 193
—, oceans 201
—, of elements in Sun 13
—, solar wind 128–134
Oscillation, atmospheric muon neutrinos 57
—, neutrino 54, 57, 60, 61
—, solar electron neutrinos 60, 61
OSO 7 112, 142
Oxygen cycle 195
Ozone 207, 255
—, ground-level 208
—, hole 207–210, 255
—, layer 207, 256
—, production by solar ultraviolet radiation 207
Ozone depletion 210, 211
—, and cataracts 211
—, and skin cancer 211
—, *Montreal Protocol* 212
—, preventive legislation 212
Ozone hole 207–210, 255
Ozone layer 207, 256

P78-1 satellite 113
Pair annihilation 256
Penumbra 256
Photon 256
Photon energy 19, 256
Photons 18
Photosphere 81, 256
—, differential rotation 74
Photosphere, faculae 91
—, five-minute oscillations 63
—, granulation 34, 35, 84, 248
—, magnetic fields 81–88
—, magnetic network 88
—, magnetograms 88, 89
—, most abundant elements 15
—, oscillations 63, 66
—, sunspots 81–88, 264
—, temperature 24
Photosynthesis 6, 256
Plage 91, 256
Planetary nebula 38
Planets, magnetic fields 166
Plasma 25, 256
Polarity 256
Polarization 257
Poloidal magnetic field 104
Positron 30, 257
—, solar flares 147, 148
Post-flare loops 158, 159, 257
Power grids, space weather disable 190
Prediction, solar flares 159–161
Project Poltergeist 43
Prominence 109, 257
—, erupting 152–155
Proton 11, 26, 257
—, from solar flares 142–146
—, mass 26
Proton-proton chain 29, 31, 257
Proton-proton reaction 30, 31
—, neutrinos 44, 45

Quadrupole moment, Sun 77
Quantum mechanics 257
Quantum theory 28
Quarks 54
Quiescent prominence 109, 110, 257

Radiation, *See* electromagnetic radiation.
Radiation belts 171–173, 258
Radiative zone 32, 33, 258
Radio communication, disrupted by space weather 187
—, ionosphere 205
Radio radiation 19, 20, 258
Radio waves 17
Radioactive dating 25
Radioactivity 11, 28, 41, 258
Radius, coronal mass ejection 156
—, Sun 24
Rain forests 195
Ramaty High Energy Solar Spectroscopic Imager 142, 144, 147, 151, 258
Recurrent geomagnetic storms 183
Resolution, optical telescope 10
RHESSI 142, 144, 147, 151, 258
Rotation 258
—, Sun 24, 74, 75

SAGE, *see* Soviet American Gallium Experiment
Satellites, disabled by space weather 188, 189
—, threat of South Atlantic Anomaly 189
—, threat of Van Allen radiation belts 189
Scintillation 258
Sea level, rising 221, 223
Seasons 198, 199, 259
Seeing 10, 259
Seventh Orbiting Solar Observatory, see OSO 7
Shintoism 3
Shock wave 259
—, coronal mass ejection 182
Sigmoid 160
Skin cancer, and ozone depletion 211
Sky, origin 193
Skylab 112, 117, 119, 121, 141, 142, 188, 259
Slow solar wind 126, 128, 130
SMM, See *Solar Maximum Mission*
Smog, ground-level zone 208
SNO, *see* Sudbury Neutrino Observatory
SNU 47
Soft X-rays 259
SOHO 53 69, 70, 72, 75, 87, 97, 112, 113, 121, 122, 131, 133, 138, 139, 142, 156, 158, 214, 259
Solar activity cycle 259
—, variable atmospheric heating 215
SOlar and Heliospheric Observatory 53, 69. 70, 75, 87, 97, 112, 113, 121, 122, 131, 133, 138, 139, 142, 156, 158, 214, 259
Solar atmosphere 81, 259
Solar constant 24, 212, 214, 259
—, variations 214
Solar core 260
Solar cycle 260
Solar disk, temperature 26
Solar dynamo model 103, 104
Solar eclipse 107, 108, 260
Solar energetic particle 260
Solar flares 95, 137, 138, 140–142, 260
—, accelerated particles 142–146
—, and space weather 184
—, bremsstrahlung 143, 144
—, electron-positron annihilation line 147
—, electrons 142–146
—, gamma rays 146–148
—, gamma-ray image 150, 151
—, gradual phase 144, 145
—, hydrogen-alpha ribbons 141
—, impulsive phase 144, 145
—, magnetic reconnection 148–152
—, model 151, 152
—, neutron-capture line 147
—, neutrons 147, 148
—, non-thermal radiation 143, 144
—, nuclear reactions 146–148
—, post-flare loops 146
—, prediction 159–161
—, protons 142–146
Solar flares, radio radiation 141, 143
—, synchrotron radiation 143
—, temperature 137
—, thermal radiation 143, 144
—, white light 140
—, X-rays 141, 145
Solar limb 260
Solar mass 260
Solar maximum 260
Solar Maximum Mission 113, 148, 188, 213, 260
Solar minimum 261
Solar neutrino detector, Gallium experiment 51, 52
—, Homestake 47, 48, 52
—, Kamiokande 50, 52
—, Soviet American Gallium Experiment 51, 52
—, Sudbury Neutrino Observatory 58–61
—, Super-Kamiokande 53, 54, 264
Solar Neutrino Problem 49, 261
—, solution 52–61
Solar Neutrino Unit 27, 261
Solar neutrinos 44–47
—, measured flux 49–52
Solar System, edge 134
Solar wind 123, 126, 261
—, and Van Allen radiation belts 173
—, comets 123, 125
—, density 126, 127
—, fast, uniform wind 126
—, origin 128–134
—, slow, gusty wind 126
—, spiral magnetic field 127
—, temperature 126, 127
—, termination shock 134, 135
—, velocity 126
Solar-A 259
Solar-B 132, 260
Solar-TErrestrial Relations Observatory 162, 191
Sound speed, inside Sun 73, 74
Sound waves 237
—, and heating of chromosphere 117
—, generation in convective zone 68
—, in Sun 63–68
South Atlantic Anomaly 261
—, threat to satellites 189
Southern lights 174, 176
Soviet-American Gallium Experiment, solar neutrino detector 51, 52
Space Shuttle 176
Space Shuttle Challenger 186
Space Shuttle Columbia 197
Space Shuttle Discovery 177
Space weather 139, 140, 183–191, 261
—, astronaut risk 185
—, communication satellites 188, 189
—, coronal mass ejections 184
—, disable power grids 190
—, disable satellites 188, 189
—, disrupt radio communications 187

Space weather, forecasts 160, 161, 191
—, solar flares 184
—, weather satellites 188, 189
Spectral line 261
Spectrograph 11, 262
Spectroheliogram 262
Spectroheliograph 90, 262
Spectroscope 262
Spectroscopy 262
Spectrum 262
—, electromagnetic 17
Spicules 92, 117, 262
Spörer minimum 216, 217, 262
Spörer's law 262
SST, *see* Swedish Solar Telescope
Stalks 110, 111
Standard Solar Model 26, 52, 53, 262
Starfish 189
Stars, composition 14
Stellar bow shock 135
STEREO 162, 191
Storms, space 183–191
Stratosphere 204, 205, 207, 263
—, ozone 207
Subatomic energy 28
Sudbury Neutrino Observatory 58–61, 263
Sun 263
—, abundance of elements 13, 74
—, age 23, 24, 36
—, anatomy 33
—, central temperature 53
—, convective zone 32, 33
—, core 26
—, differential rotation 74
—, dipolar magnetic field 100, 103, 104
—, distance 24
—, eleven-year magnetic activity cycle 99–102
—, energy-generating core 32, 33
—, fate 36–38
—, generation of sound waves in convective zone 68
—, hydrogen-alpha 91
—, inactivity 216, 217
—, ingredients 11
—, internal constitution 71
—, internal flows 76, 78
—, internal magnetic fields 104
—, internal sound waves 63–68
—, luminosity 23, 24
—, mass 24
—, neutrinos 44–47
—, oblateness 76, 77
—, origin of elements 13
—, origin of sunspots 103
—, photosphere oscillations 63
—, production of ozone 207
—, quadrupole moment 77
—, radiative zone 32, 33
—, radius 24
Sun, rotation 24
—, rotation period 24
—, sunspot cycle 99, 100
—, temperature 24
—, total eclipses 107, 108
—, variable X-ray radiation 215
—, X-rays 119, 120, 215
Sunspot 81–88, 263
Sunspot belts 264
Sunspot cycle 99, 100, 102, 264
Sunspot, temperature 82
Sunspots 81–88, 263
—, and converging flows 87
—, and differential rotation 82
—, and Earth temperature 218
—, and zonal bands 77
—, belts 264
—, bipolar pairs 94
—, cycle 99, 100, 102, 264
—, depth 87
—, Evershed effect 87
—, filaments 84
—, groups 83
—, magnetic field strength 24, 86
—, magnetic polarity 100, 101
—, missing 216, 217
—, number 99, 100
—, origin 103
—, position 99, 100
—, umbra and penumbra 83
Supergranulation 35, 36, 88, 264
Supergranule 35
Super-Kamiokande neutrino detector 53, 54, 264
Supernova explosion SN 1987A, neutrinos 50
Supernovae, and cosmic rays 165
Supersonic 264
Surge 98
Swedish Solar Telescope 84, 92
Synchrotron radiation 264
—, solar flares 143

Tacholine 76, 103
Tau neutrino 54
Temperature 264
Temperature, center of Sun 24, 33, 53
—, chromosphere 24, 115
—, corona 24, 114
—, Earth 217, 218
—, Earth atmosphere 204
—, photosphere 24
—, Sun 24
—, sunspots 82
—, thermal radiation 19
—, visible solar disk 26
Termination shock 264
The King's Mirror 177
Thermal bremsstrahlung 265
Thermal energy 19
Thermal equilibrium 265

Thermal gas 265
Thermal radiation 265
Thermonuclear fusion 265
TOMS 212
Toroidal magnetic field 103
Torsional oscillations 76, 265
Total eclipses of the Sun 107, 108
Total Ozone Mapping Spectrometer 212
TRACE 96–98, 121, 132, 133, 137, 142, 146, 153, 159, 265
Transition region 115
Transition Region And Coronal Explorer 96–98, 121, 132, 133, 137, 142, 146, 153, 159, 265
Troposphere 204, 265
Tunnel effect 29, 265

UARS 214
Ultraviolet radiation 17, 265
Ulysses Mission 128–130, 266
Umbra 266
Universe, most abundant element 14
Upper Atmosphere Research Satellite 214

Van Allen radiation belts 171–173, 266
—, and cosmic rays 173
—, and solar wind 173
—, threat to satellites 189
Velocity of light 16, 266
Velocity of sound 266
Venus 7
Very Large Array 20, 266
Visible light 266
Visible solar spectrum 12
Visible sunlight 8
Voyager 1 134, 135, 267
Voyager 2 134, 135, 267
Vulcan 98

Water 267
Water cycle 195, 196
Water vapor, greenhouse gas 200
Wavelength 16, 17, 267
Weather 204
Weather satellites, disabled by space weather 189
White dwarf star 38
White light 267
White light flare 140, 267
White sunlight 7
Wind, interstellar 135
Wind, solar 123, 126, 261
Wolf minimum 216, 217

X-ray cusp 160, 161
X-ray radiation 19, 267
X-ray sigmoid 160, 161
X-rays 16, 19
—, solar flares 141, 145
—, Sun 119, 120

Yohkoh Mission 120, 121, 131, 142, 149, 150, 161, 267

Zeeman components 267
Zeeman effect 84–86, 267
Zeeman splitting 268
Zonal flow bands 76, 78